MATHEMATICAL NONLINEAR IMAGE PROCESSING

A Special Issue of the Journal of Mathematical Imaging and Vision

Edited by:

Edward R. Dougherty
Rochester Institute of Technology

Jaakko Astola
University of Tampere

Reprinted from the Journal of Mathematical Imaging and Vision
Vol. 2, Nos. 2/3 (Nov. 1992)

KLUWER ACADEMIC PUBLISHERS
Boston/Dordrecht/London

Distributors for North America:
Kluwer Academic Publishers
101 Philip Drive
Assinippi Park
Norwell, Massachusetts 02061 USA

Distributors for all other countries:
Kluwer Academic Publishers Group
Distribution Centre
Post Office Box 322
3300 AH Dordrecht, THE NETHERLANDS

Library of Congress Cataloging-in-Publication Data

Mathematical nonlinear image processing : a special issue of the Journal of mathematical imaging and vision / edited by Edward R. Dougherty, Jaakko Astola.

p. cm.

"Reprinted from the Journal of mathematical imaging and vision. vol. 2, nos. 2/3 (Nov. 1992)."

ISBN 0-7923-9314-7 (acid free paper).

1. Image processing--Mathematics. 2. Filters (Mathematics) I. Dougherty, Edward R. II. Astola, Jaakko.

TA1632.M364 1993
621.36"7--dc20 92-40136
CIP

Printed on acid-free paper.

Printed in the United States of America

Contents

Special Issue of
Journal of Mathematical Imaging and Vision:
Nonlinear Image Processing
Guest Editors: Edward R. Dougherty and Jaakko Astola

Journal of Mathematical Imaging and Vision 2, 97 (1992).

Introduction

The special issue of *Mathematical Imaging and Vision* on nonlinear image processing springs from two factors: (1) the great expansion of nonlinear methods applied to problems in imaging and vision, and (2) the degree to which nonlinear approaches are both using and fostering new developments in diverse areas of mathematics.

Strictly speaking, digital image processing only involves nonlinear methods since the computer processing does not take place in a linear space; nevertheless, practical image modeling has often assumed a linear-space structure. This modeling has facilitated application of fundamental linear theory such as spectral representation and projection theory. Given the success of linear methods in classical signal processing, the adaptation of such methods to image processing, especially at the outset, is a natural way to proceed.

Yet linear-space theory represents only a corner of mathematics. Semantically, at least, *nonlinear image processing* encompasses all the rest of image processing and it is defined by theory and methods that utilize all areas of mathematics besides linear mathematics. And to a great extent this is true, since its domain includes probability, statistics, logic, topology, set theory, algebra, measure theory, geometry, and analysis. Indeed, one of the most active areas of nonlinear image processing is mathematical morphology, and it alone crosses all of the aforementioned areas.

As practiced, however, the thrust of nonlinear image processing, as it is usually understood, is more directed. One thinks more of morphological, neural, and nonlinear statistical methods. These three areas are not exhaustive: consider fuzzy logic. Nor are they mutually exclusive: consider the roles of neural and statistical methods in the design of morphological filters. Nevertheless there tends to be a current emphasis of interest.

There also tends to be an application focus. Nonlinear methods are widely employed to solve all the basic problems of image processing, including segmentation, restoration, recognition, compression, and analysis. Nonetheless, when the term "nonlinear" is employed, one often sees an emphasis on filtering. Perhaps the term is emphasized when nonlinear methods are used for filtering to point out the distinction with linear filtering, perhaps because so much research has focused on nonlinear filtering methods. In any event, the emphasis on filtering is reflected in the papers composing this special issue.

We hope the papers in this issue prove beneficial to people working in the area and also to those looking for new avenues of approach to solving their problems. We offer our appreciation to all who submitted papers and also to the referees, who returned their reviews within a very short time frame.

Edward R. Dougherty
Jaakko Astola

Journal of Mathematical Imaging and Vision 2, 99–116 (1992).

Statistical Properties, Fixed Points, and Decomposition with WMMR Filters

HAROLD LONGBOTHAM
Nonlinear Signal Processing Group, Department of Engineering, University of Texas at San Antonio, San Antonio, TX 78285-0665

DAVID EBERLY
Nonlinear Signal Processing Group, Department of Mathematics, University of Texas at San Antonio, San Antonio, TX 78285-0665

Abstract. WMMR^m filters weight the m ordered values in the window with minimum range. If m is not specified, it is assumed to be $N + 1$ for a window of length $2N + 1$. Previous work has demonstrated a subclass of these filters that may be optimized for edge enhancement in that their output converges to the closest perfect edge. In this work it is shown that normalized WMMR^m filters, whose weights sum to unity, are affine equivariant. The concept of the breakpoint of a filter is discussed, and the optimality of median and WMMR filters under the breakpoint concept is demonstrated. The optimality of a WMMR^m filter and of a similar generalized-order-statistic (GOS) filter is demonstrated for various non-L^P criterion, which we call closeness measures. Fixed-point results similar to those derived by Gallagher and Wise (see N.C. Gallagher and G.L. Wise, *IEEE Trans. Acoust., Speech, Signal Process.*, vol. ASSP-29, 1981, pp. 1136–1141) for the median filter are derived for order-statistic (OS) and WMMR filters with convex weights (weights that sum to unity and are nonnegative), i.e., we completely classify the fixed points under the assumption of a finite-length signal with constant boundaries. These fixed points are shown to be almost always the class of piecewise-constant (PICO) signals. The use of WMMR filters for signal decomposition and filtering based on the Haar basis is discussed. WMMR filters with window width $2N + 1$ are shown to be linear over the PICO($N + 1$) signals (minimum constant length $N + 1$). Concepts similar to lowpass, highpass, and bandpass for filtering PICO signals are introduced. Application of the filters to 1-dimensional biological data (non-PICO) and images of printed-circuit boards is then demonstrated, as is application to images in general.

Key words. WMMR filters, LMS regression, Haar basis, piecewise constant signals

1 Introduction

The median filter [1], [2] has gained recognition as a filter that will edit impulses and retain locally monotonic trends. Order-statistic filters [3], [4], such as the α-trimmed mean filter [5] have also proven useful in providing smoothers that eliminate impulses. If a filter length of $2N+1$ is assumed, the class of filters introduced here will select m of the windowed values with the smallest range and will weight the resulting ordered values. To implement the class of filters it is easiest first to order the data values $x_{(1)}, x_{(2)}, \ldots, x_{(2N+1)}$, where $x_{(i)} \le x_{(i+1)}$, then to calculate the range of each consecutive set of m values by using $r_i = |x_{(i+m-1)} - x_{(i)}|$, and then to take the dot product of the vector of ordered values computed with the m weights that determine the specific filter in the class. For example, if $m = 5$, $2N + 1 = 8$, the weight vector is $\mathbf{w} = (1/2, 1/4, 1/8, 0, 1/8)$, and the windowed values are $(10, 1, 0, 1, 2, -12, 1, 13)$, the output is $0(1/2) + 1(1/4) + 1(1/8) + 1(0) + 2(1/8)$. We call this class of filters WMMR^m filters, which stands for "weighted majority of m values with minimum range." We will restrict m to be greater than N, and if $m = N + 1$, we will refer to the filters as the WMMR filters.

Figure 1 illustrates the relationship between the finite-impulse-response (FIR), order-statistic (OS), and WMMR filters.

The class of WMMR filters includes the least-median-of-squares (LMS) [6] and the Shorth [7] estimators. The LMS filter has as output one-half the range of the $N+1$ closest clustered values, whereas the Shorth filter has as output the average of the $N+1$ closest clustered values in a window of size $2N+1$. As an example, the output of either filter for $\{5,0,10,0,6,0,0\}$ would be zero (the same as the median), whereas the output for $\{1,1,1,2,4,4,4\}$ would be 3/2, 5/4, and 2 for the LMS, Shorth, and median filters, respectively. It is shown that each member of the class of WMMR filters eliminates impulses and preserves perfect edges as filters exactly as does the median filter. In addition, the entire class of WMMR filters has been shown in previous works [8], [9] to enhance nonperfect edges (unlike the median) in that outputs of these filters converge to the closest perfect edge.

The breakdown point [10], [11] is used as a measure of the robustness for the comparison of the WMMR^m filter to existing OS filters. The breakdown point of a filter, which is defined to be the percentage of aberrant values allowed in the window before the output can be made to take on an arbitrarily large value, has been shown to be an accurate estimate of the ability of an estimator to withstand outliers. It is easily shown that for the averager the breakdown point is 0%, for the median, Shorth, LMS, and WMMR filters it is 50% (asymptotically), and for the α-trimmed mean and WMMR^m filters it can vary from 0 to 50%. Therefore the median, shorth, and LMS filters obtain the optimal breakdown point, as does the entire class of WMMR filters.

Using the breakdown point, we then look at the appropriateness of the norms currently in use (L^P norms). One object of filtering is to eliminate noise so that the input signal approaches the most probable noiseless signal or closest fixed point with respect to some measure. The sum of squares (SS or L^2) is used in most linear applications in which the noise is additive. We note that the OS and generalized-order-statistic (GOS) filters are editing filters (in that sparsely distributed aberrant values do not lead to large deviations in the output), and they demonstrate the usefulness of new measures of closeness when editing GOS filters are used for removal of impulsive noise.

The measures of closeness we will examine are used by Rousseeuw [6] in robust regression. Essentially, the squared residuals (square of the distance of the estimator from the data value) are ordered before weighting. The measures are considered robust in that a weight can be zero. More specifically, one can weight the larger squared residuals with zero, thus eliminating the effect of an outlier on the measure of closeness since sparsely distributed outliers would have no effect on an editing filter.

Some assumptions that are made herein are the use of odd length ($2N + 1$) windows, output taken at the center of the filter, and the use of sliding overlapping windows. One should note that the output of WMMR^m filters must be specified in an unambiguous manner. Essentially, we choose to average the class of possible outputs when there is a finite probability of two or more sets of values with the minimum range.

The fixed points of WMMR filters (window length $2N + 1$) are shown to be piecewise constant [PICO($N + 1$)] if the weights are convex (nonnegative and normalized, i.e., sum to one) and if the first and last weights are not equal. If the first and last weights are equal, then the set of possible fixed points also includes constant regions separated by a single point (with amplitude the average of the amplitudes of the two constant regions) and possibly a constant region with length only N. These two exceptions to a PICO($N + 1$) region occur with a very low probability, as is shown in section 3.

Wavelets [12] have recently been introduced and provide an alternative basis for the decomposition of signals to that of the Fourier series. The next step in this process is obviously the design of filtering mechanisms that take advantage of these decompositions, i.e., filtering paradigms such as lowpass, highpass, and bandpass filters based on separating noise from a signal that has a natural basis other than the Fourier series. One common wavelet basis is the Haar basis with the orthogonal space of PICO sig-

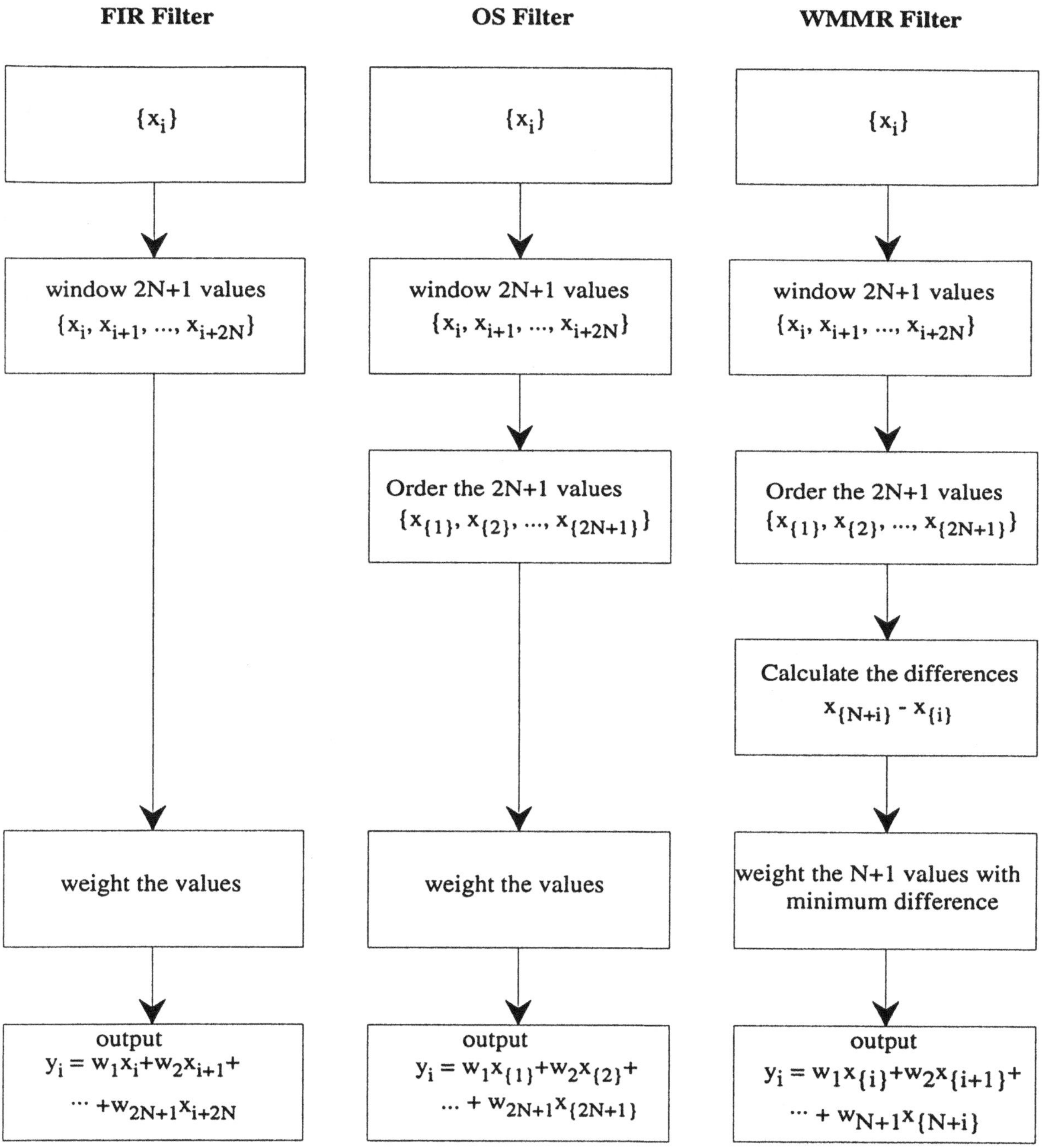

Fig. 1. Comparison of FIR, OS, and WMMR filters.

nals. In section 4 WMMR filters are shown to be linear over the class of PICO(2^M) signals for $2M \geq N+1$. WMMR filters could therefore be used for robust decomposition in the presence of impulsive noise, edge degradation [8], [9], and baseline shifts as is detailed in sections 4 and 5. In addition, WMMR filters preserve phase in that the filter can be used to decompose and filter on the different resolutions (scalings) of the Haar basis without prior knowledge of the origin (translates) in the space from which the signal was composed.

2 Statistical Properties and Optimality of a Subclass of WMMR^m Filters

This section demonstrates some of the statistical properties of WMMR^m filters. First, we show that the normalized WMMR^m filters are affine equivariant. Second, we discuss the concept of the breakdown point of a filter and classify the breakdown point of the WMMR^m and similar filters. Third, we demonstrate the optimality of the median and WMMR filters under the breakdown-point concept. Last, we discuss the optimality of a WMMR^m filter and a GOS filter for various non-L^P norms, which we call closeness measures.

Let T be an operator with domain R^{2N+1} and range R. In digital filtering two properties of nonlinear filters we may examine are linearity with respect to a dc bias and a change of scale, i.e., for a constant scalar c and a vector $\mathbf{c} = (c, c, c, \ldots, c)$, and $T\{\mathbf{x_i} + \mathbf{c}\} = T\{\mathbf{x_i}\} + \mathbf{c}$ and $T\{c\mathbf{x_i}\} = cT\{\mathbf{x_i}\}$. In the statistical literature location estimators with these properties are called location equivariant or scale equivarient, respectively. Estimators with both properties are called affine equivariant.

THEOREM 2.1. Normalized WMMR^m filters are affine equivariant estimators.

Proof. Let T be a normalized WMMR^m filter. $T(\mathbf{x_i} + \mathbf{c}\} = \text{AVE}\{\mathbf{w} * \mathbf{a} : a = \{x_{(1)}, x_{(2)}, \ldots, x_{(m)}\}$ is a subset of m values with minimum range in $\mathbf{x_i} + \mathbf{c}\}$. We note that adding a constant vector to a set does not change the range and that if $x_i \leq x_{i+1}$, then $x_i + c \leq x_{i+1} + c$. Therefore in the first sentence an $\mathbf{a}$ with minimum range in $\mathbf{x_i} + \mathbf{c}$ implies $\mathbf{a} - \mathbf{c}$ has minimum range in $\mathbf{x_i}$. Considering location equivariance, we have $T(\mathbf{x_i} + \mathbf{c}\} = \text{AVE}\{\mathbf{w} * \mathbf{a} : \mathbf{a}$ is an ordered subset of m values with minimum range in $\mathbf{x_i} + \mathbf{c}\} = \text{AVE}\{\mathbf{w} * ((\mathbf{a} - \mathbf{c}) + \mathbf{c}) : \mathbf{a} - \mathbf{c}$ has minimum range in $\mathbf{x_i}\} = c + \text{AVE}\{\mathbf{w} * (\mathbf{a} - \mathbf{c}) : \mathbf{a} - \mathbf{c}$ has minimum range in $\mathbf{x_i}\} = \mathbf{c} + T(\mathbf{x_i})$. Now for scale invariance, $T(c\mathbf{x_i}) = \text{AVE}\{\mathbf{w} * \mathbf{a} : \mathbf{a}$ has minimum range in $c\mathbf{x_i}\}$. But $\mathbf{a}$ with minimum range in $c\mathbf{x_i}$ implies $(1/c)\mathbf{a}$ has minimum range in $\mathbf{x_i}$. Therefore $T(c\mathbf{x_i}) = \text{AVE}\{(\mathbf{w} * \mathbf{a}/c)c : \mathbf{a}/c$ has minimum range in $\mathbf{x_i}\} = cT(\mathbf{x_i})$.

Historically [6], [10], [11], the breakdown point has been used in the statistical literature as a measure of the robustness of an estimator or a minimization criterion. We now describe its use as a measure of the robustness of an estimator in statistics, extend this to filters, and then use it to compare the robustness of median, averager, α-trimmed-mean, OS, and WMMR^m filters. The last part of this section will describe its use as a measure of the robustness of a minimization criterion.

Intuitively, the breakdown point is the smallest amount of contaminants that may cause an estimator to take on an arbitrarily large aberrant value. The optimal value of the breakdown point is usually accepted to be 50% since if more than 50% of the values are perturbed they could be perturbed in such a way as to mask the original distribution/sample. We note that the breakdown point of the mean is 0 and that the breakdown point of the median is 50%, providing us with examples at each extreme.

Hampel [11] used a Monte Carlo technique to study estimators of location in the presence of outliers. Six estimators were chosen, each of which rejected outliers by some criterion and then averaged the remaining values. He also calculated the breakdown point of each estimator and showed in a simple summary and theoretical explanation that the results of the Monte Carlo study could be explained by the breakdown point. These and other results [10] have led us to expect the breakdown point to be in direct correlation with the robustness of an estimator.

We wish to define a similar concept for filtering. The same concept is not applicable since in an infinite-length sequence we would need to perturb only a segment of length $2N + 1$, and if the filter depended on the signal values, at that point the value would be infinite. This would imply a breakdown of 0% for any filter. Therefore we will define the breakdown point of a filter to be the fraction of aberrant values in a particular window that will cause the filter to take on arbitrarily large values. Therefore, considering filters, we note the breakdown point for the median filter is asymptotically 50% and for the averager, it is asymptotically 0%. If the α-trimmed-mean filter is defined so as to (asymptotically) trim a total of $\alpha\%$ of the values (the $\alpha/2$ largest and the $\alpha/2$ smallest), then it has a breakdown point of $(\alpha/2)\%$. The breakdown point of the OS filter is the same as that of the α-trimmed-mean filter if the latter trims equivalent ordered values. Therefore for the optimal breakdown point of 50% (asymptotically) to be reached, the α-trimmed-mean and the OS filters must be the median. The breakdown point of a WMMR filter can asymptotically approach 50% and we can still have $N + 1$ values to weight for some other reason (e.g., edge enhancement or optimization of combinations of residuals). These considerations lead us to the following theorem.

THEOREM 2.2. The median and WMMR filters are optimal filters in that they obtain the maximum breakdown point.

Theorem 2.2 is one reason we call the median and WMMR^m filters optimal. To see the significance of this we note that if we compare the α-trimmed-mean filter that averages $N + 1$ values to the WMMR–AVE we would expect them to have similar smoothing properties except that the WMMR filter would be capable of trimming twice as many outliers if the amplitude of the impulsive noise were from one-tailed distributions.

In signal processing, norms are used in several ways. Before filtering, a norm is used to find the closest desired signal, so that after filtering the appropriateness of the filter used can be estimated. After filtering, a norm is used to measure the closeness of the output signal to the previously determined desired signal. Lastly, in many applications, such as adaptive filtering, a norm is used as part of the filtering process. Historically, the sum-of-squares (SS) or some other L^P norm has been used in all three instances. We suggest that it may be appropriate to use a different criterion for editing filters, such as the median, OS, and WMMR^m filters, when estimating the closest desired signal before filtering. But first we will discuss some parallel research in the statistical literature.

Rousseeuw [6] notes "the method of LSE (least squares error) is being criticized more and more for its dramatic lack of robustness." He suggests that instead we replace the mean with the median and use the LMS error. Indeed, he uses the breakdown point to measure the robustness of LMS error and the LSE and shows them to have breakdown points (defined similarly for norms/measures as for estimators) of 50% and 0%, respectively. Rousseeuw also shows that just as the average is the optimal least-squares estimate, the WMMR with first and last weights of 1/2 is the optimal LMS estimate. This section will extend these concepts.

DEFINITION 2.1. If $\{x_i\}, 1 \leq i \leq N$, is a set of data values, the LMS error in approximating the data set by A is MEDIAN$\{(x_i - A)^2, 1 \leq i \leq N\}$. The LMS estimate for the data set $\{x_i\}$ is the quantity that minimizes LMS error.

There are many ways to make the L^P norms more robust. The approach used in this paper is to order the distance of the location estimator from each of the data samples (these distances are the absolute value of residuals in the statistical literature) and to consider minimizing only over the squared residuals that are small. Leaving off larger squared residuals means that the error from aberrant values is not summed. This makes sense because editing filters, such as WMMR and median filters, easily eliminate sparsely distributed aberrant values, regardless of their magnitude.

Example 2.1. Let a constant signal s =

$\{s_i\}, 1 \le i \le N$, be the emitted signal, and let $r = \{r_i\}, 1 \le i \le N$, be the received signal. Then assume $r_i = s_i$ if $i \neq J$ and $r_J = s_J + d$. The L^P distance from s to r is $|d|$ for d any real number and p any positive integer. Yet s is recoverable from r on one pass of any number of editing filters. Similarly, consider $r_i = s_i + d$ for every i. The L^P distance from this signal to s_i is $(N)^{1/p} \cdot |d|$ but the emitted signal s is totally masked by the received signal r.

Therefore the L^P norm is in no way indicative of the recoverability of a signal when editing filters are used. The object in this section is to develop a measure of closeness of two signals that accurately reflects the ability of a class of filters to recover the emitted signal from the received signal in the presence of impulsive noise. We note in the example above that for each i, $R_i = (r_i - s_i)^2$ is the square of the ith residual and is the square of the distance of the ith component of r to the ith component of s. Therefore if we order these squared residuals, i.e., in the usual notation $R_{(i)} \le R_{(i+1)}$ for every i, we then ignore $N - k$ aberrant values by minimizing over only the first k ordered squared residuals. When considering location estimates, we will show that this can be accomplished by several methods. We will mention only those for which we can find an optimal estimator (filter).

Method 1. For the data values $\{r_i\}$, $1 \le i \le N$, $r_i = s + n_i$, we will use as an estimate of location the value Q_k of Q that minimizes the kth ordered squared residual $R_{(k)}$, $k > N/2$. We note that Q_k is the value that puts the minimal bound on the first k ordered squared residuals. In the statistical literature this method corresponds to the LMS measure of error introduced by Rousseeuw if the restrictions $N = 2M + 1$ and $k = M + 1$ are applied. We note that a convex hull of samples is compact and that $R_{(k)}$ is continuous in Q; therefore there is a Q that minimizes $R_{(k)}$. This means, however, that for the techniques developed here and in Method 2 below, it must be assumed the data are in the form of a discrete signal. Therefore it is assumed that there are no ties for the desired output and that with probability one a final averaging operation is not needed. This leads to the following definition.

DEFINITION 2.2. Given a data sample $\{r_i\}$, $1 \le i \le N, Q_k$ is defined as the real number that minimizes the kth-ranked squared residual.

Note that k is restricted so that $R_{(k)}$ is well defined. For example, if $N = 10$, there is no k such that $R_{(k)}$ is the median of residuals. The following example illustrates why discrete and not digital signals are assumed in this section.

Example 2.2. Consider $p = 1$, $N = 9$, $k = 5$, and the data set $\{0, 0, 0, 0, 1/2, 1, 1, 1, 1\}$. Without loss of generality, the possible Q values are between 0 and 1. If $Q = 1/2$, the residuals are 1/2, 1/2, 1/2, 1/2, 0, 1/2, 1/2, 1/2, 1/2 and therefore $R_{(5)} = 1/4$. If $0 < Q < 1/2$, the absolute residuals are Q, Q, Q, Q, $(1/2) - Q$, $1 - Q$, $1 - Q$, $1 - Q$, $1 - Q$, which implies that $R_{(5)} = \text{MAX}\{Q^2, [(1/2) - Q]^2\}$. But $\text{MAX}\{a^2, b^2\} = [\text{MAX}\{a, b\}]^2$ for a and b nonnegative. Therefore the minimal value of $R_{(5)}$ is $(1/4)^2$ when $Q = 1/4$. Similarly, if $1/2 < Q < 1$, $R_{(5)} = \text{MAX}\{(Q - 1/2)^2, (1 - Q)^2\}$, and by a similar argument $Q = 3/4$ yields a minimal $R_{(5)}$ of $(1/4)^2$. Therefore $Q = 1/2$ cannot lead to a minimal $R_{(5)}$. The average of the two possible values for Q is 1/2, and therefore we see that if a multiplicity of values is allowed for Q, an optimal value may not be obtained.

THEOREM 2.3. Let the window width be $2N + 1$. Then if the windowed values are considered as data samples, Q_k is equivalent to the output of the WMMR^k filter with coefficients $(1/2, 0, 0, \ldots, 0, 1/2)$.

Proof. To prove this theorem we need to show that if a value Q minimizes a given kth-ranked squared residual, then Q is the average of the first and last ordered data values used in calculating the first k ranked squared residuals. Now, if Q_k is the minimal bound on the kth-ranked squared residual, then it is the minimum value such that $Q_k + [R_{(k)}]^{1/2}$ or $Q_k - [R_{(k)}]^{1/2}$ defines two horizontal lines with k values from the data set between them. To prove the theorem we need show only that at least one data value lies on each of the horizontal lines. Since Q_k is the real number that minimizes $R_{(k)}$, it is obvious that there is one data value on either $Q_k + [R_{(k)}]^{1/2}$ or $Q_k - [R_{(k)}]^{1/2}$, namely, the data value with absolute residual $[R_{(k)}]^{1/2}$ from Q_k. Without loss of generality, assume that Q_k is such that the horizontal lines enclose k data values, one or more of which lie on $Q_k + [R_{(k)}]^{1/2}$ and none of which lie on $Q_k - [R_{(k)}]^{1/2}$. In this case we may increase Q_k (and thereby decrease the kth-ranked squared residual). Therefore we have a contradiction unless Q_k cannot be increased, in which case there must be at least one data value on each of the horizontal lines. Since a data value lies on each of the horizontal lines, Q_k is the average of the minimum and maximum in the set of data values used in calculating the kth-ranked squared residual.

Method 2. Assume that there are N data values, and minimize over the sum of the k smallest squared residuals; i.e., if the squared residuals of the data values from the location estimate t are $\{d_1{}^2, d_2{}^2, d_3{}^2, \ldots, d_n{}^2\}$, then consider the ordered set of squared residuals $\{d_{(1)}{}^2, d_{(2)}{}^2, \ldots, d_{(n)}{}^2\}$. Then let t_k be the location estimate that minimizes the sum of the first k ordered squared residuals. It is obvious that once the k smallest squared residuals that we are to minimize over are selected, we need only to average the corresponding data values (by comparison to the L^2 case). Assume that the sum of the R squared residuals does not correspond to R consecutive ordered data values. Then the sum may be lowered by substituting one of the skipped ordered values for the maximum or the minimum of the set of data values used. Therefore we see that R consecutive data values must be used. These observations lead to the following theorem.

THEOREM 2.4. To calculate t_k such that t_k minimizes the sum of the k smallest residuals of the data values from t_k we use the following procedure:

(a) Form the following subsets of data values:

$$\begin{aligned} s_1 &= \{x_{(1)}, x_{(2)}, \ldots, x_{(k)}\}, \\ s_2 &= \{x_{(2)}, x_{(3)}, \ldots, x_{(k+1)}\}, \\ \ldots, &\quad \vdots \\ s_{N-k+1} &= \{x_{(N-k+1)}, x_{(N-k+2)}, \ldots, x_{(N)}\}. \end{aligned}$$

(b) For each i let $a^i = \mathrm{AVE}(s_i)$,

$$\begin{aligned} D^i &= (x_{(i)} - a^i)^2 \\ &+ (x_{(i+1)} - a^i)^2 \\ &+ \cdots + (x_{(k+i-1)} - a^i)^2. \end{aligned}$$

(c) Then $t_k = a^j$ if $D^j = \mathrm{MIN}\{D^i\}$.

Theorem 2.4 is an extension of the median case in Rousseeuw's book *Robust Regression and Outlier Detection* [13], where a recursive algorithm for the calculation of the LTS estimator (t_{N+1} for our case) is presented.

3 Fixed-Point Analysis and Nonlinear Filtering

In signal processing the filter is chosen either to eliminate or to enhance certain attributes of the input signal. The design of the filter involves characterizing those signals that we want to be changed by the filter or, alternatively, characterizing those signals that are not changed by the filter. In linear filtering the set of such signals is called the passband of the filter, but in nonlinear system theory these signals are called the fixed points for the filter since they are the solutions to $F(x) = x$. For linear filters this is a matrix equation and can be solved by using standard linear algebraic methods. For nonlinear filters one must resort to other techniques to understand the structure of the fixed points [14]–[16].

Although the fixed points represent the signals that are unchanged by the filters, they also may represent limit points on iteration by the filter. To represent multiple iterates (passes) of the filter we will use superscripts. In many cases (but not always) the iterates $x^{(k+1)} = F(x^{(k)})$, $x^{(0)}$ given, converge to a fixed point. In this sense the fixed points tend to influence the shape of the filtered signal after one or more iterations.

Example 3.1. A simple example illustrating the

above point is the sign filter $f(w_i) = \text{Sign}(x_i)$, where $\text{Sign}(v) = v/|v|$ if $v \neq 0$, but $\text{Sign}(v) = 0$ if $v = 0$. For any boundary conditions **a** and **b** the fixed points **x** are those vectors whose components satisfy $x_i \in \{-1, 0, 1\}$. One pass of the filter on any input yields a fixed point.

Example 3.2. Let $N = 2$, $M = 10$, $\mathbf{a} = (3,3)$, and $\mathbf{b} = (6,6)$. The linear averager of window width 5 is $\mathbf{y} = F(\mathbf{x})$, given by

$$y_i = \frac{1}{5}\sum_{j=i-2}^{i+2} x_j \quad \text{for } 1 \leq i \leq 10.$$

The input signal is $\mathbf{x} = (2,1,6,6,0,2,5,2,9,0)$, the output signal is $\mathbf{y} = 1/5(15,18,15,15,19,15,18,18,22,23)$, and a fixed point reached after infinitely many iterations is $\mathbf{z} = 1/248(833,888,956,1019,1084,1148,1213,1276,1344,1399)$. Here the fixed point represents a smoothing of the original signal.

Example 3.3. Finally, consider the median filter applied to the input signal of Example 3.2. The output is given by $y_i = \text{median}(x_{i-2},\ldots,x_{i+2})$ for $1 \leq i \leq 10$. The output signal after one iteration is $\mathbf{y} = (3,3,2,2,5,2,2,2,5,6)$, and a fixed point reached after two iterations is $\mathbf{z} = (3,3,3,2,2,2,2,2,5,6)$. In general, the fixed points for a window length $2N+1$ median filter are those signals such that each segment of length $N+1$ from $(\mathbf{a},\mathbf{x},\mathbf{b})$ is monotonic.

Any information about the fixed points is helpful for understanding the effects of a filter on the signal. In both the linear averager and median examples, multiple iterations of the filter will smooth the signal. If the boundary conditions represent two constant segments in the signal, then the values between the boundaries may represent a distortion of an edge. If one is interested in enhancing the edge, then the above filters are not appropriate; more appropriate ones are the WMMR filters as discussed in [8] and [9].

The first result shows how to construct classes of fixed points, given a single fixed point. Define R to be the permutation matrix that reverses the components of a vector: $R(z_1,\ldots,z_L) = (z_L,\ldots,z_1)$.

LEMMA 3.1. Let $\mathbf{y} = (\mathbf{a},\mathbf{x},\mathbf{b})$ be a fixed point for the WMMR^m filter F. The following transformations produce fixed points:

(a) If $\lambda \bullet 1 = 1$, i.e., if λ is normalized or a convex combination, then $\mathbf{y} + a\mathbf{1}$ is a fixed point for any constant a.
(b) If $a > 0$ then $a\mathbf{y}$ is a fixed point. If $a < 0$ and the WMMR^m filter is symmetric, i.e., if $R\lambda = \lambda$, then $a\mathbf{y}$ is a fixed point.
(c) The reversed vector $R\mathbf{y}$ is a fixed point.

Proof. Let $\mathbf{w_i} = (\mathbf{x_{i-n}},\ldots,\mathbf{x_{i+n}})$ be the windows for $\mathbf{y}$. Since $\mathbf{y}$ is a fixed point, $\mathbf{y_i} = f(\mathbf{w_i}) = g(\mathbf{w_i}) \bullet P(\mathbf{w_i})\mathbf{w_i}$, where $P(\mathbf{w_i})$ is the permutation that orders $\mathbf{w_i}$ and where $g(\mathbf{w_i})$ are the appropriate weights, which are a function of $\mathbf{w_i}$. In each of the cases let $\mathbf{z}$ be the vector transformed from $\mathbf{y}$ with windows $\mathbf{v_i} = (\mathbf{z_{i-n}},\ldots,\mathbf{z_{i+n}})$. We need show only $\mathbf{z_i} = f(\mathbf{v_i})$ for all i.

(a) Let $\mathbf{z} = \mathbf{y} + a\mathbf{1}$; then $\mathbf{v_i} = \mathbf{w_i} + a\mathbf{1}$. The addition of $a\mathbf{1}$ changes neither the relative positions nor the clustering of the ordered window values, so $P(\mathbf{v_i}) = P(\mathbf{w_i})$ and $g(\mathbf{v_i}) = g(\mathbf{w_i})$. Therefore

$$\begin{aligned}\mathbf{z_i} &= \mathbf{y_i} + a\mathbf{1}\\ &= g(\mathbf{w_i}) \bullet P(\mathbf{w_i})\mathbf{w_i} + a(g(\mathbf{w_i}) \bullet \mathbf{1})\mathbf{1}\\ &= g(\mathbf{w_i}) \bullet P(\mathbf{w_i} + a\mathbf{1})(\mathbf{w_i} + a\mathbf{1})\\ &= g(\mathbf{v_i}) \bullet P(\mathbf{v_i})\mathbf{v_i}\\ &= f(\mathbf{v_i}),\end{aligned}$$

where we have used $\lambda \bullet \mathbf{1} = 1$ (which implies $\mathbf{g} \bullet \mathbf{1} = 1$ for any g) and $P\mathbf{1} = \mathbf{1}$ for any permutation.

(b) Let $\mathbf{z} = a\mathbf{y}$; then $\mathbf{v_i} = a\mathbf{w_i}$. If $a > 0$, the relative positions of the ordered window values do not change, and so $P(\mathbf{v_i}) = P(\mathbf{w_i})$. If $a < 0$ the ranges are scaled by $|a|$, and so the same subwindows attain minimum range and $g(\mathbf{v_i}) = g(\mathbf{w_i})$. Therefore

$$\begin{aligned}\mathbf{z_i} &= a\mathbf{y_i}\\ &= g(\mathbf{w_i}) \bullet P(\mathbf{w_i})(a\mathbf{w_i})\\ &= Rg(\mathbf{v_i}) \bullet RP(\mathbf{v_i})\mathbf{v_i}\\ &= (R^t Rg(\mathbf{v_i})) \bullet P(\mathbf{v_i})\mathbf{v_i}\\ &= g(\mathbf{v_i}) \bullet P(\mathbf{v_i})\mathbf{v_i}\\ &= f(\mathbf{v_i}),\end{aligned}$$

where we have used $R^tR = I$.

(c) Let $\mathbf{z} = R\mathbf{y}$; then $\mathbf{v_i} = R\mathbf{w_{m-i+1}}$. The permutation to order $\mathbf{v_i}$ satisfies $P(\mathbf{w_{m-i+1}}) = P(\mathbf{v_i})R$ since $\mathbf{w_{m-i+1}}$ is obtained by reversing the order of $\mathbf{v_i}$ by permutation R. The weight vector satisfies $g(\mathbf{w_{m-i+1}}) = g(\mathbf{v_i})$ since $\mathbf{w'_{m-i+1}} = \mathbf{v_i}$. Therefore

$$\begin{aligned}\mathbf{z_i} &= \mathbf{y_{m-i+1}}\\ &= g(\mathbf{w_{m-i+1}}) \bullet P(\mathbf{w_{m-i+1}})\mathbf{w_{m-i+1}}\\ &= g(\mathbf{v_i}) \bullet P(\mathbf{v_i})R\mathbf{w_{m-i+1}}\\ &= g(\mathbf{v_i}) \bullet P(\mathbf{v_i})\mathbf{v_i}\\ &= f(\mathbf{v_i}).\end{aligned}$$

We now restrict our attention to those filters with convex weights λ and with constant boundary conditions $\mathbf{a} = A\mathbf{1}$ and $\mathbf{b} = B\mathbf{1}$ for some constants A and B. The fixed points and convergence properties are qualitatively different, depending on whether the cluster size c satisfies $m < N+1$, $m > N+1$, or $m = N+1$. In most ordering filters (median and rank-order) oscillating fixed points may be excluded by the use of constant boundaries. But WMMR^m filters with cluster sizes $m < n+1$ have oscillatory behavior, even when constant boundaries are used. For instance, there are signals z such that $F^2(z) = z$ but $F(z) \neq z$, i.e., if the filter is iterated on input signals that are near these cycle-2 points, the iterations oscillate between two differently shaped signals. For example, let $N = 2$, $M = 2$, $\mathbf{a} = (0,0)$, $\mathbf{b} = (3,3)$, and $\lambda = (0.5, 0.5)$, and let $\mathbf{x} = (0,3)$ and $\mathbf{y} = (1,2)$. It is easily shown that $\mathbf{y} = F(\mathbf{x})$, $\mathbf{x} = F(\mathbf{y})$. The cluster sizes $m > N+1$ appear to produce smooth fixed points. Consider the padded signal $A^{[N]}, \mathbf{x}, B^{[N]}$. If x contains a constant segment $C^{[P]}$ with $p \neq N+1$, then this segment is unchanged by the filter since each window containing a segment value must contain at least $N+1$ equal values. The closest clustered values are the equal values, and the output is the common values. Therefore we need analyze only $A^{[N]}, \mathbf{x}, C^{[N]}$ or $C^{[N]}, \mathbf{x}, B^{[N]}$. If $\lambda_1 = \lambda_{N+1}$, we will see that a fixed point may have a constant segment of length N, but the values immediately adjacent to this segment must satisfy very stringent conditions (and are therefore unlikely to occur). Therefore, without loss of generality, we will examine fixed points x such that $x_1 \neq A$, $x_M \neq B$, and any constant segment of $\mathbf{x}$ has length at most N. Such signals can be pieced together with constant segments of length $N+1$ to form the entire set of fixed points.

LEMMA 3.2. Let F be a convex WMMR filter. In addition, assume that $\lambda_1 > 0$, that $x_1 \neq A$, $A^{[N]}, \mathbf{x}, B^{[N]}$ is a fixed point, and that $x_1 \neq A$, $x_M \neq B$. If there is a smallest index $I \neq 1$ such that $x_I = \max\{x_1, x_2, \ldots, x_M\}$, then $x_I = \cdots = x_{I+N}$. Similarly, if $\lambda_{N+1} > 0$ (instead of $\lambda_1 > 0$) and if there is a smallest index $I \neq 1$ such that $x_I = \min\{x_1, x_2, \ldots, x_M\}$, then $x_I = \ldots = x_{I+N}$.

Proof. First, let us note that the boundaries (i.e., the first $N+1$ or last $N+1$ values of $\mathbf{x}$) cannot be equal to the maximum in any window in which they are included. Consider, for example, if $A = x_I = \text{Max}\{x_1, x_2, \ldots, x_M\}$ and $1 \leq I \leq N+1$. Then $\mathbf{x}$ is not a fixed point since $x_1 \neq x_I$ and the output when centered at x_1 must be A if A is equal to the maximum in the first window. It is obvious that a similar result holds for the last window.

The output of the filter at window w_I is

$$x_i = \sum_{j=1}^{N+1} \lambda_j x'_{i+j-1}$$

where x'_k are the components of the ordered window w'_I. By our hypotheses, $x'_{I+2N} = \max w_I = x_I$. Since $\lambda_1 > 0$, the only way the maximum can be the output is for the $N+1$ (at least) largest values in the window to be all equivalent to x_I; therefore window w_I must contain at least $N+1$ occurrences of x_I. Since $x_I > x_j$ for $j < I$, the occurrences of x_I must all be in the right-half window, that is, $x_I = \cdots = x_{I+N}$ and w_I contains a constant segment of length $N+1$. A similar proof holds if x_I is the minimum in window w_I and $\lambda_{N+1} > 0$.

COROLLARY 3.1. Lemma 3.2 holds with $\max\{x_1, x_2, \ldots, x_M\}$ replaced by $\max\{x_i : x_i \in s_j(I)$, $s_j(I)$ is the jth set with minimum range in w_I, and with $\min\{x_1, x_2, \ldots, x_M\}$ replaced by $\min\{x_i : x_i \in s_j(I)$, $s_j(I)$ is the jth set with minimum range in $w_I\}$.

Proof. The proof is contained in the proof of Lemma 3.2.

COROLLARY 3.2. If the sequence $A^{[N]}, \mathbf{x}, B^{[N]}$ is a fixed point with no constant segment in $\mathbf{x}$ of length $N + 1$, then $A \leq x_i \leq B$.

Proof. The proof is result of the Lemma 3.2 and Corollary 3.1.

LEMMA 3.3. Let F be a proper convex WMMR filter, $\lambda_1 > 0$, and $\lambda_{N+1} > 0$, and let a, b be real values. Let $A^{[N]}, \mathbf{x}, B^{[N]}$ be a fixed point with no constant segment of length $N + 1$, $x_1 \neq A$, and $x_M \neq B$. Then the following conditions on the first and last window hold:

(a) If $\lambda_1 \neq \lambda_{N+1}$, then either $w_1 = A\mathbf{1} + a(0^{[N]}, \mathbf{1}, 2^{[N]})$ for some a or $x_1, \ldots, x_{N+1}$ are the closest clustered values of $\mathbf{w_M}$.
(b) If $\lambda_{N+1} \neq \lambda_1$, then either $w_M = B\mathbf{1} + b(2^{[N]}, \mathbf{1}, 0^{[N]})$ for some b or $x_{M-N}, \ldots, x_M$ are the closest clustered values of $\mathbf{w_M}$.

Proof. The following argument applies to condition (a); a similar argument applies to condition (b). In view of Lemma 3.1, part (a), we need only illustrate the case of $A = 0$. Consider window $\mathbf{w_1} = (0^{[N]}, x_1, \ldots, x_{N+1})$, where $x_1 \neq 0$. Define $u = (x_1, \ldots, x_{N+1})$. If $x_i = 0$ for some $1 < i < N+1$, then window w_1 contains $N+1$ zeros, which forces the output 0, a contradiction to $0 \neq x_1 = f(\mathbf{w_1})$. Let $\mathbf{u}'$ be the ordered values for $\mathbf{u}$. Using Corollary 3.2, we see that $A \leq x_i \leq B$ for all i, the ordered window $\mathbf{w_1}$ is $(0^{[N]}, \mathbf{u}')$, and the subwindow ranges are $\mathbf{u}'_1, \ldots, \mathbf{u}'_N$ and $\mathbf{u}'_{N+1} - \mathbf{u}'_1$; so $d_1 = \min\{\mathbf{u}'_1 \mathbf{u}'_{N+1} - \mathbf{u}'_1\}$. If $\mathbf{u}'_1 > \mathbf{u}'_{n+1} - \mathbf{u}'_1$ then $x_1, \ldots, x_{N+1}$ are the closest clustered values in $\mathbf{w_1}$ and we have satisfied the conclusion of the lemma. If $\mathbf{u}'_1 < \mathbf{u}'_{N+1} - \mathbf{u}'_1$, then $\mathbf{u_1}$ is a proper convex combination of $0^{[N]}$ and $\mathbf{u}'_1$. This implies $0 < \mathbf{u_1} < \mathbf{u}'_1$, a contradiction since $\mathbf{u}'_1 = \min \mathbf{u}$.

The remaining case is $\mathbf{u}'_1 = \mathbf{u}'_{N+1} - \mathbf{u}'_1$. Define $\mathbf{a} = \mathbf{u}'_1 = \cdots = \mathbf{u}'_J = \mathbf{u}'_{N+1} - \mathbf{u}'_1$ for some $J \neq 1$. The output is

$$u_1 = [1/(J+1)]\lambda \cdot \left[\left(\sum_{i=1,J} s_i^{(1)} \right) + s_{N+1}^{(1)} \right]$$

$$= [1/(J+1)]\lambda \cdot [(0^{[N+1-J]}, a, 2a, \ldots, Ja) + (a^{[J]}, \mathbf{u}'_{J+1}, \ldots, \mathbf{u}'_N, 2\mathbf{a})]$$

$$\leq [a/(J+1)]\lambda \cdot (1, (J+1)^{[n-1]}, J+2)$$

$$= a[1 + (\lambda_{N+1} - J\lambda_1)/(J+1)], \qquad (1)$$

where we have used $\mathbf{u}'_k \leq \mathbf{u}'_{n+1} = 2\mathbf{a}$ for $k \neq J+1$. If $\lambda_{n+1} < J\lambda_1$, then $u_1 < a$, which contradicts $a = \min u$. Otherwise, $\lambda_1 \neq \lambda_{n+1} \neq J\lambda_1$, which forces $J = 1$ and $\lambda_1 = \lambda_{n+1}$ and the inequality in (1) must be an equality. Consequently, $\mathbf{u}'_2 = \cdots = \mathbf{u}'_n = \mathbf{u}'_{n+1} = 2\mathbf{a}$ and $\mathbf{u}'_1 = \mathbf{u_1} = \mathbf{a}$, and so $\mathbf{u} = a(0^{[n]}, 1, 2^{[n]})$.

The above conditions are not the only ones under which Lemma 3.3 can be proved. Next, we state a lemma similar to Lemma 3.3 that allows λ_1 and λ_{N+1} to be zero but that requires $\lambda_N = \lambda_2 > 0$. In addition, Theorem 3.1 below could be restated to hold for the new conditions on the weights. We do not believe that the most general conditions on λ for Lemma 3.3 (and Theorem 3.1) to hold have been found.

LEMMA 3.4. Let F be a convex WMMR filter, let $\lambda_2 = \lambda_N > 0$, let a, b be real values, and let $\lambda_1 = \lambda_{N+1}$. Let $A^{[N]}, \mathbf{x}, B^{[N]}$ be a fixed point with no constant segment of length $N+1, x_1 \neq A$, and $x_M \neq B$. Then the following conditions on the first and last window hold:

(a) Either $\mathbf{w_1} = A\mathbf{1} + a(0^{[N]}, 1, 2^{[N]}$ for some a or $x_1, \ldots, x_{N+1}$ are the closest clustered values of $\mathbf{w_1}$.
(b) Either $\mathbf{w_M} = B\mathbf{1} + b(2^{[N]}, 1, 0^{[N]})$ for some b or $x_{M-N}, \ldots, x_M$ are the closest clustered values of $\mathbf{w_M}$.

Proof. All of the first part of Lemma 3.3 holds where dependency on a proper convex combination is replaced by the fact $\lambda_2 = \lambda_N > 0$. The proof starting with the inequality in (1) is slightly different.

We define a signal $A^{[n]}, \mathbf{x}, B^{[n]}$ to be piecewise constant of length at least p, dPICO(p), if it contains only constant segments of length at least p. We say that the signal is degenerate piecewise constant of length at least p, say dPICO(p), if it contains constant segments of length at least p and constant segments of length $p-1$, with isolated points subject to the following conditions:

- If b is an isolated point, then it must occur in a segment of the form $a_1^{[q_1]}, b, a_2^{[q_2]}$, with $q_1 > p-1, q_2 > p-1$, and $b = (a_1 + a_2)/2$.
- If a segment $g^{[p-1]}$ occurs, then it must be embedded in a segment of the form $a_1^{[q_1]}, b, g^{[p-1]}, d, a_2^{[q_2]}$, with $q_1 > p-1$, $q_2 > p-1$, $b = (a_1 + g)/2$, $g = (b + d)/2$, and $d = (g + a_2)/2$.

THEOREM 3.1. Let F be a proper convex WMMR filter with $c = n+1$, $\lambda_1 > 0$, $\lambda_{n+1} > 0$, and constant boundaries $A\mathbf{1}$ and $B\mathbf{1}$. If $\lambda_1 \neq \lambda_{n+1}$, then the fixed points are PICO$(n+1)$. If $\lambda_1 = \lambda_{n+1}$, then the fixed points are dPICO$(n+1)$.

Proof. Consider the case $\lambda_1 > \lambda_{n+1}$. Let $A^{[n]}, \mathbf{x}, B^{[n]}$ be a fixed point such that $x_1 \neq A$, $x_m \neq B$, and $\mathbf{x}$ contains no constant segments of length $N+1$. In view of the proof of Lemma 3.3, we can assume without loss of generality that $x_i > A$ for $i = 1, \ldots, N+1$. By the same lemma, the output of $\mathbf{w_1}$ does not include the boundary value A. Moreover, the subwindows used in the outputs of $\mathbf{w_2}, \ldots, \mathbf{w_n}$ do not contain A from the boundary conditions. This is shown by the fact that each of these windows contains the values $\mathbf{x'}_1, \ldots, \mathbf{x'}_{n+1}$ from $\mathbf{w_1}$ and at least one A from the boundary conditions. Any subwindow that contains a boundary A must also contain $\mathbf{x'}_1$. That subwindow's range is $D, \mathbf{x'}_1 - A > \mathbf{x'}_{n+1} - \mathbf{x'}_1$, and so clearly the subwindow cannot provide the minimum range.

Let $\mathbf{x'}$ be the ordered values for $\mathbf{x}$. We can write the outputs as convex combinations

$$x_i = \sum_{i-N}^{i+N} g_{ij} x_j, \quad i = 1, 2, \ldots, M,$$

where $x_j = A$ for $j < 1$ and $x_j = B$ for $j > m$. Since A is not included in the sums, $g_{ij} = 0$ for $j < 1$. Let I be the smallest index such that $x_I = \min \mathbf{x}$. Since x_I is the minimum for the entire signal $\mathbf{x}, x_I < x_j$ for $j < I$. Moreover, x_I is the minimum of the values used in calculating the output of window $\mathbf{w_I}$. By Corollary 3.2, $\mathbf{w_I}$ contains a constant segment of length $n+1$, a contradiction to our hypotheses.

Now consider the case for which $\lambda_1 = \lambda_{N+1} > 0$, and assume that $\mathbf{x}$ contains no constant segments of length N. Then by Lemma 3.3 there are two choices: (a) $x_1, x_2, \ldots, x_{N+1}$ form the minimum range in $\mathbf{w_1}$, in which case we have the same contradiction (a constant segment of length $N+1$ must exist in $\mathbf{x}$) or (b) the first window is of the form $(A^{[N]}, (A+B)/2, B^{[N]})$ since there is no sequence of length N in $\mathbf{x}$. Therefore we have that the fixed point is either PICO$(N+1)$ or dPICO$(N+1)$.

The case for $0 < \lambda_1 \leq \lambda_{N+1}$ is similar.

4 Linearity, Decomposition, and Filtering

Wavelets have been proposed as alternatives to signal decomposition [12]. One of the simpler examples of an orthonormal wavelet decomposition is the Haar basis which is based on scalings and translations of the function $h(x)$, which is 1 for the interval $0 \leq x < 1/2, -1$ for the interval $1/2 \leq x < 1$, and 0 otherwise. The dual to this basis is formed by translations of $g(x)$, where $g(x)$ is 1 on $0 \leq x < 1$ and 0 elsewhere. By suitable scalings, either forms an unconditional basis for $L^P(R)$, $1 < P < \infty$, and together they form a method of efficient multiresolution decomposition of a signal [17]. This paper deals with filtering on comparable decompositions of a sequence of real numbers, usually called a discrete signal [18]. The resolution will be at a certain level or resolution m, where m is an integer, as we now outline.

DEFINITION 4.1. For a given m we define B^{2^m} to be the set of sequences such that each sequence is piecewise constant with the length of each constant sequence a multiple of 2^m, and for each sequence in B^{2^m} the values at times $0, 1, \ldots, 2^m - 1$ are constant.

We will refer to these sequences as the PICO$'(2^m)$ sequences and note they are a subset of PICO(2^m) restricted in phase and pulse width, i.e., they must have a constant sequence of length a multiple of 2^m, starting at 0, and each point is embedded in a constant region that must have length a multiple of 2^m. A basis for this set of sequences is easily seen to be the set of all translations (by $k2^M, k$ an integer) of the sequence that is one on the interval $0 \le i < 2^M$ and 0 elsewhere. Any PICO$'(2^m)$ signal may obviously be written as a weighted sum of these basis elements, and there is an obvious one-to-one transformation from the set of all sequences to B^{2^m}. The following results are now easily shown.

LEMMA 4.1. A PICO(M) signal is a fixed point to WMMR filtering if the weights sum to unity and $M > N$.

LEMMA 4.2. For scalars a and α, if $\{x_i\}$ is PICO(M), $M \ge N + 1$, and the filter weights sum to α, then WMMR$\{a\{x_i\}\} = a\alpha\{x_i\}$, i.e., a PICO($M$) signal is an eigenvector to a WMMR filter with eigenvalue a if the filter weights sum to a.

LEMMA 4.3. If $M_1 > M_2$, then the sum of a PICO$'(2^{M_1})$ sequence and a PICO$'(2^{M_2})$ sequence is a PICO$'(2^{M_2})$ sequence.

LEMMA 4.4. Let $\{x_i\}$ and $\{y_i\}$ be PICO$'(2^{M_1})$ and (2^{M_2}) respectively, and let $2^{M_1}, 2^{M_2} \ge N + 1$. Then WMMR$\{x_i\}$ + WMMR$\{y_i\}$ = WMMR$\{x_i + y_i\}$.

THEOREM 4.1. The class of WMMR filters is linear over the class of PICO$'(2^M)$ signals if $2^M \ge N + 1$.

It is the existence of a suitable decomposition basis (complex exponentials) over which the linear time-invariant operator are linear, realistic filtering criterion, and suitable norms (L^P) that have made the development of the digital signal processing theory possible. Similar properties are now described for the WMMR filters on the Haar basis and PICO signals. The comparison technique is the ranked squared residual technique of section 3.

A commonality of the OS and WMMR filters is their stopband (impulse rejection), but an important difference in the filters is their passband (fixed points or eigenfunctions). Fixed points for the median and other OS filters are not suitable for orthogonal decomposition of signals [1], [19], but the WMMR filters have PICO signals for their fixed points and, as noted above, a subset of these, the PICO$'$ signals, or the orthogonal space with Haar basis, form a suitable basis for orthonormal decomposition of a signal. To show how one may develop a realistic criterion for filtering using WMMR filters one could consider passbands–stopbands based on pulsewidths, filtering of corrupted edges, and minimization under a ranked squared residual criterion.

First, how do we design lowpass, highpass, and bandpass filters for PICO signals? Here lowpass, highpass, and bandpass have nothing to do with the frequency domain; instead lowpass means the filter will pass the elements of the Haar basis (or PICO) signals that are of a certain width or wider, and similarly for highpass and bandpass. As shown above, a PICO(M) signal is a fixed point for a WMMR filter of window width $2N + 1$ if $M \ge N + 1$. It is easily shown that a burst of noise of length $K \le N$ is totally eliminated by using the linearity property if the phase restrictions hold. If the phase restrictions do not hold, then a corrupted edge results and Theorem 4.2 below states how the WMMR filters will restore the signal to the closest PICO($N+1$) signal. A highpass filter can be designed by simply designing the lowpass filter that passes all PICO components greater than the desired length and then subtracting the result from the original signal. This procedure is used in the biological example in section 5. A combination of the two methods would result in a bandpass filter.

In reconstructing a perfect edge from a nonperfect edge, it is natural to extend the constant regions by the number of intermediate points closest to them. If all the intermediate values lie halfway between the two regions, it would be only natural (a) in the case of an even number of values to extend each of the constant regions by half the number of values in the intermediate region or (b) in the case of an odd number of values, to leave at more one value halfway between the constant regions and to extend each constant region by half the number of the remaining intermediate values closer to that

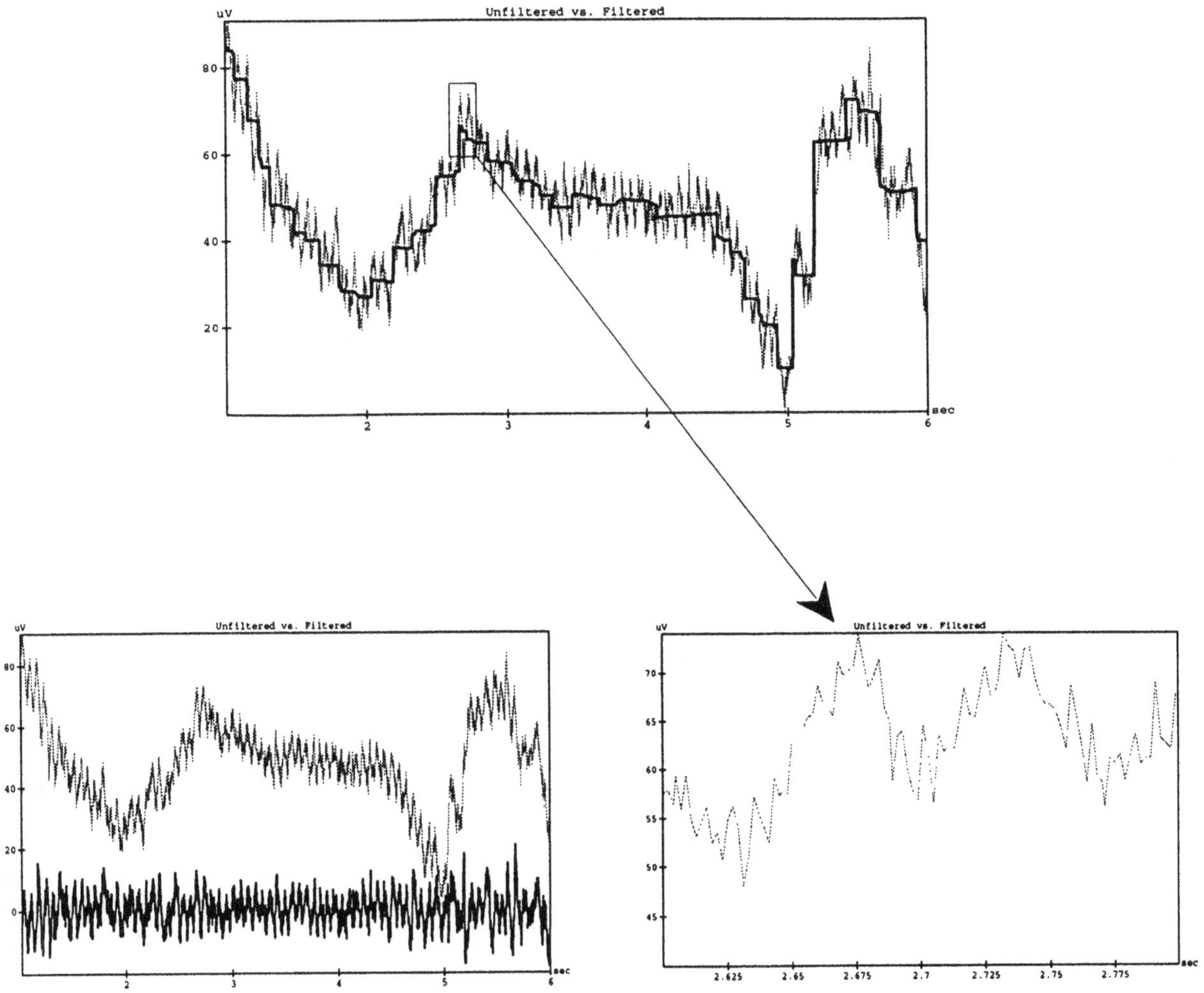

Fig. 2(a).

constant region. This is exactly what each filter in the class of symmetric normalized WMMR filters will do [8] if the selected $N+1$ values are ordered before weighting. If the edge is corrupted but the result is a LOMO signal, there is no need to order the selected values before weighting, and if the median weights for the WMMR filter are selected, the edge is reconstructed in one pass.

THEOREM 4.2. When one uses WMMR filters with weights that are normalized, nonnegative, and symmetric, then nonperfect edges asymptotically approach the closest perfect edge on iterative application of the filter.

5 Applications

The ability of the WMMR filter to extract the dc drift (assumed to be a PICO signal) from a biological signal is illustrated in figure 2. Then 2-dimensional applications are discussed.

The result of this paper concerning the WMMR are very similar to those of the median in the following two respects. The theoretical results are in one dimension, but many applications will be in two dimensions. In the

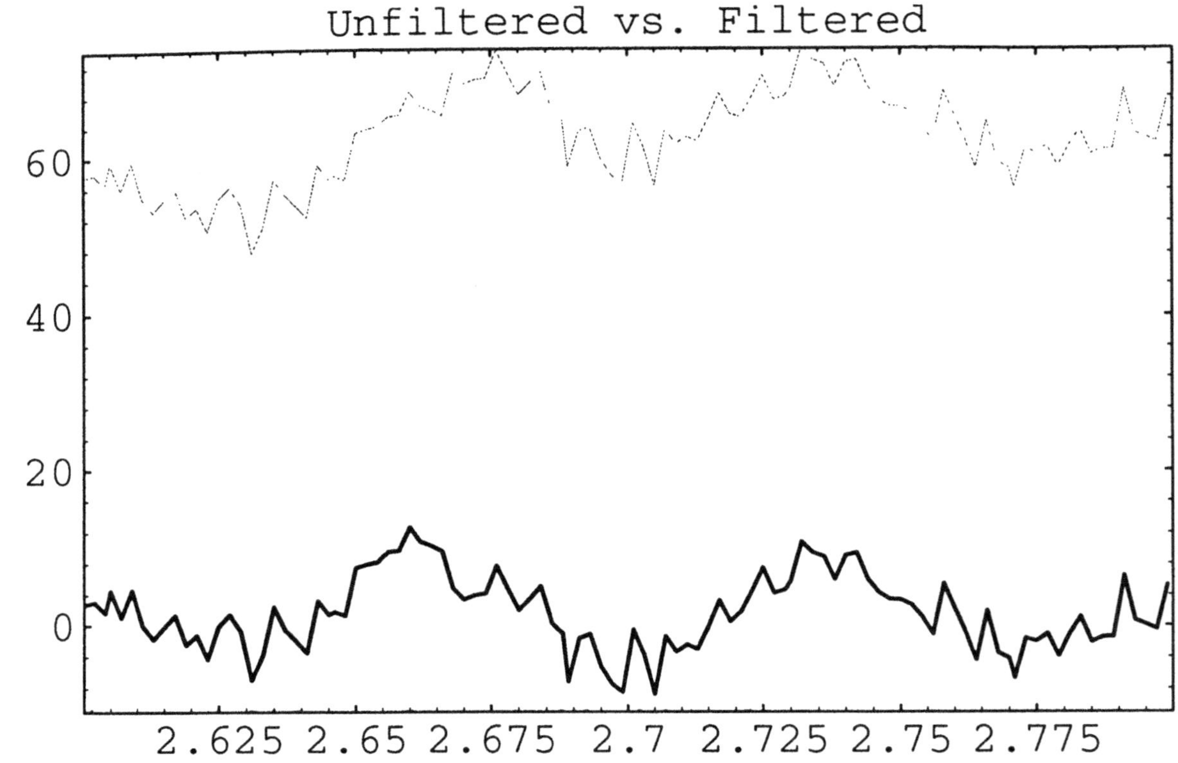

Fig. 2(b)

Fig. 2. Ability to WMMR filter to extract dc drift from a biological signal. (a) Top graph, raw visual evoked potential (VEP) signal (light curve) and PICO signal (bold curve) after appropriate WMMR filtering; lower left, the original biological signal and the resultant signal after the PICO signal is subtracted; two periods of the original signal. In the graph in the lower right the filter width is 72 (note that $N + 1 = 37$), and there are approximately 36 points in the cycle of the resultant signal. (b) Segments of the VEP signals before and after filtering. Note that the muscle artifact was eliminated without significant loss of detail. Approaches now being investigated preserve even more detail.

application of the filters to two dimensions the shape of the window must also be taken into consideration. In the following we discuss some results of applications of the WMMR filter to real data and synthetic data, and we discuss the design criteria used.

In an application of a 1-dimensional WMMR filter, Wang [20] used the LMS estimate in designing an algorithm for a robot wall follower. The problem was to design, a robot that would follow corridor walls, turn into the first corridor possible (following the wall around the corner), and ignore plants, doorways, water fountains, windows, and other obstacles. He compared the LMS estimate with several other robust estimates and found it to be preferable. In [21] Shelton and Longbotham applied the WMMR–MED to sinusoidal signals (which were shown to converge to rectangular pulses). They demonstrated that the process yielded a practical alternative for spatial localization of a frequency within one-half period in the presence of impulsive noise, independent and identically distributed (i.i.d.) noise, and baseline shifts. The process was applied to biological signals (the visual evoked potential) to demonstrate identification of a response within one period, whereas previous methods needed a minimum of 15 periods.

In two dimensions the WMMR–MED has been applied to x rays with a high degree of success. Glatt has applied the WMMR–MED to mammograms [22] for the selection of tumors based on size. The tumors were circular in nature. The median and WMMR filters with a hexagonal window (binary image hexagonally

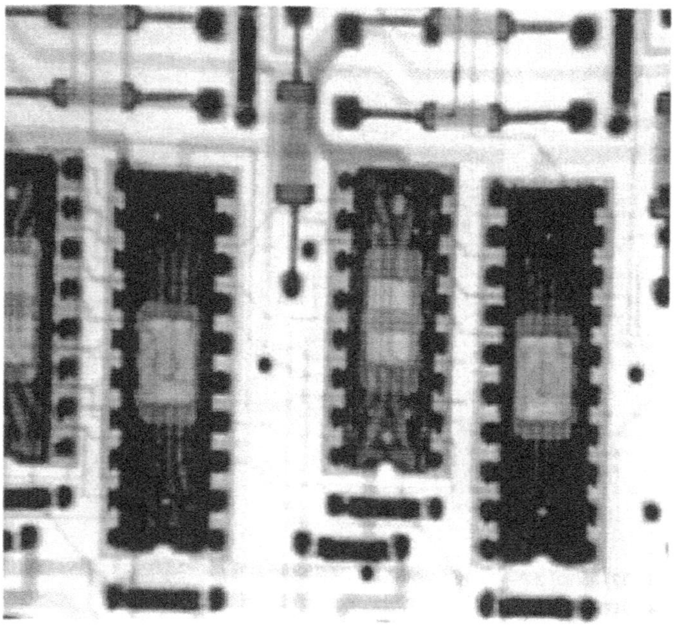

(a)

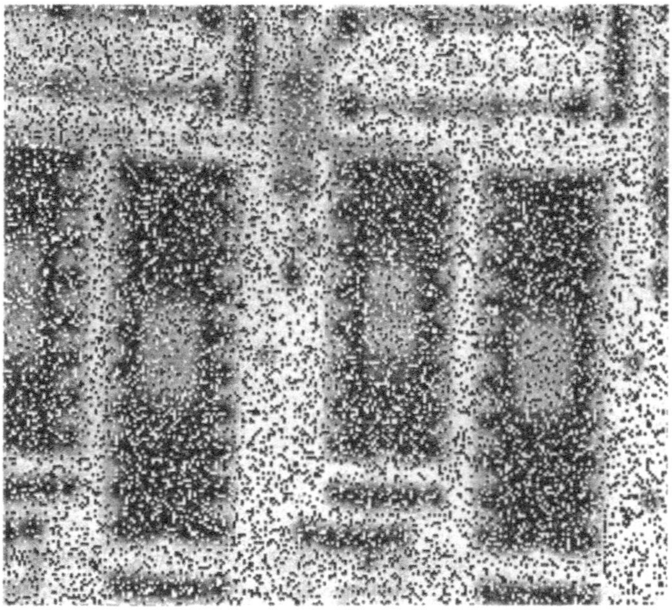

(b)

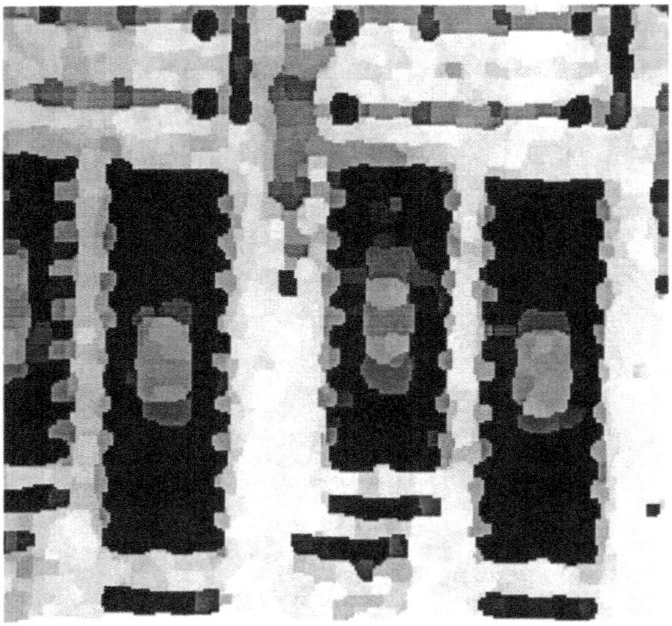

(c)

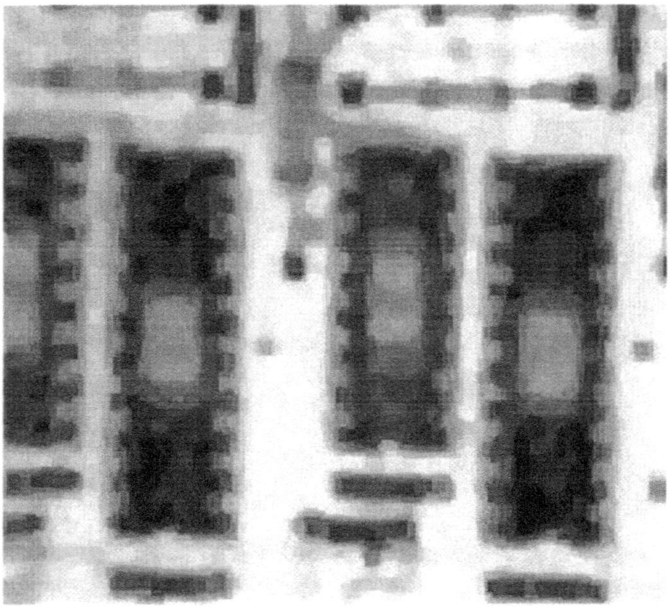

(d)

Fig. 3. Ability of the WMMR filter to enhance edges and smooth noise in an image: (a) original image; (b) original image with edge corruption by a 5 × 5 square averager, the addition of i.i.d. noice [$N(0, 15)$], and the addition of 20% impulsive noise (20% of the pixels set to white or black); (c) result of applying the WMMR filter of length 7 three times to the noisy image; (d) noisy image filtered three times by the 7 × 7 (square window) median filter. In (d), as is expected, smoothing occurs, but the edges are not enhanced. Dithering may be seen because of the use of a NeXT printer with only two bits of gray scale.

sampled) were shown to have fixed points that were circular in nature, and therefore the hexagonal window was applied. In the application to dental x rays [23] the goal was to teach artificial neural networks (ANN's) to recognize dental decay from digitized x rays. The results show that an ANN preceded by a WMMR–MED filter (with a square window) results in a 20% improvement in the classification of caries over that of a trained dentist. Other filters were also applied as precursors to ANN training but with less improvement.

In figure 3 the WMMR is applied to noisy x-ray images of printed circuit boards for enhancement of integrated-circuit boundaries. In this case corner preservation is important. It is easily shown that the median or WMMR filter with the cross window will preserve corners but will not enhance edges perpendicular to the direction of travel. This fact is also apparent in table 1, where the WMMR–MED and WMMR–AVE, when implemented with a cross window, are shown to lower the error of the edge-corrupted image only initially (because of corner correction) in comparison with the 1-dimensional WMMR that perfectly enhances edges. In general, the 1-dimensional WMMR–MED is recommended if the primary concern is to enhance edges and to smooth. In figure 3 we assumed that the boundary was white for filtering. The order of the noise application was edge corruption, addition of i.i.d. $[N(0, 15)]$, and impulsive noise (created by setting a pixel to one extreme if it is in the other half of the gray-scale values).

Acknowledgments

We thank Jerome Keating, Statistics Department, the University of Texas at San Antonio, for introducing us to the LMS filter through reference [20]. We also thank the U.S. Air Force for sponsoring Dr. Longbotham's research in part through contract AFOSR-89-0490, and we thank the National Science Foundation for sponsoring Dr. Eberly's research under contract DMS-9003037. The U.S. Government is authorized to reproduce reprints for government purposes not withstanding any copyright notice herein.

References

1. N.C. Gallagher and G.L. Wise, "A theoretical analysis of the properties of the median filter," *IEEE Trans. Acoust., Speech, Signal Process.*, vol. ASSP-29, pp. 1136–1141, 1981.
2. S.G. Tyan, "Median filtering: deterministic properties," in *Two Dimensional Signal Processing: Transforms and Median Filters*, T.S. Huang, ed., Springer-Verlag: New York, 1981.
3. A.C. Bovik, T.S. Huang, and D.C. Munson, "A generalization of median filtering using linear combinations of order statistics, *IEEE Trans. Acoust., Speech, Signal Process.*, vol. ASSP-31, pp. 1342–1350, 1983.
4. H.G. Longbotham and A.C. Bovik, "Theory of order statistic filters and their relationship to FIR filters," *IEEE Trans. Acoust., Speech, Signal Process.*, vol. ASSP-37, pp. 275–287, 1989.
5. J.D. Bednar and T.L. Watt, "Alpha-trimmed means and their relationship to median filters," *IEEE Trans. Acoust., Speech, Signal Process.*, vol. ASSP-32, pp. 145–153, 1984.
6. P.J. Rousseeuw, "Least median of squares regression," *J. Am. Statist. Assoc.*, vol. 79, pp. 871–880, 1984.
7. J.W. Tukey, *Exploratory Data Analysis,* Addison-Wesley: Reading, MA, 1971.
8. H. Longbotham and D. Eberly, "The WMMR filters: a class of robust edge enhancers," *IEEE Trans. Acoust., Speech, Signal Process.*, to be published.
9. H. Longbotham and N. Barsalou, "A class of optimal generalized order statistic filters that are edge enhancers," in *Nonlinear Image Processing*, E.J. Delp. ed., Proc. Soc. Photo-Opt. Instrum. Eng., vol. 1247, 1991.
10. D. Donoho and P.J. Huber, "The notion of breakdown point," in *A Festschrift for Erich L. Lehman*, P.J. Bickel, K. Doksum, and J.L. Hodges, eds., pp. 157–184, Wadsworth: Belmont, CA, 1983.
11. F.R. Hampel, "The breakdown points of the mean combined with some rejection rules," *Technometrics,* vol. 27, pp. 95–107, 1985.
12. I. Daubechies, "Orthonormal bases of compactly supported wavelets," *Commun. Pure Appl. Math.*, vol. 41, pp. 909–996, 1988.
13. P.J. Rousseeuw and A.M. Leroy, *Robust Regression and Outlier Detection*, John Wiley: New York, 1987.
14. V.I. Istratescu, *Fixed Point Theory,* Reidel: Dordrecht, The Netherlands, 1981.
15. S. Karamardian, *Fixed Points: Algorithms and Applications,* Academic: San Diego, CA, 1977.
16. F.F. Bonsall, *Lectures on Some Fixed Point Theorems of Functional Analysis*, Tata Institute: Bombay, 1962.
17. S.G. Mallet, "Multifrequency channel decompositions of images and wavelet models," *IEEE Trans. Acoust., Speech, Signal Process.*, vol. ASSP-37, pp. 2091–2110, 1989.
18. A.V. Oppenheim and R.S. Schafer, *Disrete-Time Signal Processing*, Prentice-Hall: New York, 1989.

19. D. Eberly and H. Longbotham, "Fixed points of order statistic filters," *IEEE Trans. Acoust., Speech, Signal Process.*, to be published.
20. C.M. Wang, "A robust estimator for wall following," *Commun. Statist. Theory Methods*, vol. 17, pp. 411–422, 1988.
21. D. Shelton and H.G. Longbotham, "Robust time domain frequency analysis," in *Nonlinear Signal Processing III*, J. Astola, C.G. Boncelet, and E.R. Dougherty, eds., Proc. Soc. Photo-Opt. Instrum. Eng., vol. 1658, 1992.
22. L.A. Glatt, H.G. Longbotham, T.L. Arnow, D.J. Shelton, and P. Ravdin, "An application of WMRR filters to detection and sizing of tumors in mammograms," in *Medical Imaging VI: Image Processing*, ed., Proc. Soc. Photo-Opt. Instrum. Eng., vol. 1652, 1992.
23. B. Dove, D. McDavid, L. Gay, and H.G. Longbotham, "Applications of neural networks for the detection of proximal caries," *J. Dent. Res.*, vol. 71, p. 241, 1992.
24. Y.H. Lee and T.F. Adly, "An edge gradient enhacing adaptive order statistic filter," *IEEE Trans. Acoust., Speech, Signal Process.*, vol. ASSP-35, pp. 680–695, 1987.
25. Y.H. Lee and T.F. Adly, "A study of convex/concave edges and edge enhancing operators based on the Laplacian," *IEEE Trans. Circuits Syst.*, vol. 37, pp. 940–946.
26. H.G. Longbotham, A.C. Bovik, and A.P. Restrepo, "Generalized order statistic filters," in *Proc. IEEE Internat. Conf. on Acoust., Speech, Signal Process.*, Glasgow, Scotland, 1989.
27. D. Eberly, H.G. Longbotham, and J. Aragon, "Complete classification of roots to 1-dimensional median and rank-order filters," *IEEE Trans. Acoust., Speech, Signal Process.*, vol. ASSP-39, 1991.
28. P.D. Wendt, E.J. Coyle, and N.C. Gallagher, "Stack filers," *IEEE Trans. Acoust., Speech, Signal Process.*, vol. ASSP-34, pp. 898–911, 1986.
29. C.M. Wang, "A robust estimator for wall following," *Commun. Statist. Theory Methods*, vol. 17, pp. 411–422, 1988.
30. D. Regan, *Evoked Potentials in Psychology, Sensory Physiology and Clinical Medicine*, Chapman and Hall: London, 1972.

Harold Gene Longbotham received the B.S. degree in mathematics and physics from Stephen F. Austin University, Nacadoches, Texas, in 1968, the M.S. degree in physics from the University of New Mexico in 1971, the M.S. degree in statistics from the University of Texas at San Antonio in 1983, and the M.S. and Ph.D. degrees in electrical engineering from the University of Texas at Austin in 1984 and 1987, respectively.

From 1968 and 1970 he was at the Houston Manned Spacecraft Center working for Lockheed Electronics Company in the area of Fourier-transform holography. From 1971 to 1983 he taught at several institutions, including the Pine Ridge Indian Reservation, the University of New Mexico, Texas Lutheran College, and the University of Texas at San Antonio, where he is currently an assistant professor. His current research interests include digital signal processing, digital-filter design with emphasis on nonlinear filters, and digital control theory.

Dr. Longbotham was elected to *Who's Who in American Colleges and Universities* and is a member of Alpha Chi and Sigma Phi Sigma.

David Eberly received the B.A. degree in mathematics from Bloomsburg University in Bloomsburg, Pennsylvania, in 1979, the M.S. degree in applied mathematics from the University of Colorado in 1981, and the Ph.D. degree in mathematics from the University of Colorado in 1984. From 1984 to 1992 he was a faculty member in the Department of Mathematics, Computer Science, and Statistics at the University of Texas at San Antonio. From 1990 to 1992 he was an adjunct faculty member in the Department of Radiology at the University of Texas Health Science Center. He is currently a graduate student in the Department of Computer Science at the University of North Carolina, Chapel Hill.

Journal of Mathematical Imaging and Vision 2, 117–135 (1992).

Asymptotic Behavior of Morphological Filters

LASSE KOSKINEN AND JAAKKO ASTOLA
Department of Mathematical Sciences, Tampere University, Tampere, Finland

Abstract. The connection between morphological and stack filters is used in the analysis of the statistical properties of morphological filters. Closed-form expressions for the output distributions of morphological filters are given, and their statistical symmetry properties are analyzed. Asytotically tight bounds on the expectations of two-dimensional morphological filters, and asymptotic formulas for the variances of one-dimensional morphological filters are derived. These results form the basis for analyzing general asymptotic properties of morphological filters.

Key words. morphological filters, stack filters, statistical properties, asymptotic analysis

1 Introduction

In most image-processing applications the performance of the filter depends on how well it can suppress the noise and retain the desired information. Obtaining quantitative information on how much the filters reduce noise and how biased they are requires statistical analysis. Since morphological filters are nonlinear and are based on geometrical concepts, standard statistical methods cannot be directly applied to their analysis. Thus it is not surprising that the statistical theory of morphological filters is still far from mature. Certain statistical properties have been analyzed in Stevenson and Arce [1], and the restoration and representation properties of noisy images with use of morphological filters have been studied, e.g., in Schonfeld and Goutsias [2]. The connection between stack and morphological filters, studied by Maragos and Schafer [3] and Koskinen et al. [4], [5], can be used in the derivation of the output-distribution formulas for morphological filters, and the purpose of this paper is to apply the stack-filter method both to the derivation of output distributions and to the asymptotic analysis of statistical properties of morphological filters. The asymptotic analysis is important because it shows us the general behavior of morphological filters.

This paper is organized as follows. In section 2 basic definitions and some basic properties of morphological and stack filters are given. To obtain an adequate understanding of the filtering of noisy signals, it is desirable to determine the output distribution of the filter in terms of the input distributions. In section 3 formulas for the output distributions of morphological filters are derived for the case of independent inputs. These formulas apply to any input distribution and can be used when the structuring set is small to moderately large. Certain symmetry properties stemming from the duality properties of morphological filters are also presented. In section 4 the asymptotic behavior of dilation and erosion is studied, and in section 5 bounds (that are asymptotically tight for Laplace distributions) on the expectations of a large class of two-dimensional morphological filters are derived. In section 6 analytical formulas for output distributions are derived for the case of one-dimensional morphological filters. The simulation results in Rustanius et al. [6] indicate that these formulas approximate the two-dimensional case well. The bias and the noise attenuation of the filter depend on the noise distribution. The analytical formulas of output distributions make it possible to derive asymptotic expressions for the output expectations and variances for uniform (short-tailed) and Laplace (heavy-tailed) input distributions.

2 Basic Definitions

In this section we recall the definitions of morphological filters that process discrete signals by sets (see, e.g., Dougherty and Ciardiana [7], Serra [8], Maragos and Schafer [3], [9], Matheron [10], and Chu and Del. [11] (one-dimensional application)) and the definitions of stack filters (see, e.g., Wendt et al. [12] and Yli-Harja et al. [13]).

In this paper the set of natural numbers is denoted by **N**, the set of integers by **Z**, and the set of real numbers by **R**.

The *structuring set* B is a finite subset of $\mathbf{Z}^{(m \in \mathbf{N})}$. If B is one dimensional, *i.e.*, if $B \subset \mathbf{Z}$, and if for all $k, n \in \mathbf{Z}(k > 0)$ the property $n, n+k \in B$ implies that $\{n+1, n+2, \dots, n+k-1\} \subseteq B$, then the structuring set B is called *convex*. The *symmetric set* B^s of B is defined by $B^s = \{-x : x \in B\}$, the *translated set* B_x, where the set B is translated by $x \in \mathbf{Z}^m$, is defined by $B_x = \{x+y : y \in B\}$, and the *Minkowski sum* of structuring sets, $A, B \subset \mathbf{Z}^m$, is defined by $A + B = \{x_1 + x_2 : x_1 \in A, x_2 \in B\}$.

The operations dilation, erosion, closing, and opening by $B, B \subset \mathbf{Z}^m$, transform a signal $f, f : \mathbf{Z}^m \to \mathbf{R}$, to another signal by the following rules: The *dilation* of f by B is denoted by $f \oplus B^s$ and is defined by

$$(f \oplus B^s)(x) = \max_{y \in B_x}\{f(y)\}, \quad x \in \mathbf{Z}^m.$$

The *erosion* of f by B is denoted by $f \ominus B^s$ and is defined by

$$(f \ominus B^s)(x) = \min_{y \in B^s_x}\{f(y)\}, \quad x \in \mathbf{Z}^m.$$

The *closing* of f by B is denoted by f^B and is defined by

$$f^B(x) = [(f \oplus B^s) \ominus B](x), \quad x \in \mathbf{Z}^m.$$

The *opening* of f by B is denoted by f_B and is defined by

$$f_B(x) = [(f \ominus B^s) \oplus B](x), \quad x \in \mathbf{Z}^m.$$

In the same way that closing and opening were defined as dilation followed by erosion and erosion followed by dilation, clos-opening by the structuring set B is defined as closing by B followed by opening by B, and open-closing by B is defined as opening by B followed by closing by B. The *clos-opening* of f by B is denoted by $(f^B)_B$ and is defined by

$$(f^B)_B(x) = [(((f \oplus B^s) \ominus B) \ominus B^s) \oplus B](x), \quad x \in \mathbf{Z}^m.$$

The *open-closing* of f by B is denoted by $(f_B)^B$ and is defined by

$$(f_B)^B(x) = [(((f \ominus B^s) \oplus B) \oplus B^s) \ominus B](x), \quad x \in \mathbf{Z}^m.$$

In Boolean expressions we use $x \wedge y$ for "x AND y," $x \vee y$ for "x OR y," and $\overline{x}$ for "NOT x," where x and y are Boolean variables. In some formulas binary values are to be understood as being real 1's and 0's. The number of elements in a finite set A is denoted by $|A|$.

The relation "$\geq$" of binary vectors $\underline{x} = (x_1, x_2, \cdots, x_n)$ and $\underline{y} = (y_1, y_2, \cdots, y_n)$ is defined as $\underline{x} \geq \underline{y}$ if and only if $x_i \geq y_i$ for all $i \in \{1, 2, \cdots, n\}$. Because this relation is reflexive, antisymmetric, and transitive, it defines a partial ordering on the set of binary vectors. This order property is known as the *stacking property*, and it is said that $\underline{x}$ and $\underline{y}$ stack if $\underline{x} \geq \underline{y}$. There is also a natural ordering on Boolean functions. Let f and g be two Boolean functions. We write $f \geq g$ if and only if $f(\underline{x}) \geq g(\underline{x})$ for all $\underline{x}$. A Boolean function f is said to be *increasing* if the relation $\underline{x} \geq \underline{y}$ implies the relation $f(\underline{x}) \geq f(\underline{y})$. Filters that are defined by increasing Boolean functions are called *stack filters*. The Boolean function g_D is a *dual* of g if and only if $g_D(\underline{x}) = \overline{g(\overline{\underline{x}})}$ for all $\underline{x}$.

Let A be a finite subset of $\mathbf{Z}^m$. Then the *Boolean function g indexed by the set A* is a Boolean expression of variables $z_a, a \in A$, denoted by $g(\underline{z}) = g(z_a : a \in A)$.

Let $u(t)$ denote the real *unit-step function*

$$u(t) = \begin{cases} 1 & \text{if } t \geq 0, \\ 0 & \text{otherwise.} \end{cases}$$

Then we use a Boolean function with its variables indexed by a set to define the corresponding continuous stack filter in the following way. Let A be a finite subset of $\mathbf{Z}^m$, let $g(\underline{z})$ be an increasing Boolean function indexed by A, and let $f : \mathbf{Z}^m \to \mathbf{R}$ be a signal. Then the *continuous stack filter* S corresponding to $g(\underline{z})$ is defined by

$$S(f)(x) = \max\{t \in \mathbf{R}| \\ g(u(f(x+a)-t) : a \in A) = 1\}.$$

3 General Statistical Properties

The objective of this section is to derive output distributions for morphological filters and to study certain symmetry properties of dual filters. Knowledge of the properties discussed in this section is important when we apply morphological filters to noisy signals since it will give us an idea of the noise-suppression capability and the biasing effects of the filters.

An attractive property of stack filters is that it is possible to derive analytical results for their statistical properties. For example, the output distribution of a continuous stack filter can be expressed by using the following proposition (Yli-Harja et al [13]).

PROPOSITION 1. Let the input values $X_b (b \in B)$ in the window B of a stack filter S be independent random variables having distribution functions $F_b(t)$, respectively. Then the output distribution function $G(t)$ of the stack filter S is

$$G(t) = \sum_{\underline{z} \in g^{-1}(0)} \prod_{b \in B} (1 - F_b(t))^{z_b} F_b(t)^{1-z_b}, \quad (1)$$

where $g(\underline{z})$ is an increasing Boolean function that corresponds to the stack filter S and $g^{-1}(0) = \{\underline{z} | g(\underline{z}) = 0\}$.

Example. The Boolean function $g(\underline{z}) = x_{-1}x_0 + x_{-1}x_1 + x_0x_1$ corresponds to a median filter whose window is $B = \{-1, 0, 1\}$. If the input at -1 is a random variable having a distribution function $F_1(t)$ and inputs at 0 and 1 are random variables having a common distribution function $F_2(t)$, then by Proposition 1 the output distribution $G_m(t)$ of the median filter is $G_m(t) = F_1(t)F_2(t)^2 + (1 - F_1(t))F_2(t)^2 + 2F_1(t)(1 - F_2(t))F_2(t)$.

In the following we derive the output distributions of morphological filters when the values of the input signal are independent random variables. The following simple proposition gives explicit stack-filter expressions for morphological filters (Koskinen *et al.* [4], [14]). These expressions make it possible to calculate the output distributions of morphological filters by using Proposition 1.

PROPOSITION 2. Let B be a structuring set. Then the positive Boolean function that corresponds to stack-filter expression of

(a) dilation by B is $g_d(\underline{z}) = \vee_{b \in B^{z_b}}$,
(b) erosion by B is $g_e(\underline{z}) = \wedge_{b \in B^{z_b}}$,
(c) closing by B is $g_c(\underline{z}) = \wedge_{a \in B^s} (\vee_{b \in B_a z_b})$,
(d) opening by B is $g_o(\underline{z}) = \vee_{a \in B^s} (\wedge_{b \in B_a z_b})$,
(e) clos-opening by B is $g_{co}(\underline{z}) = \vee_{a \in B^s} (\wedge_{b \in (B+B^s)_a} (\vee_{c \in B_b z_c}))$,
(f) open-closing by B is $g_{oc}(\underline{z}) = \wedge_{a \in B^s} (\vee_{b \in (B+B^s)_a} (\wedge_{c \in B_b z_c}))$.

Proof. We will prove only case (e); the other cases can be proved in a similar way. Let f denote the signal to be clos-opened. We obtain the equations

$$\begin{aligned}(f^B)_B(x) &= [(((f \oplus B^s) \ominus B) \ominus B^s) \oplus B](x) \\ &= \max_{a \in B^s}(\min_{b \in B_a}(\min_{c \in B_b^s}(\max_{d \in B_c} f(x+d)))) \\ &= \max_{a \in B^s}(\min_{b \in (B+B^s)_a}(\max_{c \in B_b} f(x+c))) \\ &= \max\{t \in \mathbf{R}| \vee_{a \in B^s} (\wedge_{b \in (B+B^s)_a} \\ &\quad (\vee_{c \in B_b} u(f(x+c)-t))) = 1\}. \quad (2)\end{aligned}$$

So, by the definition of the continuous stack filter, the Boolean function that corresponds to clos-opening by B is $g_{co}(\underline{z}) = \vee_{a \in B^s} (\wedge_{b \in (B+B^s)_a} (\vee_{c \in B_b} z_c))$.

Because dilations and erosions are, in fact, local extremes, their statistical properties are extensively studied in the theory of order statistics; see e.g., Castillo [15] and Galambos [16]. Proposition 3 is a simple result of order statistics and is here formulated in terms of morphology

(Koskinen et al. [4, 14]). It shows that the statistical properties of dilation and erosion can be analyzed easily and do not depend on the shape of the structuring set.

PROPOSITION 3. Consider a discrete signal f and dilation and erosion by structuring set B of size n at point x_0. Let the values $f(x_0 + b)(b \in B)$ be independent random variables having the distribution functions $F_1(t), F_2(t), \cdots, F_n(t)$. Then the distribution function G_d of the value $(f \oplus B^s)(x_0)$ is

$$G_d(t) = \prod_{i=1}^{n} F_i(t), \tag{3}$$

and the distribution function G_e of the value $(f \ominus B^s)(x_0)$ is

$$G_e(t) = 1 - \prod_{i=1}^{n}(1 - F_i(t)). \tag{4}$$

Proof. Let the structuring set of the dilation and the erosion be B, where the size of B is n. Then by Proposition 2 the positive Boolean function that corresponds to the stack-filter expression is $g(\underline{z}) = \vee_{b \in B} z_b$ for dilation and is $p(\underline{z}) = \wedge_{b \in B} z_b$ for erosion. Since $g^{-1}(0) = \{\underline{0}\}$, Proposition 1 implies that

$$G_d(t) = \sum_{\underline{z}=\underline{0}} \prod_{i=1}^{n} (1 - F_i(t))^{z_i} F_i(t)^{1-z_i}$$
$$= \prod_{i=1}^{n} F_i(t). \tag{5}$$

Similarly, since $p^{-1}(0) = \{0,1\}^n - \{\underline{1}\}$ (all components of 1's are 1's), by using Proposition 1 we obtain the result

$$G_e(t) = \sum_{\underline{z} \neq \underline{1}} \prod_{i=1}^{n} (1 - F_i(t))^{z_i} F_i(t)^{1-z_i}$$
$$= 1 - \prod_{i=1}^{n}(1 - F_i(t)). \tag{6}$$

Propositions 4 and 5 (Koskinen et al. [4, 14]) are direct consequences of Propositions 1 and 2.

PROPOSITION 4. Consider a discrete signal f and the closing of f by a structuring set B at point x_0. Let the values $f(x_0 + b)(b \in B + B^s)$ be independent random variables having the distribution functions $F_b(t)$, respectively. Then the distribution function $G_c(t)$ of the value $f^B(x_0)$ of the closed signal is

$$G_c(t) = \sum_{\underline{z} \in g^{-1}(0)} \prod_{b \in B + B^s} (1 - F_b(t))^{z_b} F_b(t)^{1-z_b}, \tag{7}$$

where $g(\underline{z}) = \wedge_{a \in B^s}(\vee_{b \in B_a} z_b)$.

PROPOSITION 5. Consider a discrete signal f and the clos-opening of f by a structuring set B at point x_0. Let the values $f(x_0 + b)(b \in (B + B^s) + (B + B^s))$ be independent random variables having the distribution functions $F_b(t)$, respectively. Then the distribution function $G_{co}(t)$ of the value $(f^B)_B(x_0)$ of the clos-opened signal is

$$G_{co}(t) = \sum_{\underline{z} \in g^{-1}(0)} \prod_{b \in (B+B^s)+(B+B^s)} (1 - F_b(t))^{z_b} F_b(t)^{1-z_b}, \tag{8}$$

where $g(\underline{z}) = \vee_{a \in B^s}(\wedge_{b \in (B+B^s)_a}(\vee_{c \in B_b} z_c))$.

Later we will have to know exactly what is meant by the term "same shape." The mathematical definition is given by the following equivalence relation.

DEFINITION Let A and B be structuring sets. We say that A and B are *congruent* if A can be transformed to B by applying translations, reflections, and rotations by $\pi/2$. If A and B are congruent, then we write $A \sim B$.

If $A \sim B$, then Proposition 4 shows that closing by A and closing by B have the same statistical behavior in the case of independently and identically distributed inputs since the same Boolean function corresponds to the stack-filter expressions of closing by A and closing by B. Obviously, this conclusion is also valid for clos-opening.

Propositions 4 and 5 offer a straightforward method for calculating output distributions for closing and clos-opening. On the other hand, if we know the distribution function of closing by B or clos-opening by B, Proposition 6 (Koskinen et al. [5]) gives us an easy way to find the output distributions of opening by B or open-closing by B. The reason for this is that the Boolean functions corresponding to the stack-filter expressions of opening and open-closing are the duals of those that correspond to closing and clos-opening.

PROPOSITION 6. Let g be a positive Boolean function, let g_D be the dual of g, let S be the stack filter defined by g, and let S_D be the stack filter defined by g_D. Consider the filtering of a discrete signal f by the stack filter S at point x_0 where the values in the moving window $B(|B| = n)$ of the stack filter S are independent random variables having the distribution functions $F_1(t), F_2(t), \cdots, F_n(t)$. If the distribution function of the value $S(f)(x_0)$ is $G(F_1(t), F_2(t), \cdots, F_n(t))$, then the distribution function of the value $S_D(f)(x_0)$ is

$$G_D(F_1(t), F_2(t), \cdots, F_n(t)) = 1 - G((1 - F_1(t)), (1 - F_2(t)), \cdots, (1 - F_n(t))). \quad (9)$$

Proof. Because the filtering by S corresponds to the positive Boolean function $g(\underline{z})$, Proposition 1 implies that the distribution function of the value $S(f)(x_0)$ is

$$G(F_1(t), F_2(t), \cdots, F_n(t)) = \sum_{\underline{z} \in G^{-1}(0)} \prod_{i=1}^{n} (1 - F_i(t))^{z_i} F(t)^{1-z_i}, \quad (10)$$

where n is the size of the moving window A that corresponds to the stack filter S. Since g_D is the dual of g, then $g_D(\underline{z}) = 0$ if and only if $g(\overline{\underline{z}}) = 1$. Thus Proposition 1 implies that

$$G_D(F_1(t), F_2(t), \cdots, F_n(t)) = \sum_{\underline{z} \in g_D^{-1}(0)} \prod_{i=1}^{n} (1 - F_i(t))^{z_i} F_i(t)^{1-z_i}$$
$$= 1 - \sum_{\underline{z} \in g^{-1}(0)} \prod_{i=1}^{n} (1 - F_i(t))^{1-z_i} F_i(t)^{z_i}$$
$$= 1 - G((1 - F_1(t)), (1 - F_2(t)), \cdots, (1 - F_n(t))). \quad (11)$$

Example. Consider an image f where the gray-level values $f(i,j)$ of the pixels are independent random variables having a common distribution function $F(t)$. Then by Proposition 4 the distribution function $G_c(t)$ of the pixel values after closing by $B = \{(0,0), (0,1), (1,0)\}$ is $G_c(t) = F(t)^7 + 6(1 - F(t))F(t)^6 + 15(1 - F(t))^2F(t)^5 + 12(1 - F(t))^3F(t)^4 + 3(1 - F(t))^4F(t)^3$. In addition, by Proposition 6, after opening by B the distribution function is $G_o(t) = 1 - (1 - F(t))^7 - 6(1 - F(t))^6F(t) - 15(1 - F(t))^5F(t)^2 - 12(1 - F(t))^4F(t)^3 - 3(1 - F(t))^3F(t)^4$.

Proposition 2 shows that the basic morphological filters that are duals (in the morphological sense) of each other are duals also in a stack-filter sense. This implies the following statistical symmetry properties (Koskinen et al. [5]).

PROPOSITION 7. Consider the filtering of a discrete signal f whose values are independent, identically and symmetrically distributed random variables having a common distribution function $F(t)$ and the expectation μ. Let g be a positive Boolean function, let g_D be the dual of g, let S be the stack filter defined by g, and let S_D be the stack filter defined by g_D. If the expectation of the values of f after filtering by S is $E\{S(f)\} = \mu + \xi$, then the expectation of the values of f after filtering by S_D is $E\{S_D(f)\} = \mu - \xi$. Moreover, the output variances of S and S_D are equal.

Proof. Proposition 1 implies that the distribution function of the values of the signal after filtering by S is ($w(z)$ is the Hamming weight of z)

$$G(F(t)) = \sum_{\underline{z} \in g^{-1}(0)} (1 - F(t))^{w(\underline{z})} F(t)^{n-w(\underline{z})} \quad (12)$$

and that the distribution function of the values of the signal after filtering by S_D is

$$G_D(F(t)) = \sum_{\underline{z} \in g_D^{-1}(0)} (1-F(t))^{w(\underline{z})} F(t)^{n-w(\underline{z})}. \quad (13)$$

Now, Proposition 6 implies that

$$\sum_{\underline{z}\in g^{-1}(0)} (1-F(t))^{w(\underline{z})}F(t)^{n-w(\underline{z})} = 1 - \sum_{\underline{z}\in g_D^{-1}(0)} (1-F(t))^{n-w(\underline{z})}F(t)^{w(\underline{z})}. \quad (14)$$

Since $F(t)$ is symmetric, $F(\mu+\xi) = 1-F(\mu-\xi)$, implying

$$\begin{aligned} G(F(\mu+\xi)) &= \sum_{\underline{z}\in g^{-1}(0)} (1-F(\mu+\xi))^{w(\underline{z})} F(\mu+\xi)^{n-w(\underline{z})} \\ &= 1 - \sum_{\underline{z}\in g_D^{-1}(0)} (1-F(\mu+\xi))^{n-w(\underline{z})} F(\mu+\xi)^{w(\underline{z})} \\ &= 1 - \sum_{\underline{z}\in g_D^{-1}(0)} (1-F(\mu-\xi))^{w(\underline{z})} F(\mu-\xi)^{n-w(\underline{z})} \\ &= 1 - G_D(F(\mu-\xi)). \end{aligned} \quad (15)$$

Thus the output distributions of S and S_D are mirror images with respect to μ. This proves Proposition 7.

COROLLARY 7.1. Consider the filtering of a discrete signal f whose values are independent, identically and symmetrically distributed random variables having the expectation μ. Let the expectation of the output after dilation by B be $E\{f \oplus B\} = \mu + \xi_1$, let the expectation of the output after closing by B be $E\{f^B\} = \mu + \xi_2$, and let the expectation of the output after clos-opening by B be $E\{(F^B)_B\} = \mu + \xi_3$. Then

(a) after erosion by B the expectation is $E\{f \ominus B\} = \mu - \xi_1$,
(b) after opening by B the expectation is $E\{f_B\} = \mu - \xi_2$,
(c) after open-closing by B the expectation is $E\{(f_B)^B\} = \mu - \xi_3$.

COROLLARY 7.2. Consider the filtering of a discrete signal f whose values are independent, identically and symmetrically distributed random variables. Then

(a) the variances of the dilated signal $f \oplus B$ and the eroded signal $f \ominus B$ are equal;
(b) the variances of the closed signal f^B and the opened signal f_B are equal;
(c) the variances of the clos-opened signal $(f^B)_B$ and the open-closed signal $(f_B)^B$ are equal.

4 Asymptotic Behavior of Dilation and Erosion

In this section we analyze the asymptotic behavior of dilation and erosion in the cases of uniform and Laplace distributions. The results that we derive here are important because they are useful in the analysis of compound morphological operations. First, we recall one concept. Let $f : \mathbf{N} \to \mathbf{R}$ and $g : \mathbf{N} \to \mathbf{R}$ be two functions. Then f is *asymptotically dominated* by g if and only if there exist real numbers A and n_0 such that $|f(n)| \le A|g(n)|$ for all $n > n_0$. If f is asymptotically dominated by g, we write $f(n) = O(g(n))$. The following proposition is a direct consequence of Proposition 3.

PROPOSITION 8. Consider the filtering of a discrete signal f whose values are independent and identically uniformly distributed on [0, 1]. Let B be a structuring set of size n. Then after dilation by B, for the expectation

$$E\{f \oplus B\} = 1 - \frac{1}{n+1} \quad (16)$$

and for the variance

$$V\{f \oplus B\} = \frac{1}{n^2} + O\left(\frac{1}{n^3}\right), \quad (17)$$

and after erosion by B, for the expectation

$$E\{f \ominus B\} = \frac{1}{n+1} \quad (18)$$

and for the variance

$$V\{f \ominus B\} = \frac{1}{n^2} + O\left(\frac{1}{n^3}\right). \quad (19)$$

Henceforth, we denote by $L(\alpha)$ the *Laplace*

distribution whose distribution function $F(t)$ is

$$F(t) = \int_{-\infty}^{t} \frac{\alpha}{2} e^{-\alpha|x|} dx. \tag{20}$$

When we study the asymptotic behavior of morphological filters in the case of the Laplace distribution, the following two lemmas are very useful.

LEMMA 1. Let $F(t)$ be the distribution function of the Laplace distribution $L(\alpha)$. Then it holds

$$\int_{-\infty}^{\infty} x \left(\frac{d}{dx}F(x)^n\right) dx = -\frac{1}{\alpha n 2^n} + \frac{1}{\alpha}\sum_{k=1}^{n}\frac{1-2^{-k}}{k}. \tag{21}$$

Proof. First we show that

$$\int_{0}^{\infty} x \left(\frac{d}{dx}F(x)^n\right) dx = \frac{1}{\alpha}\sum_{k=1}^{n}\frac{1-2^{-k}}{k}. \tag{22}$$

Substituting $e^{-\alpha x} = 2t$, we obtain

$$\begin{aligned}\int_{0}^{\infty} x \left(\frac{d}{dx}F(x)^n\right) dx &= \frac{n}{2}\int_{0}^{\infty} x\left(1-\frac{e^{-\alpha x}}{2}\right)^{n-1} \alpha e^{-\alpha x} dx \\ &= \frac{n}{2}\int_{1/2}^{0}\left(-\frac{\ln 2t}{\alpha}\right)(1-t)^{n-1}(-2)dt \\ &= \frac{n}{\alpha}\int_{0}^{1/2}(-\ln 2t)(1-t)^{n-1}dt. \end{aligned}\tag{23}$$

We write

$$A(n) = \int_{0}^{1/2}(-\ln 2t)(1-t)^{n-1}dt, \tag{24}$$

and so

$$\begin{aligned}A(n) &= \int_{0}^{1/2}(-\ln 2t)(1-t)^{n-2} - t(-\ln 2t)(1-t)^{n-2}dt \\ &= A(n-1) + \int_{0}^{1/2} t\ln 2t(1-t)^{n-2}dt \\ &= A(n-1) + \int_{0}^{1/2}\ln 2t\frac{(1-t)^{n-1}}{n-1}dt + \int_{0}^{1/2}\frac{(1-t)^{n-1}}{n-1}dt \\ &= A(n-1) - \frac{1}{n-1}A(n) - \left(\frac{(1-t)^n}{n(n-1)}\right)\Bigg|_{0}^{1/2} \\ &= A(n-1) - \frac{1}{n-1}A(n) + \frac{1-2^{-n}}{n(n-1)}. \end{aligned}\tag{25}$$

Writing (25) then in form

$$nA(n) = (n-1)A(n-1) + \frac{1-2^{-n}}{n} \tag{26}$$

and solving $nA(n)$ gives

$$A(n) = \frac{1}{n}\sum_{k=1}^{n}\frac{1-2^{-k}}{k}, \tag{27}$$

implying equation (22). Since

$$\int_{-\infty}^{0} x \left(\frac{d}{dx}F(x)^n\right) dx = -\frac{1}{\alpha n 2^n}, \tag{28}$$

Lemma 1 follows.

LEMMA 2. Let $F(t)$ be the distribution function of the Laplace distribution $L(\alpha)$. Then it holds that

$$\int_{-\infty}^{\infty} x^2 \left(\frac{d}{dx}F(x)^n\right) dx = \frac{1}{\alpha^2 n^2 2^{n-1}} + \frac{1}{\alpha^2}\left(\sum_{k=1}^{n}\frac{2}{k}\sum_{l=1}^{k}\frac{1-2^{-l}}{l}\right). \tag{29}$$

Proof. First we show that

$$\int_{0}^{\infty} x^2 \left(\frac{d}{dx}F(x)^n\right) dx = \frac{1}{\alpha^2}\left(\sum_{k=1}^{n}\frac{2}{k}\sum_{l=1}^{k}\frac{1-2^{-l}}{l}\right). \tag{30}$$

Substituting $e^{-\alpha x} = 2t$, we obtain

$$\int_0^\infty x^2 \left(\frac{d}{dx}F(x)^n\right) dx$$
$$= \frac{n}{2}\int_0^\infty x^2 \left(1 - \frac{e^{-\alpha x}}{2}\right)^{n-1} \alpha e^{-\alpha x} dx$$
$$= \frac{n}{2}\int_{1/2}^0 \left(-\frac{\ln 2t}{\alpha}\right)^2 (1-t)^{n-1}(-2)dt$$
$$= \frac{n}{\alpha^2}\int_0^{1/2} (\ln 2t)^2(1-t)^{n-1}dt. \quad (31)$$

We write

$$B(n) = \int_0^{1/2} (\ln 2t)^2(1-t)^{n-1}dt, \quad (32)$$

and so

$$B(n) = \int_0^{1/2} (\ln 2t)^2(1-t)^{n-2}$$
$$-t(\ln 2t)^2(1-t)^{n-2}dt$$
$$= B(n-1)$$
$$- \int_0^{1/2} t(\ln 2t)^2(1-t)^{n-2}dt$$
$$= B(n-1)$$
$$+2\int_0^{1/2} (-\ln 2t)\frac{(1-t)^{n-1}}{n-1}dt$$
$$- \int_0^{1/2} (\ln 2t)^2\frac{(1-t)^{n-1}}{n-1}dt$$
$$= B(n-1) + \frac{1}{n-1}B(n)$$
$$+2\int_0^{1/2} (-\ln 2t)\frac{(1-t)^{n-1}}{n-1}dt. \quad (33)$$

Since by equation (27)

$$2\int_0^{1/2} (-\ln 2t)\frac{(1-t)^{n-1}}{n-1}dt$$
$$= \frac{2}{n^2-n}\sum_{k=1}^{n}\frac{1-2^{-k}}{k}, \quad (34)$$

we obtain

$$B(n) = B(n-1) - \frac{1}{n-1}B(n)$$
$$+\frac{2}{n^2-n}\sum_{k=1}^{n}\frac{1-2^{-k}}{k}, \quad (34)$$

and solving

$$B(n) = \frac{1}{n}\left(\sum_{k=1}^{n}\frac{2}{k}\sum_{l=1}^{k}\frac{1-2^{-l}}{l}\right) \quad (36)$$

as before implies equation (29). Since

$$\int_{-\infty}^0 x^2 \left(\frac{d}{dx}F(x)^n\right) dx = \frac{1}{\alpha^2 n^2 2^{n-1}}, \quad (37)$$

Lemma 2 follows.

LEMMA 3. Let $F(t)$ be the distribution function of the Laplace distribution $L(\alpha)$, and let the distribution function of a random variable Y be $F(t)^n$ Then for the expectation of Y

$$E\{Y\} = \frac{1}{\alpha}\ln n + O(1), \quad (38)$$

and for the variance of Y

$$V\{Y\} = \eta \ln n + O(1), \quad (39)$$

where

$$\eta = \frac{1}{\alpha^2}\sum_{k=1}^{\infty}\frac{2^{-k}}{k}. \quad (40)$$

Proof. Since the output distribution function of Y is $F(t)^n$, by using Lemma 1 we obtain the asymptotic formula for the expectation

$$E\{Y\} = -\frac{1}{\alpha n 2^n} + \frac{1}{\alpha}\sum_{k=1}^{n}\frac{1-2^{-k}}{k}$$
$$= \frac{1}{\alpha}\ln n + O(1). \quad (41)$$

By Lemma 2 we also obtain the second moment

$$E\{Y^2\} = \frac{1}{\alpha^2 n^2 2^{n-1}}$$
$$+ \frac{1}{\alpha^2}\left(\sum_{k=1}^{n}\frac{2}{k}\sum_{l=1}^{k}\frac{1-2^{-l}}{l}\right). \quad (42)$$

Now, using equations (41) and (42), we obtain the asymptotic formula for the variance

$$
\begin{aligned}
V\{Y\} &= E\{Y^2\} - (E\{Y\})^2 \\
&= \frac{1}{\alpha^2 n^2 2^{n-1}} + \frac{1}{\alpha^2}\sum_{k=1}^{n}\frac{2}{k}\sum_{l=1}^{k}\frac{1-2^{-l}}{l} \\
&\quad - \left(-\frac{1}{\alpha n 2^n} + \frac{1}{\alpha}\sum_{k=1}^{n}\frac{1-2^{-k}}{k}\right)^2 \\
&= \frac{1}{\alpha^2}\sum_{k=1}^{n}\frac{2}{k}\sum_{l=1}^{k}\frac{1-2^{-l}}{l} \\
&\quad - \left(\frac{1}{\alpha}\sum_{k=1}^{n}\frac{1-2^{-k}}{k}\right)^2 + O(1) \\
&= \frac{1}{\alpha^2}\sum_{k=1}^{n}\frac{1}{k}\sum_{l=1}^{k}\frac{1-2^{-l}}{l} \\
&\quad - \frac{1}{\alpha^2}\sum_{k=1}^{n}\frac{1}{k}\sum_{l=k+1}^{n}\frac{1-2^{-l}}{l} \\
&\quad - \frac{1}{\alpha^2}\sum_{k=1}^{n}\frac{2^{-k}}{k}\sum_{l=1}^{n}\frac{1-2^{-l}}{l} + O(1) \\
&= \frac{1}{\alpha^2}\sum_{k=1}^{n}\frac{1}{k}\sum_{l=1}^{k}\frac{1-2^{-l}}{l} \\
&\quad - \frac{1}{\alpha^2}\sum_{k=1}^{n}\frac{1}{k}\sum_{l=1}^{k-1}\frac{1-2^{-l}}{l} \\
&\quad + \frac{1}{\alpha^2}\sum_{k=1}^{n}\frac{2^{-k}}{k}\sum_{l=1}^{n}\frac{1-2^{-l}}{l} + O(1) \\
&= \frac{1}{\alpha^2}\sum_{k=1}^{n}\frac{2^{-k}}{k}\sum_{l=1}^{n}\frac{1}{l} + O(1). \qquad (43)
\end{aligned}
$$

Thus

$$V\{Y\} = \frac{1}{\alpha^2}\sum_{k=1}^{n}\frac{2^{-k}}{k}\ln n + O(1). \qquad (44)$$

PROPOSITION 9. Consider the filtering of a discrete signal f whose values are independent and identically distributed random variables having the Laplace distribution $L(\alpha)$. Let B be a structuring set of size n, and let

$$\eta = \frac{1}{\alpha^2}\sum_{k=1}^{\infty}\frac{2^{-k}}{k}. \qquad (45)$$

Then after dilation by B, for the expectation

$$E\{f \oplus B\} = \frac{1}{\alpha}\ln n + O(1) \qquad (46)$$

and for the variance

$$V\{f \oplus B\} = \eta \ln n + O(1), \qquad (47)$$

and after erosion by B, for the expectation

$$E\{f \ominus B\} = -\frac{1}{\alpha}\ln n + O(1) \qquad (48)$$

and for the variance

$$V\{f \ominus B\} = \eta \ln n + O(1). \qquad (49)$$

Proof. Let $F(t)$ be the distribution function of the Laplace distribution $L(\alpha)$. Since by Proposition 3 the distribution function of a random variable $f \oplus B$ is $F(t)^n$, Lemma 3 implies that

$$E\{f \oplus B\} = \frac{1}{\alpha}\ln n + O(1) \qquad (50)$$

and

$$V\{f \oplus B\} = \eta \ln n + O(1). \qquad (51)$$

Now, Corollary 7.1 implies equation (48), and Corollary 7.2 implies equation (49).

5 Bounds on the Expectations of Two-Dimensional Closing and Opening

The nonlinear nature of closings and openings makes it difficult to carry out an accurate analysis of the effects of noise in the case of two-dimensional closing and opening since their statistical properties depend on both the shape and the size of the structuring set. However, we will show that by using the stack-filter method one can also derive bounds on the output expectations for two-dimensional morphological filters when the values of the signal to be filtered are independent and identically distributed.

In the following we derive a lower bound, which is asymptotically tight for the Laplace distribution, on the expectation of closing by an $n \times n$ square structuring set. This bound can then be used for more general structuring sets containing a square. Moreover, the method of this section can also be used for other types of structuring sets whose shapes are "symmetric enough." The following lemma is a central tool in what follows.

LEMMA 4. *Consider a random variable Y whose distribution function is $u(F(t))$, where $F(t)$ is a strictly monotonous and piecewise differentiable distribution function and $u : [0,1] \to [0,1]$ is a differentiable function. Let the integrals $\int_{-\infty}^{\infty} tF'(t)dt$ and $\int_{-\infty}^{\infty} tu'(F(t))F'(t)dt$ be finite, and let $v(t) : [0,1] \to [0,\infty)$ be a differentiable function that satisfies $v(0) = 0, v(1) = 1$ and $v(t) \geq u(t)$ for all $0 < t < 1$. Then for the expectation of Y*

$$E\{Y\} \geq \int_{-\infty}^{\infty} tv'(F(t))F'(t)dt. \quad (52)$$

Proof. Because $\int_{-\infty}^{\infty} tu'(F(t))F'(t)dt$ is finite, the expectation of Y exists and satisfies

$$\begin{aligned} E\{Y\} &= \int_{-\infty}^{\infty} tu'(F(t))F'(t)dt \\ &= \int_{-\infty}^{0} tu'(F(t))F'(t)dt \\ &\quad + \int_{0}^{\infty} tu'(F(t))F'(t)dt \\ &= \Big(tu(F(t))\Big)\Big|_{-\infty}^{0} \\ &\quad - \int_{-\infty}^{0} u(F(t))dt \\ &\quad - \Big(t(1-u(F(t)))\Big)\Big|_{0}^{\infty} \\ &\quad + \int_{0}^{\infty} (1-u(F(t)))dt \\ &= -\int_{-\infty}^{0} u(F(t))dt \\ &\quad + \int_{0}^{\infty} (1-u(F(t)))dt. \quad (53) \end{aligned}$$

Because $\int_{-\infty}^{\infty} tF'(t)dt$ is finite, we have

$$\Big(tF(t)\Big)\Big|_{-\infty}^{0} = \Big(t(1-F(t))\Big)\Big|_{0}^{\infty} = 0 \quad (54)$$

Thus the assumption that $v(t) : [0,1] \to [0,\infty)$ is a differentiable function and satisfies $v(0) = 0, v(1) = 1$ implies that

$$\Big(tv(F(t))\Big)\Big|_{-\infty}^{0} = \Big(t(1-v(F(t)))\Big)\Big|_{0}^{\infty} = 0. \quad (55)$$

Now, a manipulation similar to the one above leads to the equation

$$\begin{aligned} &\int_{-\infty}^{\infty} tv'(F(t))F'(t)dt \\ &= -\int_{-\infty}^{0} v(F(t))dt \\ &\quad + \int_{0}^{\infty} (1-v(F(t)))dt, \quad (56) \end{aligned}$$

from which Lemma 4 follows since $v(t) \geq u(t)$ for all $t \in [0,1]$.

To bound the distribution function of closing we need the following combinatorial lemma (Koskinen and Astola [17]).

LEMMA 5. *Consider an $m \times m$ square $\mathrm{Sq}(m) = \{(i,j) \in \mathbf{Z}^2 | -n < i < n, -n < j < n\} (m = 2n-1)$, and let $\mathrm{Sq}_n(i,j)$ denote the $n \times n$ subsquare having its northeast corner at $(i,j), 1 \leq i \leq n, 1 \leq j \leq n$. For $l = 0, 1, \ldots, m^2$ let $\nu(l)$ denote the number of subsets $S \subseteq \mathrm{Sq}(m)$, with $|S| = l$, such that $S \cap \mathrm{Sq}_n(i,j) = \emptyset$ for some $i = 1, \ldots, n$ and $j = 1, \ldots, n$, i.e.,*

$$\begin{aligned} \nu(l) = |\{S \subseteq \mathrm{Sq}(m) \mid |S| \\ = l, S \cap \mathrm{Sq}_n(i,j) = \emptyset \\ \text{for some } i,j = 1,\cdots,n\}|. \quad (57) \end{aligned}$$

Then

$$\begin{aligned} \nu(l) \leq \binom{m-n^2}{l} \\ +(2n-2)\left(\binom{m-n^2}{l} - \binom{m-n^2-n}{l}\right) \\ +(n-1)^2\left(\binom{m-n^2}{l} - 2\binom{m-n^2-n}{l}\right. \end{aligned}$$

$$+\binom{m-n^2-2n}{l}\Big)\Big). \qquad (58)$$

Proof. Let $1 \le l \le m - n^2$. If we fix (i,j), the number of S, $(S \subseteq \mathrm{Sq}(m)), |S| = l$ and $S \cap \mathrm{Sq}_n(i,j) = \emptyset$, is obviously

$$\binom{m-n^2}{l}.$$

If we just let (i,j) run through points $1 \le i,j \le n$, we obtain

$$\nu(l) \le n^2\binom{m-n^2}{l}, \qquad (59)$$

which is too crude. To obtain a sharper bound, we let (i,j) run through the points in the following order:

$$\begin{array}{c}(1,1),(1,2),\cdots,(1,k),\\(2,1),(3,1),\cdots,(k,1),\\(2,2),(2,3),\cdots,(2,k),\\(3,2),(3,3),\cdots,(3,k),\\\vdots\\(k,2),(k,3),\cdots,(k,k)\end{array}$$

(top border from left to right, left border from top to bottom, and the rest row-wise from left to right).

Let us now count for each $\mathrm{Sq}_n(i,j)$ the sets $S, |S| = l$ and $S \cap \mathrm{Sq}_n(i,j) = \emptyset$, so that the sets that were counted for $\mathrm{Sq}_n(i-1,j)$ or $\mathrm{Sq}_n(i,j-1)$ are not counted (they have been counted before). This gives

$$\begin{aligned}\nu(l) \le \binom{m-n^2}{l} &+(2n-2)\left(\binom{m-n^2}{l} - \binom{m-n^2-n}{l}\right)\\ &+(n-1)^2\left(\binom{m-n^2}{l} -2\binom{m-n^2-n}{l} + \binom{m-n^2-2n}{l}\right). \qquad (60)\end{aligned}$$

The last term in (60) results from the fact that the contribution of both points $(i-1,j)$ and $(i,j+1)$ would otherwise be subtracted twice.

Astola and Neuvo [18] have shown that whenever the inputs are independent and identically distributed random variables that can attain arbitrarily large values, the expectation of closing (opening) tends to infinity as the size of the structuring set increases. The following proposition (Koskinen and Astola [7]) gives an asymptotically tight bound on the expectation in the case of the Laplace distribution.

PROPOSITION 10. Let $f : \mathbf{Z}^2 \to \mathbf{R}$ be a signal whose values are independent and identically distributed random variables having a common strictly monotonous and piecewise-differentiable distribution function $F(t)$. Consider the closing of f by a square structuring set

$$\mathrm{Sq}(n) = \{(i,j) \in \mathbf{Z}^2 \mid -k < i < k, -k < j < k\}, \qquad (61)$$

where $n = 2k - 1$. Then after closing by $\mathrm{Sq}(n)$, for the expectation

$$E\{f^{\mathrm{Sq}(n)}\} \ge \int_{-\infty}^{\infty} t\left(\frac{\partial}{\partial t}G(F(t),n)\right)dt, \qquad (62)$$

where

$$G(x,n) = n^2x^{n^2} - 2(n^2-n)x^{n^2+n} + (n^2-2n+1)x^{n^2+2n}. \qquad (63)$$

Proof. Let $M = \mathrm{Sq}(n) + \mathrm{Sq}(n)$, and consider the characteristic function of each subset $S \subseteq M$ as a Boolean vector of dimension $|M| = m = n^2$. For $S \subseteq M$, denote by $f^{\mathrm{Sq}(n)}(S)$ the Boolean function determined by the output of the closing of S by $\mathrm{Sq}(n)$. For $l = 0,1,\cdots,m$ we write

$$\nu(l) = |\{S \subseteq M \mid |S| = l, f^{\mathrm{Sq}(n)}(S) = 0\}|. \qquad (64)$$

Now Proposition 1 implies that the distribution function $H(t)$ of the values of the closed signal f^B is

$$H(t) = u(F(t)), \qquad (65)$$

where

$$u(x) = \sum_{l=0}^{m} \nu(l)(1-x)^l x^{m-1}. \quad (66)$$

Substituting the bound of Lemma 5 in (66) and performing the summations, we get

$$\begin{aligned} u(x) \leq x^{n^2} &+ (2n-2)(x^{n^2} - x^{n^2+n}) \\ &+(n-1)^2(x^{n^2} - 2x^{n^2+n} \\ &+x^{n^2+2n}), \end{aligned} \quad (67)$$

and the result now follows from Lemma 4.

COROLLARY 10.1. Consider a gray-level image whose pixel values are white noise uniformly distributed in [0, 1], and perform closing by an $n \times n$ square structuring set. Then the expectation E of the gray-level of the closed image satisfies

$$E \geq 1 - 5/n^2 + O(1). \quad (68)$$

This shows that on the constant regions of a noisy image the output will follow the upper tail of the noise distribution, giving a smooth but biased output.

The following proposition (Koskinen and Astola [7]) shows that for the Laplace distribution the above bound (62) is asymptotically tight. First we need the following lemma.

LEMMA 6. Let $F(t)$ be the distribution function of the Laplace distribution $L(\alpha)$, and let $G(x, n)$ be defined by (63). Then

$$\int_{-\infty}^{\infty} t\left(\frac{\partial}{\partial t}G(F(t), n)\right) dt = \frac{2}{\alpha}\ln n + O(1). \quad (69)$$

Proof. Applying Lemma 3 to each term of the right-hand side of (62), we obtain

$$\begin{aligned} &\int_{-\infty}^{\infty} t\left(\frac{\partial}{\partial t}G(F(t), n)\right) dt \\ &= \frac{n^2}{\alpha}\Big(\ln(n^2) - 2\ln(n^2+n) + \ln(n^2+2n)\Big) \\ &+\frac{2n}{\alpha}\Big(\ln(n^2+n) - \ln(n^2+2n)\Big) \\ &+\frac{1}{\alpha}\ln(n^2+2n) + O(1) \\ &= \frac{2}{\alpha}\ln n + O(1). \end{aligned} \quad (70)$$

PROPOSITION 11. Let $f : \mathbf{Z}^2 \to \mathbf{R}$ be a signal whose values are independent and identically distributed random variables having the Laplace distribution $L(\alpha)$. Consider the closing of f by an $n \times n$ square structuring set $\mathrm{Sq}(n)$. Then after closing by $\mathrm{Sq}(n)$, for the expectation

$$E\{f^{\mathrm{Sq}(n)}\} = \frac{2}{\alpha}\ln n + O(1). \quad (71)$$

Proof. From Lemma 6 we obtain an asymptotic lower bound for $E\{f^{\mathrm{Sq}(n)}\}$, and, surprisingly, the same upper bound results from comparing closing by $\mathrm{Sq}(n)$ to dilation by $\mathrm{Sq}(n)$. The relation

$$f^{\mathrm{Sq}(n)} \leq f \oplus \mathrm{Sq}(n) \quad (72)$$

implies that

$$E\{f^{\mathrm{Sq}(n)}\} \leq E\{f \oplus \mathrm{Sq}(n)\}, \quad (73)$$

and Proposition 9 gives

$$E\{f \oplus \mathrm{Sq}(n)\} = \frac{2}{\alpha}\ln n + O(1). \quad (74)$$

Proposition 11 now follows from Proposition 10 and Lemma 6.

Propositions 10 and 11 were formulated for square structuring sets, but they imply similar results for a large class of structuring sets that can be suitably approximated by a square structuring set.

DEFINITION Let A and B be structuring sets. Then A *is shrouded* by B if there exist C such that $C \sim B$ and for all $c \in C^s$ there exists an $a \in A^s$ such that $A_a \subseteq C_c$.

LEMMA 7. Consider the closing of a signal f whose values are independent and identically distributed random variables. Let A and B be such structuring sets that A is shrouded by B. Then

$$E\{f^A\} \leq E\{f^B\}. \quad (75)$$

Proof. Since B shrouds A, there exists C such that $C \sim B$, and for all $c \in C^s$ there exists an $a \in A^s$ such that $A_a \subseteq C_c$. Then we obtain the relation $f^A \leq f^C$, which implies that $E\{f^A\} \leq E\{f^C\}$. Since closing by B and closing by C have the same statistical behavior in the case of independently and identically distributed inputs, we obtain Lemma 7.

Now we can state the asymptotic results in Proposition 12 (Koskinen and Astola [17]).

PROPOSITION 12. Let $f : \mathbf{Z}^2 \to \mathbf{R}$ be a signal whose values are independent and identically distributed random variables having the Laplace distribution $L(\alpha)$, and let ρ be a real number. Consider a sequence $B_n, n = 1, 2, \ldots,$ of structuring sets such that $\mathrm{Sq}(n)$ is shrouded by B_n and $n^2 \leq |B_n| \leq \rho n^2$ for $n = 1, 2, \ldots$. Then for the expectations it holds that

$$E\{f^{B_n}\} = \frac{2}{\alpha} \ln n + O(1) \tag{76}$$

and

$$E\{f_{B_n}\} = -\frac{2}{\alpha} \ln n + O(1). \tag{77}$$

Proof. By Lemma 7 the assumption that $\mathrm{Sq}(n)$ is shrouded by B_n implies that

$$E\{f^{B_n}\} \geq E\{f^{\mathrm{Sq}(n)}\}, \tag{78}$$

and the relation

$$f^{B_n} \leq f \oplus B_n \tag{79}$$

implies that

$$E\{f^{B_n}\} \leq E\{f \oplus B_n\}. \tag{80}$$

Now by using Propositions 9 and 10 and the relation $|B| \leq \rho n^2$ we obtain equation (76), and by using Corollary 7.1 we obtain equation (77).

6 Asymptotic Behavior of One-Dimensional Morphological Filters

In general, the statistical properties of morphological filters depend on both the shape and the size of the structuring set. As a result, it is difficult to derive analytical expressions for these properties. However, in the case of a one-dimensional convex structuring set we can derive analytical expressions for the output distributions, and we can study the second-order statistical properties of closing, opening, clos-opening, and open-closing when the values of the input signal are independent and identically distributed random variables (Koskinen *et al.* [5]). These results can also be used to approximate the statistical properties of two-dimensional morphological filters since the simulation results in Rustanius et al. [6] indicate that the statistical properties of morphological filters are mainly determined by the size of the structuring set.

PROPOSITION 13. Consider a discrete signal f whose values are independent and identically distributed random variables having a common distribution function $F(t)$. Let f be closed by using a convex one-dimensional structuring set B of length n. Then the distribution function $G_c(t)$ of the values of the closed signal f^B is

$$G_c(t) = nF(t)^n - (n-1)F(t)^{n+1}. \tag{81}$$

Proof. Consider the closing by a convex one-dimensional structuring set B of length n. By Proposition 4 we need to compute the number of the vectors $\underline{z} = (z_1, z_2, \cdots, z_{2n-1})$ of each weight in primage $g^{-1}(0)$ of the Boolean function

$$g(\underline{z}) = \wedge_{a \in B^s} (\wedge_{b \in B_a} z_b), \tag{82}$$

which corresponds to the stack-filter expression of closing.

Now the number of the vectors of weight $2n - 1 - s$ in $g^{-1}(0)$ equals the number of the vectors of weight $2n - 1 - s$ containing at least n consecutive 0's. Divide these vectors into n distinct classes in the following way. The class C_0 contains all vectors

$$b^{(0)} = (\overbrace{0, 0, \cdots, 0}^{n \text{ times}}, *, *, \cdots, *),$$

and the class C_l contains all vectors such that

$$b^{(l)} = (\overbrace{*, *, \cdots, *, 1}^{l \text{ times}}, \overbrace{0, 0, \cdots, 0}^{n \text{ times}}, *, *, \cdots, *),$$

where $*$ denotes 0 or 1. It is easy to see that C_0 consists of

$$\binom{n-1}{s-n}$$

vectors of weight $2n-1-s$ and C_l consists of

$$\binom{n-2}{s-n}$$

vectors of weight $2n-1-s$ for all $l = 1, 2, \cdots, n-1$. So there are

$$\binom{n-1}{s-n} + (n-1)\binom{n-2}{s-n}$$

vectors of weight $2n-1-s$ in $g^{-1}(0)$. Now Proposition 4 implies that the distribution function $G_c(t)$ of the closed signal is

$$\begin{aligned} G_c(t) &= \sum_{i=n}^{2n-1}\left(\binom{n-1}{i-n} + (n-1)\binom{n-2}{i-n}\right) \\ &\quad (1-F(t))^{2n-1-i}F(t)^i \\ &= \sum_{i=0}^{n-1}\left(\binom{n-1}{i} + (n-1)\binom{n-2}{i}\right) \\ &\quad (1-F(t))^{n-1-i}F(t)^{i+n} \\ &= F(t)^n \sum_{i=0}^{n-1}\binom{n-1}{i} \\ &\quad (1-F(t))^{n-1-i}F(t)^i \\ &\quad +(n-1)F(t)^n(1-F(t)) \\ &\quad \sum_{i=0}^{n-1}\binom{n-2}{i} \\ &\quad (1-F(t))^{n-2-i}F(t)^i \\ &= nF(t)^n - (n-1)F(t)^{n+1}. \end{aligned} \tag{83}$$

Using Proposition 6, we obtain the following corollary.

COROLLARY 13.1. Consider a discrete signal f whose values are independent and identically distributed random variables having a common distribution function $F(t)$. Let f be opened by using a convex one-dimensional structuring set B of length n. Then the distribution function $G_o(t)$ of the values of the opened signal f_B is

$$\begin{aligned} G_o(t) &= 1 - n(1-F(t))^n \\ &\quad +(n-1)(1-F(t))^{n+1}. \end{aligned} \tag{84}$$

Let f be a discrete signal whose values are independent and identically distributed random variables having the Laplace distribution $L(\alpha)$, and let f be closed by a convex one-dimensional structuring set B of length n. Then Proposition 13 and Lemmas 1 and 2 imply that for random variable $y = f^B$

$$\begin{aligned} E\{y\} &= -\frac{n+3}{\alpha(n+1)2^{n+1}} \\ &\quad +\frac{1}{\alpha}\left(\sum_{k=1}^{n}\frac{1-2^{-k}}{k} - \frac{n-1}{n+1}\right. \\ &\quad \left. +\frac{n-1}{n+1}2^{-n-1}\right) \end{aligned} \tag{85}$$

and

$$\begin{aligned} E\{y^2\} &= \frac{n^3+5n^2+2n}{\alpha^2 n^2 (n+1)^2 2^n} \\ &\quad +\frac{1}{\alpha^2}\left(\sum_{k=1}^{n}\frac{2}{k}\sum_{l=1}^{k}\frac{1-2^{-l}}{l}\right. \\ &\quad \left. -\frac{2n-2}{n+1}\sum_{l=1}^{n+1}\frac{1-2^{-l}}{l}\right). \end{aligned} \tag{86}$$

In the same way, using Proposition 1 and Lemmas 1 and 2, one can derive exact output expectations and variances for all stack filters whose windows are of a moderate size and whose inputs are independent random variables having a Laplace distribution.

PROPOSITION 14. Consider a discrete signal f whose values are independent and identically distributed random variables having a common distribution function $F(t)$. Let f be clos-opened by using a convex one-dimensional structuring set B of length n, where $n > 2$. Then the distribution function $G_{co}(t)$ of the values of the clos-opened signal $(f^B)_B$ is

$$G_{co}(t) = \frac{n^2-n-2}{2}F(t)^{2n+2}$$

$$+(-n^2+n+1)F(t)^{2n+1}$$
$$+\frac{n^2-n}{2}F(t)^{2n}-(n-1)F(t)^{n+1}$$
$$+nF(t)^n. \qquad (87)$$

Proof. Consider the clos-opening by a convex one-dimensional structuring set B of length $n > 2$. By Proposition 5 we need to compute the number of the vectors $\underline{z} = (z_1, z_2, \cdots, z_{4n-3})$ of each weight in the preimage $g^{-1}(0)$ of the Boolean function

$$g(\underline{z}) = \vee_{a\in B^s}(\wedge_{b\in(B+B^s)_a}(\vee_{c\in B_b}z_c)), \qquad (88)$$

which corresponds to the stack-filter expression of clos-opening.

The number of the vectors of weight $4n-3-s$ in $g^{-1}(0)$ equals the number of the vectors $\underline{z} = (z_1, z_2, \cdots, z_{4n-3})$ of weight $4n-3-s$ such that each subvector $(z_i, z_{1+i}, \cdots, z_{3n-3+i})$ contains at least n consecutive 0's for all $i \in \{1, 2, \cdots, n\}$.

First we divide $g^{-1}(0)$ into three distinct classes A, B, and C: $A = \{\underline{z} \in \{0,1\}^{4n-3} |$ for some $i \in \{0, \ldots, n-1\}, z_{n+i} = 0, \ldots, z_{2n-1+i} = 0\}$, $B = \{\underline{z} \in \{0,1\}^{4n-3} | z_{2n-1} = 0$ and for some nonnegative i, j such that $i + j \le n-2, z_{n-1-i} = 0, \ldots, z_{2n-2-i} = 0$ and $z_{2n+j} = 0, \ldots, z_{3n-1+j} = 0$ and for some nonnegative l, m such that $l + m \le n-4, z_{2n-2-l} = z_{2n+m} = 1\}$, and $C = \{\underline{z} \in \{0,1\}^{4n-3} | z_{2n-1} = 1$ and for some nonnegative i, j such that $i + j \le n-2, z_{n-1-i} = 0, \ldots, z_{2n-2-i} = 0$ and $z_{2n+j} = 0, \ldots, z_{3n-1+j} = 0\}$.

Then we divide C again into three distinct classes $C1, C2$ and $C3$: $C1 = \{\underline{z} \in C | z_{n-1} = 0, \ldots, z_{2n-2} = 0\}$, $C2 = \{\underline{z} \in C | z_{2n} = 0, \ldots, z_{3n-1} = 0$ and for some i such that $0 \le i \le n-3, z_{2n-2-i} = 1\}$, and $C3 = \{\underline{z} \in C |$ for some nonegative i, j such that $i, j \le n-4, z_{2n-2-i} = z_{2n+j} = 1\}$.

Using the same method as in the proof of Proposition 13, we see that class A consists of

$$\binom{3n-3}{s-n} + (n-1)\binom{3n-4}{s-n}$$

vectors of weight $4n-3-s$. We can calculate directly that class B consists of

$$\sum_{i=1}^{n-3} i\binom{2n-6}{s-2n-1}$$

$$= \frac{n^2-5n+6}{2}\binom{2n-6}{s-2n-1} \qquad (89)$$

and that class $C3$ consists of

$$\sum_{i=1}^{n-3} i\binom{2n-6}{s-2n} = \frac{n^2-5n+6}{2}\binom{2n-6}{s-2n} \qquad (90)$$

vectors of weight $4n-3-s$.

It is easy to see that class $C1$ consists of

$$\binom{2n-4}{s-2n} + (n-2)\binom{2n-5}{s-2n},$$

and that class $C2$ consists of

$$(n-2)\binom{2n-5}{s-2n}$$

vectors of weight $4n-3-s$.

So the number $t(s)$ of the vectors of weight $4n-3-s$ in $g^{-1}(0)$ is

$$t(s) = \binom{3n-3}{s-n} + (n-1)\binom{3n-4}{s-n}$$
$$+\frac{n^2-5n+6}{2}\binom{2n-6}{s-2n-1}$$
$$+\frac{n^2-5n+6}{2}\binom{2n-6}{s-2n}$$
$$+\binom{2n-4}{s-2n}$$
$$+(2n-4)\binom{2n-5}{s-2n}. \qquad (91)$$

Now the output distribution $G_{co}(t)$ for clos-opening is given by

$$G_{co}(t) = \sum_{i=0}^{4n-3} t(i)(1-F(t))^{4n-3-i}F(t)^i$$
$$= (2n-4)(1-F(t))^2F(t)^{2n}$$
$$+(1-F(t))F(t)^{2n}$$
$$+\frac{n^2-5n+6}{2}(1-F(t))^3F(t)^{2n}$$
$$+\frac{n^2-5n+6}{2}(1-F(t))^2F(t)^{2n+1}$$
$$+F(t)^n + (n-1)(1-F(t))F(t)^n$$
$$= \frac{n^2-n-2}{2}F(t)^{2n+2}$$

$$+(-n^2+n+1)F(t)^{2n+1} + \frac{n^2-n}{2}F(t)^{2n} - (n-1)F(t)^{n+1} + nF(t)^n. \quad (92)$$

Using Proposition 6, we obtain the following corollary.

COROLLARY 14.1. Consider a discrete signal f whose values are independent and identically distributed random variables having a common distribution function $F(t)$. Let f be open-closed by using a convex structuring set B of length n, where $n > 2$. Then the distribution function $G_{oc}(t)$ of the values of the open-closed signal $(f_B)^B$ is

$$G_{oc}(t) = 1 - \frac{n^2-n-2}{2}(1-F(t))^{2n+2} - (-n^2+n+1)(1-F(t))^{2n+1} - \frac{n^2-n}{2}(1-F(t))^{2n} + (n-1)(1-F(t))^{n+1} - n(1-F(t))^n. \quad (93)$$

Formula (87) defines the output distribution of clos-opening in terms of the input distribution. Another formula for the output distribution of clos-opening is given in Stevenson and Arce [1], but this formula does not define the correct output distribution in terms of the input distribution.

When the inputs are independent and identically distributed random variables, the analytical forms of the output distribution functions also allow us to compute the asymptotic approximations for the expectation and the variance in the cases of uniform or Laplace distributions. The tails of the uniform density function are zero-value, and morphological filters attenuate this kind of noise well. On the other hand, the density function of Laplace distribution is heavy-tailed, and morphological filtering increases the variance of this kind of noise.

Using Propositions 13 and 14, we obtain after lengthy computations (performed by using a symbolic program) the following result (Koskinen et al. [5]), which shows that the output variance of compound morphological operations decreases very fast as the length of the structuring set increases in the case of the uniform distribution.

PROPOSITION 15. Consider the filtering of a discrete signal f whose values are independent and identically uniformly distributed on [0, 1]. Let B be a convex one-dimensional structuring set of length n. Then

(a) after closing by B, for the expectation

$$E\{f^B\} = 1 - \frac{2}{n} + O\left(\frac{1}{n^2}\right) \quad (94)$$

and for the variance

$$V\{f_B\} = \frac{2}{n^2} + O\left(\frac{1}{n^3}\right); \quad (95)$$

(b) after opening by B, for the expectation

$$E\{f_B\} = \frac{2}{n} + O\left(\frac{1}{n^2}\right) \quad (96)$$

and for the variance

$$V\{f_B\} = \frac{2}{n^2} + O\left(\frac{1}{n^3}\right); \quad (97)$$

(c) after clos-opening by B, for the expectation

$$E\{(f^B)_B\} = 1 - \frac{17}{8n} + O\left(\frac{1}{n^2}\right) \quad (98)$$

and for the variance

$$V\{(f^B)_B\} = \frac{119}{64n^2} + O\left(\frac{1}{n^3}\right); \quad (99)$$

(d) after open-closing by B, for the expectation

$$E\{(f_B)^B\} = \frac{17}{8n} + O\left(\frac{1}{n^2}\right) \quad (100)$$

and for the variance

$$V\{(f_B)^B\} = \frac{119}{64n^2} + O\left(\frac{1}{n^3}\right). \quad (101)$$

The following asymptotic formulas show that the output expectations and the output variances of compound morphological operations increase logarithmically with the length of the structuring set.

PROPOSITION 16. Consider the filtering of a discrete signal f whose values are independent and identically distributed having the Laplace distribution $L(\alpha)$. Let B be a convex one-dimensional structuring set of length n, and let

$$\eta = \frac{1}{\alpha^2} \sum_{k=1}^{\infty} \frac{2^{-k}}{k}. \quad (102)$$

Then

(a) after closing or clos-opening by B, for the expectation

$$E\{f^B\} = E\{(f^B)_B\} = \frac{1}{\alpha} \ln n + O(1) \quad (103)$$

and for the variance

$$V\{f^B\} = V\{(f^B)_B\} = \eta \ln n + O(1), \quad (104)$$

(b) after opening or open-closing by B, for the expectation

$$E\{f_B\} = E\{(f_B)^B\} = -\frac{1}{\alpha} \ln n + O(1) \quad (105)$$

and for the variance

$$V\{f_B\} = V\{(f_B)^B\} = \eta \ln n + O(1). \quad (106)$$

Proof. Lemma 3 and Proposition 13 imply that

$$\begin{aligned} E\{f^B\} &= \frac{1}{\alpha}\Big(n \ln n - (n-1)\ln(n+1)\Big) \\ &\quad + O(1) \\ &= \frac{1}{\alpha}\Big(\ln n + (n-1)\ln\Big(\frac{n}{n+1}\Big)\Big) \\ &\quad + O(1) \\ &= \frac{1}{\alpha} \ln n + O(1) \end{aligned} \quad (107)$$

and

$$\begin{aligned} V\{f^B\} &= \eta\Big(n \ln n - (n-1)\ln(n+1)\Big) \\ &\quad + O(1) \\ &= \eta\Big(\ln n + (n-1)\ln\Big(\frac{n}{n+1}\Big)\Big) \\ &\quad + O(1) \\ &= \eta \ln n + O(1). \end{aligned} \quad (108)$$

Similarly, Lemma 3 and Proposition 14 imply that

$$E\{(f^B)_B\} = \frac{1}{\alpha} \ln n + O(1) \quad (109)$$

and

$$V\{(f^B)_B\} = \eta \ln n + O(1). \quad (110)$$

If the length of the structuring set is n, then the output variance of linear and median filters is $O(1/n)$ for both the uniform and the Laplace distributions. On the contrary, Propositions 8, 9, 15, and 16 show that the output variances of morphological filters are $O(1/n^2)$ for uniform distribution and the output variances of morphological filters increase logarithmically with the size of the structuring set for the Laplace distribution. Thus morphological operations are statistically unstable operations whose behavior is extremely sensitive to a change in the type of noise distribution.

Propositions 9, 12, and 16 show that the expectations of one- and two-dimensional morphological operations have the same asymptotic increasing rate in the case of the Laplace distribution. This is evidence of the major role of the size of the structuring set when we compare the effect of the size and the shape of the structuring set on the statistical properties.

The morphological filters and median filters have many similar deterministic and impulsive noise-attenuation properties; see, e.g., Maragos and Schafer [3] and Justusson [19]. However, the above results show that the statistical behaviors of these filters differ greatly: morphological filters are unstable and median filters are stable in a statistical sense, and in the case of short-tailed and heavy-tailed noise, the behaviors of median and morphological filters are quite opposite.

7 Conclusions

When the output distribution function of a filter is known, all other statistical properties of this filter can be derived from this distribution. In this paper a closed-form expression for the output distribution in terms of the input distribution has been given for morphological filters. Certain statistical symmetry properties stemming from the duality of morphological filters have also been presented.

By using the expressions of output distributions, asymptotic formulas for the output variances and for the expectations of morphological filters have been derived for the cases of uniform and Laplace distributions. These formulas show that the asymptotic behavior of morphological filters is extremely sensitive to a change in the type of noise distribution, leading to the conclusion that morphological filters may be unstable under noisy conditions.

References

1. R. Stevenson and G. Arce, "Morphological filters: statistics and further syntactic properties," *IEEE Trans. Circuits and Systems*, vol. ASSP-34, pp. 1292–1305, 1987.
2. D. Schonfeld and J. Goutsias, "Optimal morphological pattern restoration from noisy binary images," *IEEE Trans. Patt. Anal. Mach. Intell.*, vol. PAMI-13, pp. 14–29, 1990.
3. P. Maragos and R. Schafer, "Morphological filters – part II: their relations to median, order-statistics, and stack filters," *IEEE Trans. Acoust., Speech, Signal Process.* vol. ASSP-35, pp. 1170–1184, 1987.
4. L. Koskinen, J. Astola, and Y. Neuvo, "Morphological filtering of noisy images," in *Visual Communications and Image Processing*'90: *Fifth in a Series*, M. Kunt, ed., *Proc. Soc. Photo-Opt. Instrum. Eng.*, vol. 1360, pp. 155–165, 1990.
5. L. Koskinen, J. Astola, and Y. Neuvo, "Analysis of noise attenuation in morphological image processing," in *Nonlinear Image Processing II*, G.R. Arce, C.G. Boncelet, and E.R. Dougherty, eds., *Proc. Soc. Photo-Opt. Instrum. Eng.*, vol. 1451, pp. 102–113, 1991.
6. P. Rustanius, L. Koskinen, and J. Astola, "Theoretical and experimental analysis of the effects of noise in morphological image processing," in *Proc. SPIE Symp. on Image Algebra and Morphological Image Processing III*, San Diego, CA, July 1992.
7. E. Dougherty and C. Ciardiana, *Morphological Methods in Image and Signal Processing*, Prentice-Hall: Englewood Cliffs, NJ, 1988.
8. S. Serra, *Image Analysis and Mathematical Morphology*, Academic Press: London, 1988.
9. P. Maragos and R. Schafer, "Morphological filters – Part I: their set theoretic analysis and relations to linear shift-invariant filters," *IEEE Trans. Acoust., Speech, Signal Process.* vol. ASSP-35, pp. 1153–1169, 1987.
10. G. Matheron, *Random Sets and Integral Geometry*, John Wiley: New York, 1975.
11. C. Chu and E. Delp, "Impulsive noise suppression and background normalization of electrocardiogram signals using morphological operators," *IEEE Trans. Biomed. Eng.*, vol. 36, pp. 226–273, 1989.
12. P. Wendt, E. Coyle, and N. Callager, "Stack filters," *IEEE Trans. Acoust., Speech, Signal Process.* vol. ASSP-34, pp. 898–911, 1986.
13. O. Yli-Harja, J. Astola, and Y. Neuvo, "Analysis of the properties of median and weighted median filters using threshold logic and stack filter representation," *IEEE Trans. Signal Process.*, vol. 39, pp. 395–410, 1991.
14. L. Koskinen, J. Astola, and Y. Neuvo, "Statistical properties of discrete morphological filters," in *Proc. IEEE Int. Symp. on Circuits and Systems*, New Orleans, LA, May 1990, pp. 1219–1222.
15. E. Castillo, *Extreme Value Theory in Engineering*, Academic Press: London, 1988.
16. J. Galambos, *The Asymptotic Theory of Order Statistics*, John Wiley: New York, 1978.
17. L. Koskinen and J. Astola, "Statistical properties of soft morphological filters," in *Proc. SPIE Symp. on Nonlinear Image Processing III*, San Jose, CA, February 1992.
18. J. Astola and Y. Neuvo, "An efficient tool for analyzing weighted median and stack filters," submitted to *IEEE Trans. Circuits and Systems*.
19. B. Justusson, "Median filtering: statistical properties," in *Topics in Applied Physics, Two Dimensional Digital Signal Processing II*, T.S. Huang, ed., Springer-Verlag: Berlin, 1981.

Lasse Koskinen received the M.Sc., and licentiate degrees in mathematics from Tampere University, Finland, in 1989 and 1991, respectively. Currently he is finishing his Ph.D. thesis in mathematics at Tampere University. His research interests are signal and image processing.

Jaakko Astola received the B.Sc., M.Sc., licentiate, and Ph.D. degrees in mathematics from Turku University, Finland, in 1972, 1973, 1975, and 1978, respectively. From 1976 to 1977 he was a research assistant at the Research Institute for Mathematical Sciences of Kyoto University, Japan. Between 1979 and 1987 he was with the Department of Information Technology, Lappeenranta University of Technology, Finland, holding various teaching positions in mathematics, applied mathematics, and computer science. In 1987, he was a professor in the Department of Electrical Engineering, Tampere University of Technology, Finland; he is currently Associate Professor in Applied Mathematics at Tampere University. His research interests include signal processing, coding theory, and statistics.

Journal of Mathematical Imaging and Vision 2, 137–154 (1992).

NonLinear Filtering Structure for Image Smoothing in Mixed-Noise Environments

ROBERT L. STEVENSON AND SUSAN M. SCHWEIZER
Laboratory for Image and Signal Analysis, Department of Electrical Engineering, University of Notre Dame, Notre Dame, IN 46556

Abstract. This paper introduces a new nonlinear filtering structure for filtering image data that have been corrupted by both impulsive and nonimpulsive additive noise. Like other nonlinear filters, the proposed filtering structure uses order-statistic operations to remove the effects of the impulsive noise. Unlike other filters, however, nonimpulsive noise is smoothed by using a maximum *a posteriori* estimation criterion. The prior model for the image is a novel Markov random-field model that models image edges so that they are accurately estimated while additive Gaussian noise is smoothed. The Markov random-field-based prior is chosen such that the filter has desirable analytical and computational properties. The estimate of the signal value is obtained at the unique minimum of the *a posteriori* log likelihood function. This function is convex so that the output of the filter can be easily computed by using either digital or analog computational methods. The effects of the various parameters of the model will be discussed, and the choice of the predetection order statistic filter will also be examined. Example outputs under various noise conditions will be given.

Key words. image processing, nonlinear filtering, stochastic image models

1 Introduction

In many data acquisition, transmission, and storage systems noise is introduced into the data. When the data are a two-dimensional image, this noise reduces the picture quality of the original signal. Various filtering techniques have been developed to suppress the noise in the signal in order to improve the overall picture quality. For images, linear-filtering operations do not perform well because images usually contain many sharp edges and thin structures that tend to be smeared or lost in the linear-filtering process. For these reasons nonlinear filters have been examined. Nonlinear filters are designed to suppress the additive-noise component in an image while preserving the important structural information, such as edges and lines. As with any nonlinear system, the analysis of these filters is difficult.

Over the years many nonlinear-filtering ideas have been put forth. For noise environments that are very impulsive, rank-based estimators have been very successful. The median [1], morphological [2]–[4], and stack filters are three important and related rank-based filter types. For nonimpulsive noise the rank-based filters do not generally remove noise as well as does the class of linear filters, but, as was pointed out above, the linear filters suffer from poor syntactical performance. For this reason several hybrid filter designs, which incorporate both rank-based and linear-based sections in the filter, have been proposed [5]–[9]. Although these filters work better in mixed-noise environments, they suffer from some of the same limitations as the ranked-based filters, namely their designs are based on statistical or on syntactical considerations, and the other properties are difficult to determine and quantify.

This paper proposes a new nonlinear filtering structure based on a maximum *a posteriori* estimation criterion which uses a Markov random field model for the prior distribution. The form of the Markov random field is such that estimates obtained with the proposed model permit discontinuities in the signal to be accurately estimated while additive Gaussian noise is smoothed. Impulses in the image data are removed by predetecting the locations of the impulses with an order-statistic filter and by modifying the noise model used in the maximum *a posteriori* estimation procedure. Section 2 presents the necessary elements of estimation theory, discusses random-field models, and presents the Huber–Markov random field model. It ends with the statement of an estimation procedure for filtering Gaussian noise from image data. Section 3 examines the problem of predetecting the location of impulsive data and including that information in the noise model. A criterion is established so that various order-statistic filters can be ranked so as to determine which filter works best for the detection of impulse location. Section 4 examines the computational issues of computing the image estimates. In section 5 example outputs for various noise conditions are shown and discussed. Finally, section 6 summarizes our work to this point and indicates the directions that are under investigation to make this design more versatile.

2 Estimation Theory and Signal Models

The problem addressed in this paper can be stated as follows: The $N \times N$ image $\mathbf{z}$ has been corrupted with an additive-noise process $\mathbf{n}$, and the noisy $N \times N$ image $\mathbf{y}$,

$$\mathbf{y} = \mathbf{z} + \mathbf{n} \tag{1}$$

is observed. The goal is to find a perceptually high-quality estimate of $\mathbf{z}$, denoted $\hat{\mathbf{z}}$, from only knowledge of $\mathbf{y}$ and limited knowledge of the corrupting noise process $\mathbf{n}$. In this paper the noise process is assumed to be an ϵ-corrupted Guassian process. The Gaussian process is corrupted by an impulsive-noise process; the probability of an impulse occurring is given by ϵ. This noise model was chosen to simulate some of the severe degradation that can occur in the transmission of analog and digital video sequences.

2.1 Estimation Theory

Incorporation of this problem into a statistical framework requires the development of probabilistic models for the noise and source signal. First, some basic notation is defined. A random field $\mathbf{x}$ is defined on a set of $N \times N$ points, and each point, $\mathbf{x}_{m,n}$ for $1 \leq m, n \leq N$, takes on values in $\Re$. Let $\Pr(\mathbf{x})$ denote the probability that the random field takes on the value $\mathbf{x}$ (i.e., a probability density function). In order to write linear operations as vector–matrix equations, the stacked-vector notation is used for the random fields and image data. One commonly used random-field model is the Gaussian model. It is characterized by the distribution

$$\Pr(\mathbf{x}) = \frac{1}{\sqrt{(2\pi)^{N^2}|\mathbf{B}|}} \exp\left(-\frac{1}{2}\mathbf{x}^t\mathbf{B}^{-1}\mathbf{x}\right), \tag{2}$$

where $\mathbf{B}^{-1}$ is a symmetric, positive definite matrix, $|\cdot|$ denotes the matrix determinant, and $\mathbf{x}^t$ is the transpose of $\mathbf{x}$. If the field is independent and identically distributed (i.i.d.), the model becomes

$$\Pr(\mathbf{x}) = \frac{1}{\sqrt{(2\pi\sigma^2)^{N^2}}} \exp\left(-\frac{\|\mathbf{x}\|^2}{2\sigma^2}\right), \tag{3}$$

where σ^2 is the variance and $\|\cdot\|$ is the Euclidean norm.

An accurate estimate of the image $\mathbf{z}$ will have two important properties. First, the estimate must accurately fit the observed data, and second, the estimate should reflect any reasonable prior information about the signal structure. A natural measure of the accuracy of the fit is the log likelihood function

$$L(\mathbf{y} \mid \mathbf{z}) = \log \Pr(\mathbf{y} \mid \mathbf{z}), \tag{4}$$

where $\Pr(\mathbf{y} \mid \mathbf{z})$ is the density function of $\mathbf{y}$, given $\mathbf{z}$. The maximum likelihood (ML) estimate is the estimate that maximizes the probability that the data were observed,

$$\hat{\mathbf{z}} = \arg \max_{\tilde{\mathbf{z}} \in \Re^N} L(\mathbf{y} \mid \tilde{\mathbf{z}}). \tag{5}$$

The ML estimate does not provide any mechanism for incorporating prior information about the image data. In practice, there is only one set of observable data $\mathbf{y}$ for each image, and prior information is needed to reduce the effects of noise; this requires that a prior distribution for the image Pr(**z**), be defined. Then the log likelihood function of the *a posteriori* distribution can be computed by

$$L_p(\mathbf{z} \mid \mathbf{y}) = \log \Pr(\mathbf{z} \mid \mathbf{y}), \\ = L(\mathbf{y} \mid \mathbf{z}) + \log \Pr(\mathbf{z}) \\ - \log \Pr(\mathbf{y}). \quad (6)$$

This leads to a maximum *a posteriori* (MAP) estimation criterion that can be formulated as

$$\hat{\mathbf{z}} = \arg \max_{\tilde{\mathbf{z}} \in \Re^N} L_p(\mathbf{y} \mid \tilde{\mathbf{z}}) \\ = \arg \max_{\tilde{\mathbf{z}} \in \Re^N} \{L(\mathbf{y} \mid \tilde{\mathbf{z}}) + \log \Pr(\tilde{\mathbf{z}})\}. \quad (7)$$

The MAP estimate provides mechanisms for incorporating both the closeness of fit and any prior information known about the signal structure.

To use the MAP technique to estimate the desired information **z**, appropriate statistical models for the conditional distribution Pr(**y** | **z**) and the prior Pr(**z**) must be developed. Given the model of data acquisition, the conditional distribution is related to the noise distribution by

$$\Pr(\mathbf{y} \mid \mathbf{z}) = \Pr(\mathbf{n})|_{\mathbf{n}=\mathbf{y}-\mathbf{z}} \quad (8)$$

Thus it is necessary to develop accurate statistical models for the noise Pr(**n**) and for the signal Pr(**z**). The i.i.d. Gaussian is often used as a model for the additive-noise term, i.e.,

$$\Pr(\mathbf{y} \mid \mathbf{z}) = \frac{1}{\sqrt{(2\pi\sigma^2)^{N^2}}} \exp\left(-\frac{\| \mathbf{y} - \mathbf{z} \|^2}{2\sigma^2}\right). \quad (9)$$

This results in a least-squares maximum-likelihood estimate, i.e.,

$$\hat{\mathbf{z}} = \arg \max_{\tilde{\mathbf{z}} \in \Re^N} - \| \mathbf{y} - \tilde{\mathbf{z}} \|^2, \\ = \arg \min_{\tilde{\mathbf{z}} \in \Re^N} \| \mathbf{y} - \tilde{\mathbf{z}} \|^2. \quad (10)$$

The Gaussian model can also be used as a model for the source signal; this results in the MAP estimator

$$\hat{\mathbf{z}} = \arg \min_{\tilde{\mathbf{z}} \in \Re^N} \Big\{ \tilde{\mathbf{z}}^t \mathbf{B}^{-1} \tilde{\mathbf{z}} \\ + \frac{1}{\sigma^2} \| \mathbf{y} - \tilde{\mathbf{z}} \|^2 \Big\}, \quad (11)$$

where the matrix $\mathbf{B}^{-1}$ describes the relationship among a neighboring set of pixels in the image. The optimization problem is convex, and so the estimate $\hat{\mathbf{z}}$ can be easily computed. Such an estimation scheme has been used extensively in array signal processing [10], image restoration [11], and computer vision [12]. The problem for such a scheme is that in most applications (even the ones in which it is used) the Gaussian model is not an accurate model of the source signal. If the source signal has discontinuities, such an estimation scheme will smooth out discontinuities and produce smooth signal estimates.

2.2 *Markov Random Fields*

The Markov random field (MRF) model is a more versatile statistical model. A Gibbs distribution is used to write explicitly the distributions of MRF's. A Gibbs distribution is any distribution that can be expressed in the form

$$\Pr(\mathbf{x}) = \frac{1}{Z} \exp\left(-\sum_{c \in C} V_c(\mathbf{x})\right), \quad (12)$$

where Z is a normalizing constant, $V_c(\cdot)$ is any function of a local group of points c, and C is the set of all such local groups. Note that the Gaussian model is a special case of the MRF.

If an MRF model is used for the source model, the MAP estimation results in the following optimization problem:

$$\hat{\mathbf{z}} = \arg \min_{\tilde{\mathbf{z}} \in \Re^N} \Big\{ \sum_{c \in C_s} V_c(\tilde{\mathbf{z}}) \\ + \frac{1}{2\sigma^2} \| \mathbf{y} - \tilde{\mathbf{z}} \|^2 \Big\}, \quad (13)$$

where C_s denotes the set of functions that form the model for the signal MRF. A major difficulty in using such a model in MAP estimation is that it is very easy to construct a statistical model for which the resulting optimization problem is nonconvex [13]–[17]. This is especially true when the model is trying to incorporate the fact that discontinuities are present in the source signal. This poses many theoretical and practical problems since the optimization of a nonconvex functional is an ill-posed problem. Algorithms have been developed to form estimates for such nonconvex optimization problems [18]; however, the estimates are unstable. The algorithms are also very computationally complex in time or space or both, and thus they are not practical in many applications.

To understand how to include discontinuities into the statistical model, it is important to understand first what the model represents and how to define the model for a particular application. For the source signal $\mathbf{z}_1$, the value of the probability measure $\Pr(\mathbf{z}_1)$ is related to how closely $\mathbf{z}_1$ matches our prior information about the source. So a $\mathbf{z}_1$ that closely matches our prior information should have a higher probability than one that does not. For this to be true the functions $V_c(\cdot)$ should provide a measure of the consistency of a particular $\mathbf{z}$, where a $\mathbf{z}$ that is more consistent with the prior information will have smaller values of $V_c(\cdot)$. The situation that is important in this work occurs when the prior information is mostly true but a limited amount of inconsistency is allowable (e.g., a piecewise-smooth surface, that is, a surface that is mostly smooth but for which a few discontinuities are allowable).

This paper uses a special form of the MRF that has this very desirable property. The model is characterized by a special form of the Gibbs's distribution

$$\Pr(\mathbf{x}) = \frac{1}{Z}\exp\left(-\frac{1}{\lambda}\sum_{c\in C}\rho_T\left(\mathbf{d}_c^t\mathbf{x}\right)\right) \tag{14}$$

where λ is a scalar constant that is greater than zero, $\mathbf{d}_c$ is a collection of linear operators, and the function $\rho_T(\cdot)$ is given by

$$\rho_T(x) = \begin{cases} x^2, & x \le T, \\ T^2 + 2T\mid x - T\mid, & x > T; \end{cases} \tag{15}$$

see figure 1(a). Since $\rho_T(\cdot)$ is convex, this particular form of the MRF results in a convex optimization problem when used in the MAP estimation formulation in (13). Therefore such MAP estimates are unique, are stable, and can be computed efficiently. The function $\rho_T(\cdot)$ is known as the Huber minimax function [19], [20], and for that reason this statistical model is referred to as the Huber–Markov random-field (HMRF) model.

For this distribution the linear operators $\mathbf{d}_c$ provide the mechanism for incorporating what is considered consistent most of the time, whereas the function $\rho_T(\cdot)$ is the mechanism for permitting some inconsistency. The parameter T controls the amount of inconsistency allowable. The function $\rho_T(\cdot)$ permits some inconsistency by reducing the importance of the consistency measure when the value of the consistency measure exceeds some threshold T.

The fact that the difference between a pixel and its local neighbors should be small is used for the measure of consistency, that is, there should be few local variations in the image. For this assumption an appropriate set of consistency measures is

$$\begin{aligned}&\{\mathbf{d}_c\mathbf{z}\}_{c\in C_s}\\ &= \{\mathbf{z}_{m,n} - \mathbf{z}_{k,l}\}_{k,l\in\mathcal{N}_{m,n},1\le m,n\le N},\end{aligned} \tag{16}$$

where $\mathcal{N}_{m,n}$ consists of the eight nearest neighbors of the pixel located at (m,n). Across discontinuities this measure is large, but the relative importance of the measure at such a point is reduced because of the use of the Huber function.

2.3 *Filtering Algorithm for Gaussian Noise*

If the HMRF model is used as the image model and an i.i.d. Gaussian noise model is used, the MAP estimator becomes

$$\begin{aligned}\hat{\mathbf{z}} = \arg\min_{\tilde{\mathbf{z}}\in\Re^N} &\left\{\sum_{1\le m,n\le N}\ \sum_{k,l\in\mathcal{N}_{m,n}}\rho_T(\tilde{\mathbf{z}}_{m,n} - \tilde{\mathbf{z}}_{k,l})\right.\\ &\left.+\frac{\lambda}{2\sigma^2}\sum_{1\le m,n\le N}(\mathbf{y}_{m,n} - \tilde{\mathbf{z}}_{m,n})^2\right\}.\end{aligned} \tag{17}$$

The choice of the image-model parameters λ and T are, strictly speaking, signal dependent, but, as will be shown, the parameters can be chosen to work for a large class of images. The implementation of this optimization procedure is discussed in section 4, and example results are shown in section 5.

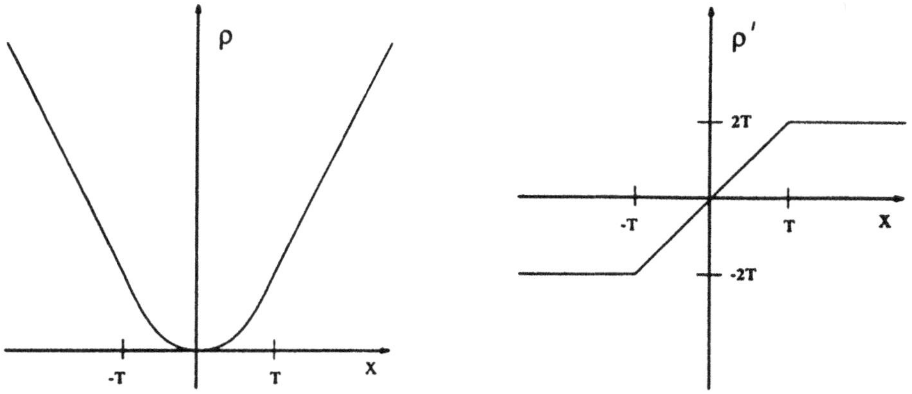

Fig. 1. (a) $\rho(x)$, (b) $\rho'(x)$.

3 Impulse Detection and Filtering

Corruption by an impulsive-noise process commonly occurs in image data. This source of noise can come from several types of physical processes, such as defective sensors or bit errors in image transmission systems. This section addresses the problem of removing these outliers from the image data.

3.1 Inclusion of Impulse Information

If a pixel in the image is corrupted by an impulse, this pixel value provides no information about the true, underlying image value at that point. This lack of information from a particular observed data value can be modeled by a uniform conditional distribution, i.e.,

$$\Pr(\mathbf{y}_{m,n} \mid \mathbf{z}) = \frac{1}{R} \tag{18}$$

if (m, n) is corrupted by an impulse, where R is the range of values for the observed data (e.g., $R = 256$ for 8-bit data). If ϵ is the percentage of the image that is corrupted by impulses, the conditional distribution in (9) becomes

$$\Pr(\mathbf{y} \mid \mathbf{z}) = \frac{1}{R^{\epsilon N^2}\sqrt{(2\pi\sigma^2)^{(1-\epsilon)N^2}}} \exp\left(-\frac{(\mathbf{y}-\mathbf{z})^t I_{\mathcal{I}}(\mathbf{y}-\mathbf{z})}{2\sigma^2}\right), \tag{19}$$

where $I_{\mathcal{I}}$ is a diagonal matrix with values along the diagonal equal to 1 if that point is corrupted with Gaussian noise and equal to 0 if that point is corrupted with impulsive noise. In Huber's work on robust estimation of one-dimensional signals corrupted with an outlier process [19], [20], he suggests modifying the i.i.d. Gaussian noise model by replacing the squared term in the exponent with the Huber function (15), i.e.,

$$\Pr(\mathbf{y} \mid \mathbf{z}) = \frac{1}{Z}\exp\left(-\frac{1}{2\sigma^2}\sum_{1\le m,n\le N} \rho_T(\mathbf{y}_{m,n} - \mathbf{z}_{m,n})\right). \tag{20}$$

Using this idea for image data does not work well. This model does not completely reduce the influence of outliers, but, more important, it also treats data along edges as outlier data and leads to increased smoothing of edges.

If the noise model in (19) is used for the noise process and the same HMRF model is used for the image model, the MAP estimator in (17) becomes

$$\hat{\mathbf{z}} = \arg\min_{\hat{\mathbf{z}}\in\Re^N} \left\{ \sum_{1\le m,n\le N}\sum_{k,l\in\mathcal{N}_{m,n}} \rho_T(\hat{\mathbf{z}}_{m,n} - \hat{\mathbf{z}}_{k,l}) + \frac{\lambda}{2\sigma^2}\sum_{(m,n)\notin\mathcal{I}} (\mathbf{y}_{m,n} - \hat{\mathbf{z}}_{m,n})^2 \right\}, \tag{21}$$

where $\mathcal{I}$ denotes the set of (m, n) that is corrupted by impulse. As should be expected, the MAP estimator essentially throws away those points that provide no information about the underlying image data. From a practical point of view, the problem that remains is how to determine which of the data points are corrupted by impulses from only the observable data $\mathbf{y}$ without knowledge of the true distribution (19).

3.2 Impulse Detection

The simple idea of comparing a data point in question with a filtered version of the image is used for performing the impulse detection [21]. If the difference between the original data point and the filter is substantial, then that location is marked as an impulse location, i.e., if

$$| \mathbf{y}_{m,n} - F(\mathbf{y})_{m,n} | \geq D, \text{ then } (m,n) \in \mathcal{I}_D. \quad (22)$$

The filtered image $F(\mathbf{y})$ can be computed by many types of linear- or nonlinear-filtering algorithms; results from several of the filters tried in this research will be described in subsection 3.3.

The ideal filter $F(\cdot)$ would barely modify data corrupted by Gaussian noise (i.e., $\mathbf{y}_{m,n} \approx F(\mathbf{y})_{m,n}$) while easily detecting and removing outliers (i.e., $\mathbf{y}_{m,n} - F(\mathbf{y})_{m,n}$ is large). The threshold D would then be picked to differentiate between these two regions. The filter should also be robust to image structures, such as edges, so that it does not misclassify edge pixels as impulse-corrupted data. These properties indicate that the class of robust image filters developed over the years, such as the various median, stack, and morphological filters, should perform well for impulse detection.

3.3 Comparison of Robust Filters for Impulse Detection

To fairly compare the many possible choices for the image filter in the impulse-detection stage, we established a criterion based on our experience. An image (different from that of the example shown in section 5) was corrupted with an ϵ-corrupted Gaussian noise source (ϵ varied from 5% to 20%), and the true locations of the impulses $\mathcal{I}$ were stored for comparison. For each filter under consideration the

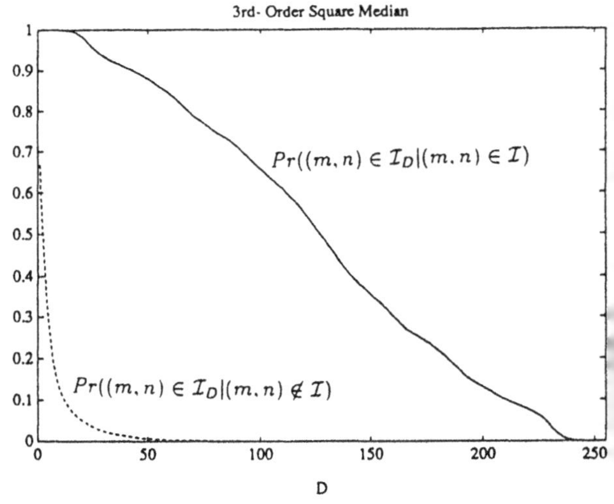

Fig. 2. Percentage of impluses correctly detected and percentage misclassified.

set of impulses $\mathcal{I}_D$ detected for each possible threshold D was computed and was compared with the true set of impulses $\mathcal{I}$. From this, the percent of impulses correctly classified (i.e., an estimate of $\Pr((m,n) \in \mathcal{I}_D \mid (m,n) \in \mathcal{I})$) and the percentage of Gaussian-corrupted points misclassified as impulses (i.e., an estimate of $\Pr((m,n) \in \mathcal{I}_D \mid (m,n) \notin \mathcal{I})$) were computed. These percentages were computed for the range of possible values of D. An example plot of these percentages for a 3rd-order square median (7×7 square window) is shown in figure 2.

It is desirable to choose a threshold D such that all impulses are detected and no locations are misclassified (i.e., $\Pr((m,n) \in \mathcal{I}_D \mid (m,n) \in \mathcal{I}) \approx 1$ while $\Pr((m,n) \in \mathcal{I}_D \mid (m,n) \notin \mathcal{I}) \approx 0$). This causes a conflict since it is not possible to meet both goals. Thus a compromise must be made. Since not detecting an impulse means that a sharp and noticeable peak will appear in the output image, whereas misclassifying a Gaussian-corrupted data point means only that some potentially useful information is thrown away, it is more important that almost all of the impulses be detected (i.e., it is more important that $\Pr((m,n) \in \mathcal{I}_D \mid (m,n) \in \mathcal{I}) \approx 1$ than that $\Pr((m,n) \in \mathcal{I}_D \mid (m,n) \notin \mathcal{I}) \approx 0$). This suggests that a Neyman–Pearson-type criterion should be used for choosing the decision threshold D, that is, choose D by

$$D = \arg \max_{D \in [0,R]} \{D : \Pr((m,n) \in \mathcal{I}_D \mid (m,n) \in \mathcal{I}) \geq \alpha\}. \quad (23)$$

Setting $\alpha = 0.98$ ensured that, for the most part, all of the most significant impulses were correctly identified. As was mentioned above, it is also desirable to have $\Pr((m,n) \in \mathcal{I}_D \mid (m,n) \notin \mathcal{I}) \approx 0$; this provides a criterion by which to rank the various filters. For each filter, the threshold D is chosen by (23) and the filters are ranked based on how close $\Pr((m,n) \in \mathcal{I}_D \mid (m,n) \notin \mathcal{I})$ is to zero. Table 1 shows results of this ranking for various types and orders of robust filters. References to the filter descriptions are also provided in the table. Based on the chosen criterion, the 3rd-order multilevel median [24] has the best ability to discern between impulsive and Gaussian contamination. Example results for a severely corrupted image are shown in section 5.

4 Implementations

The previous sections presented a mathematically well-posed technique for computing high-resolution and robust signal estimates. The implementation of the impulse detector is standard and thus is not discussed in this paper. The optimization algorithm arising from the MAP estimator is fairly complex and is discussed in this section. The technique results in a convex but nonquadratic functional-minimization problem. This section examines the computational issues of the resulting mathematical problem statement and examines two very different approaches to forming estimates. One of the advantages of this class of estimators is the inherent parallelism associated with these algorithms. This parallelism can be used to greatly reduce the total computational time. First, the implementation of the algorithm on a general-purpose digital computer is examined. Next, analog techniques for computing the estimate are discussed. The digital implementation has the advantage of being easy to program, whereas the analog implementation has the advantage of extremely fast computation. Thus, depending on the requirements of the application, either flexibility or extremely high computation speeds can be achieved.

4.1 Digital Network

The most prevalent class of algorithms for digital convex functional minimization are based on iterative techniques, in which the update at each iteration monotonically decreases the functional value. Let $M(\mathbf{z},\mathbf{y})$ be the resulting functional-minimization problem obtained from the MAP estimation procedure. Let $\mathbf{z}^k$ denote the function value at the kth iteration. At each iteration the function is updated by

$$\mathbf{z}^{k+1} = \mathbf{z}^k + \alpha^k \mathbf{p}^k, \quad (24)$$

where the vector $\mathbf{p}^k$ is the direction of the update and the scalar α^k determines the size of the step taken in that direction. Since our function is convex, any of the descent-based methods will converge to the optimal solution, given any starting vector $\mathbf{z}^0$. However, overall computation time will depend on the initial guess $\mathbf{z}^0$, the scheme for choosing the update vector $\mathbf{p}^k$, and the step size α^k. For a particular application, computation time can be dramatically reduced if some quick technique is used to form a rough estimate to be used as the initial guess [25], [26]. An intuitive choice for the update vector is the direction of steepest descent, that is, the direction for which the functional will decrease the fastest. The direction of steepest descent is the negative of the gradient vector, i.e.,

$$\begin{aligned}\mathbf{p}^k &= -\nabla M(\mathbf{z}^k,\mathbf{y}) \\ &= -\left\{ \sum_{c \in C_s} \rho'_T(\mathbf{d}_c \mathbf{z}^k)\mathbf{d}_c + \frac{\lambda}{\sigma^2}\mathbf{I}_{\mathcal{I}}(\mathbf{y} - \mathbf{z}^k) \right\},\end{aligned} \quad (25)$$

where $\rho'_T(\cdot)$ is the first derivative of $\rho_T(\cdot)$, as shown in figure 1(b). Once an update vector is chosen, the next step is to compute the size of the step α^k that will be taken in that direction. The maximum decrease for a given $\mathbf{p}^k$ occurs when α^k is chosen so as to minimize the functional along that direction,

$$\alpha^k = \arg \min_{\alpha \in \Re} M(\mathbf{z}^k + \alpha \mathbf{p}^k, \mathbf{y}). \quad (26)$$

Table 1. Comparison of impulse-detection filters.

Estimation Method	References	Order	D	$\Pr((m,n) \in \mathcal{I}_D \mid (m,n) \notin \mathcal{I})$
Square median	1,22	1	17	0.017055
		2	19	0.036018
		3	20	0.050518
		4	21	0.059532
Cross median	1,22	1	15	0.012359
		2	17	0.026103
		3	18	0.036472
		4	19	0.042945
Max/median	23	1	0	1.000000
		2	19	0.034812
		3	22	0.047900
		4	23	0.047717
Seperable median	1	1	17	0.015365
		2	18	0.035774
		3	20	0.046966
		4	21	0.056224
Multilevel median	24	1	0	1.000000
		2	6	0.019488
		3[a]	13	0.008098
		4	14	0.011051

[a] Best filter.

This results in a univariate nonquadratic minimization problem. It is not necessary to compute the optimal step size at each step; instead, this one-dimensional minimization problem can be approximated by minimization of the osculating parabola,

$$\alpha^k = \arg\min_{\alpha\in\Re} \Big\{ M(\mathbf{z}^k,\mathbf{y}) + \alpha \nabla M(\mathbf{z}^k,\mathbf{y})^t \mathbf{p}^k + \frac{1}{2}\alpha^2 \mathbf{p}^{k^t} \nabla^2 M(\mathbf{z}^k,\mathbf{y})\mathbf{p}^k \Big\}. \quad (27)$$

In this case α^k is well defined and is given by

$$\alpha^k = \frac{\nabla M(\mathbf{z}^k,\mathbf{y})^t \mathbf{p}^k}{\mathbf{p}^{k^t} \nabla^2 M(\mathbf{z}^k,\mathbf{y})\mathbf{p}^k}, \quad (28)$$

where

$$\nabla^2 M(\mathbf{z}^k,\mathbf{y}) = \sum_{c\in C_s} \rho_T''(\mathbf{d}_c\mathbf{z}^k)\mathbf{d}_c\mathbf{d}_c^t - \frac{\lambda}{\sigma^2}\mathbf{I}_{\mathcal{I}} \quad (29)$$

and ρ_T'' is the second derivative of $\rho_T(\cdot)$. For the first several iterations, while the step size is relatively large and the approximation is poor, it is thus necessary to verify that the step taken does in fact reduce the functional value. If it does not, then the step value is reduced by half until an appropriate step size that reduces the functional is found.

This process iterates until the change in the image during a particular step is small (e.g., it stops when $\| \alpha^k \mathbf{p}^k \|^2 < 10$). The number of iterations needed to converge to the solution varies with the amount of smoothing required. For little smoothing ($\lambda/2\sigma^2$ large) the process will converge in two or three iterations, whereas for large amounts of smoothing the number of iterations will increase. Note that because the consistency measure consists of only local computations, it is possible to significantly reduce the total computation time through the use of a tightly coupled mesh of parallel-processing nodes [27].

4.2 *Analog Network*

Variational principles, such as the functional-minimization problem arising from MAP estimation theory, have also been solved by using analog networks [28], [29]. This section examines solving the discontinuity-preserving functional by using analog electrical networks. The class of variational principles that can be solved by an analog electrical network is dictated by

Tellegen's theorem [30]. Basically, the properties of convexity and well-posedness ensure that the resulting network will compute the correct solution. For quadratic variational principles the resulting analog network can be made with linear components; however, for the nonquadratic variational principles presented here the resulting networks will have nonlinear components. The case of a quadratic functional-minimization problem, built with linear components, has been examined [12], [31]–[33], and at least one working VLSI chip has been designed and tested [33].

To ease the notational jumble, this section examines only the construction of a network that will solve the problem of estimating a Gaussian-corrupted one-dimensional signal with discontinuities. Extensions to the two-dimensional case should be clear. For piecewise-smooth signal reconstruction, the form of the variational principle is

$$M(\mathbf{z},\mathbf{y}) = \sum_{i=1}^{N} \rho_T(\mathbf{z}_{i+1} - \mathbf{z}_i) + \frac{\lambda}{2\sigma^2} \sum_{i \in \mathbf{I}_\mathcal{I}} (\mathbf{z}_i - \mathbf{y}_i)^2. \quad (30)$$

By taking the derivative of this functional with respect to $\mathbf{z}_i$ and setting the equation equal to zero the following N nonlinear equations are obtained:

$$0 = \rho_T'(\mathbf{z}_i - \mathbf{z}_{i-1}) - \rho_T'(\mathbf{z}_{i+1} - \mathbf{z}_i) + \frac{\lambda}{\sigma^2}(\mathbf{z}_i - \mathbf{y}_i)I_i, \quad (31)$$

where I_i is an indicator function

$$I_i = \begin{cases} 1, & i \notin \mathcal{I}, \\ 0, & i \in \mathcal{I}. \end{cases} \quad (32)$$

The equations at the boundaries are slightly different. Examine the network in figure 3. The two-terminal passive device ξ is characterized by the voltage–current relationship $I = \rho_T'(V)$. For this network the following N equations at each node can, by using Kirchhoff's current law, be written

$$0 = \rho_T'(\mathbf{z}_i - \mathbf{z}_{i-1}) - \rho_T'(\mathbf{z}_{i+1} - \mathbf{z}_i) + \frac{1}{R}(\mathbf{z}_i - \mathbf{y}_i)I_i; \quad (33)$$

this has the same form as equation (31). Therefore if the network is excited with the constraint-controlled voltage sources, the node voltages $\mathbf{z}$ represent the solution to the variational principle. Theoretically this solution will be obtained instantaneously; however, capacitance in any real implementation will cause transients when the constraints are applied. Once the network has settled, the solution to the variational principle can be obtained by measuring the node voltages. The nonlinear circuit elements can be constructed through the use of nonlinear devices, such as diodes or zener diodes, and through the use of active devices, such as operational amplifiers [28], [29], [34].

5 Examples

As a demonstration of the type of image estimates that can be obtained with this method, the noise-free image in figure 4(a) was corrupted with both i.i.d. Gaussian noise, $N(0, 20)$, and an ϵ-corrupted Gaussian noise source, $N(0, 20)$ and $\epsilon = 5\%$. The Gaussian-corrupted image is shown in figure 4(b), and the mixed-noise-corrupted image is shown in figure 4(c).

5.1 *Filtering Gaussian-Corrupted Image Data*

For images corrupted with just Gaussian noise (figure 4(b)), the algorithm in subsection 2.3 is used to form the estimate. This case will be examined so that the properties of the HMRF model can be shown. The HMRF model has two parameters, T and λ, which control the characteristics of the estimated data. The value of T depends on the degree to which discontinuities should be retained. Discontinuities below the value of T will be smoothed with a Gaussian-type smoother, whereas discontinuities above the value of T will be better retained. As a result of the smooth transition between the quadratic and linear regions of $\rho(\cdot)$ (since $\rho'(\cdot)$ is continuous),

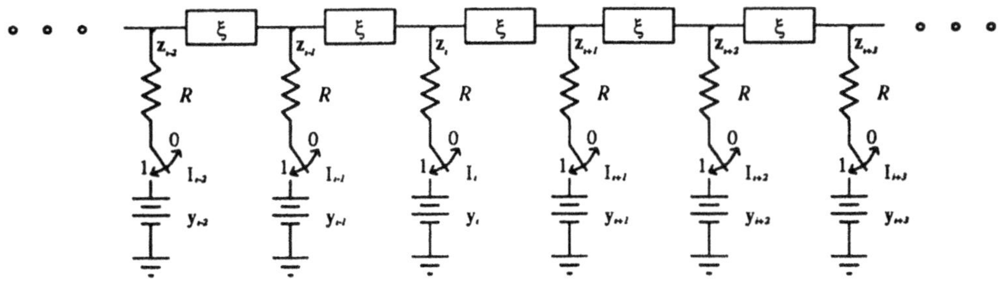

Fig. 3. Analog network for solving the one-dimensional filtering problem.

the estimate is not very sensitive to this parameter. Changes as much as ±20% in the value of T make no perceivable change in the quality of the estimate. For natural image data, values of T in the range of 2 to 10 generally produce the best results. For synthetic image data, T in the range of 0.1 to 1 produces good results. Figure 5 shows the signal estimates obtained for $\lambda = 0.5$ and T varying from 1.0 to 10.0. Notice that as the value of T increases, fewer discontinuities are retained and the estimates appear smoother.

The value of λ controls the level of smoothing within a nondiscontinuous region. For smaller values of λ the regions between discontinuities will be smoothed to a greater extent. Figure 6 shows four levels of smoothing for the case in which $T = 5.0$. Notice that λ can be varied from very little smoothing ($\lambda = 5.0$) to extreme oversmoothing ($\lambda = 0.1$). It is interesting that even in the extreme-oversmoothing case ($\lambda = 0.1, T = 5.0$), the discontinuities between regions are still maintained.

The accuracy with which this technique can estimate the original uncorrupted data was shown by computing the error between the original and estimated signals for the HMRF model and for several other popular nonlinear-filtering structures. Table 2 compares the resulting mean-squared-error (MSE) and the mean-absolute-error (MAE) computations. In each case the filter parameters that produced the smallest MSE and MAE were used and are given in the table. References to the filter descriptions are also given in the table. Notice that using the HMRF model produces image estimates that are much closer to the original uncorrupted image data. Figure 7 shows an expanded view of part of the estimated image for both the HMRF estimate and for other methods that produced the most visually pleasing artifacts. Notice that HMRF models the true underlying image data much more accurately than do any of the other methods. The other estimators leave disturbing visual structures from the noise.

5.2 *Filtering Images Corrupted by Mixed Noise*

For images contaminated by a mixed-noise process, the multilevel median was used to predetect impulse locations, and the estimator described by equation (21) was used to smooth the estimate. The decision threshold for the estimate was set to $D = 15$, and the impulse locations from figure 4(c) are shown in figure 8(a). Notice that although there are obviously Gaussian-corrupted points misclassified as impulses in figure 8(a), these locations are spread uniformly throughout the image. This is important because if such mistakes were concentrated along edge regions, throwing out that information would lead to severe degradation of the estimate in that area. The final image estimate computed from (21) is shown in figure 8(b).

As with the Gaussian noise example in subsection 5.1, the image estimate obtained with the HMRF estimator was compared with that obtained with other popular estimation techniques. Table 3 compares several filters quantitatively, and figure 9 compares the output from several estimators visually. As before, the HMRF estimator outperforms the other popular estimators both quantitatively and qualitatively.

(a) (b) (c)

Fig. 4. (a) Original image "Lena," (b) image corrupted with $N(0,20)$, and (c) image corrupted mixed noise, $N(0,20)$ and $\varepsilon = 5\%$.

(a) (b) (c) (d)

Fig. 5. Image estimates for $\lambda = 0.5$ and (a) $T = 1.0$, (b) $T = 2.0$, (c) $T = 5.0$, and (d) $T = 10.0$.

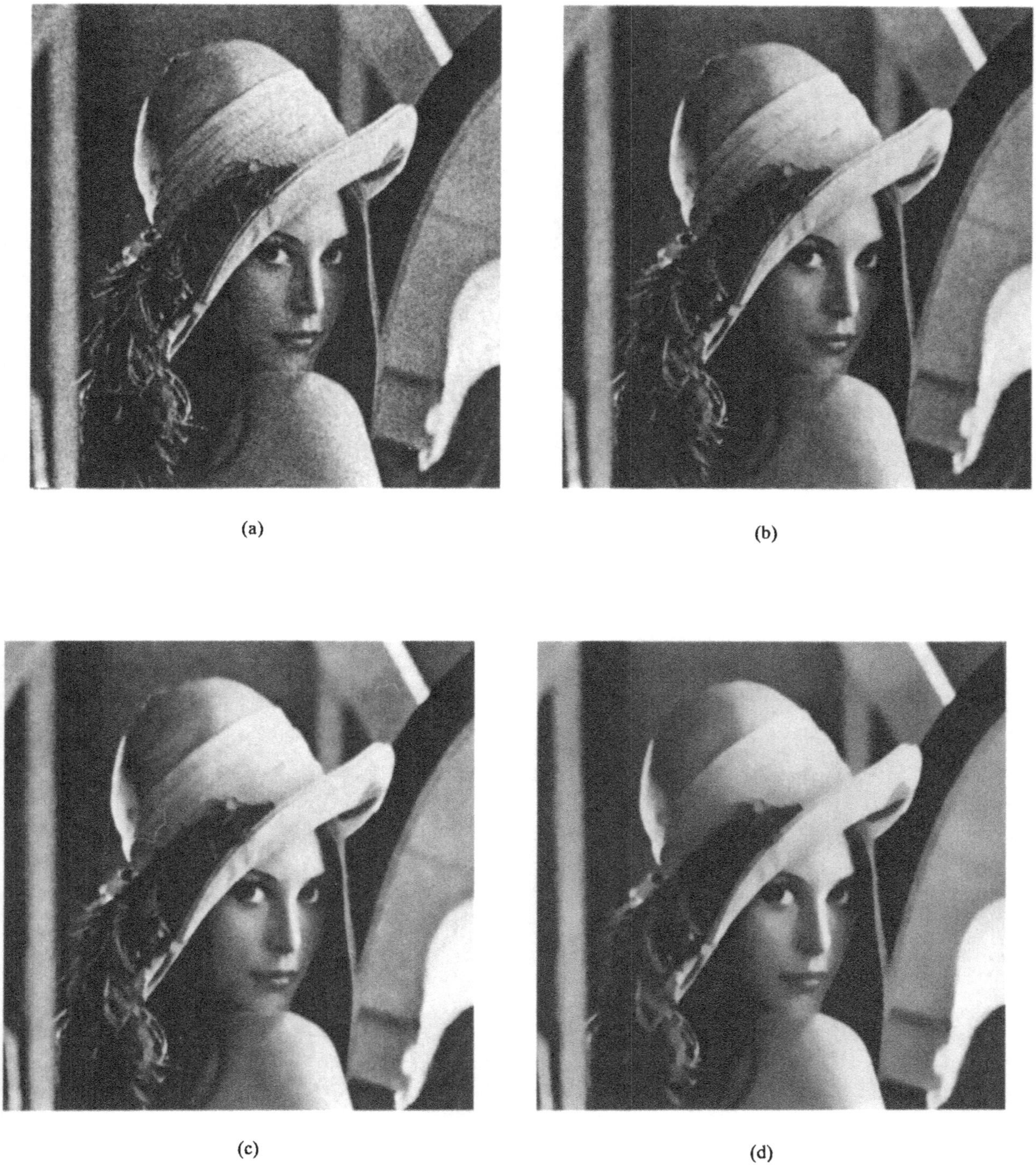

Fig. 6. Image estimates for $T = 5.0$ and (a) $\lambda = 5.0$, (b) $\lambda = 1.0$, and (c) $\lambda = 0.5$, (d) $\lambda = 0.1$.

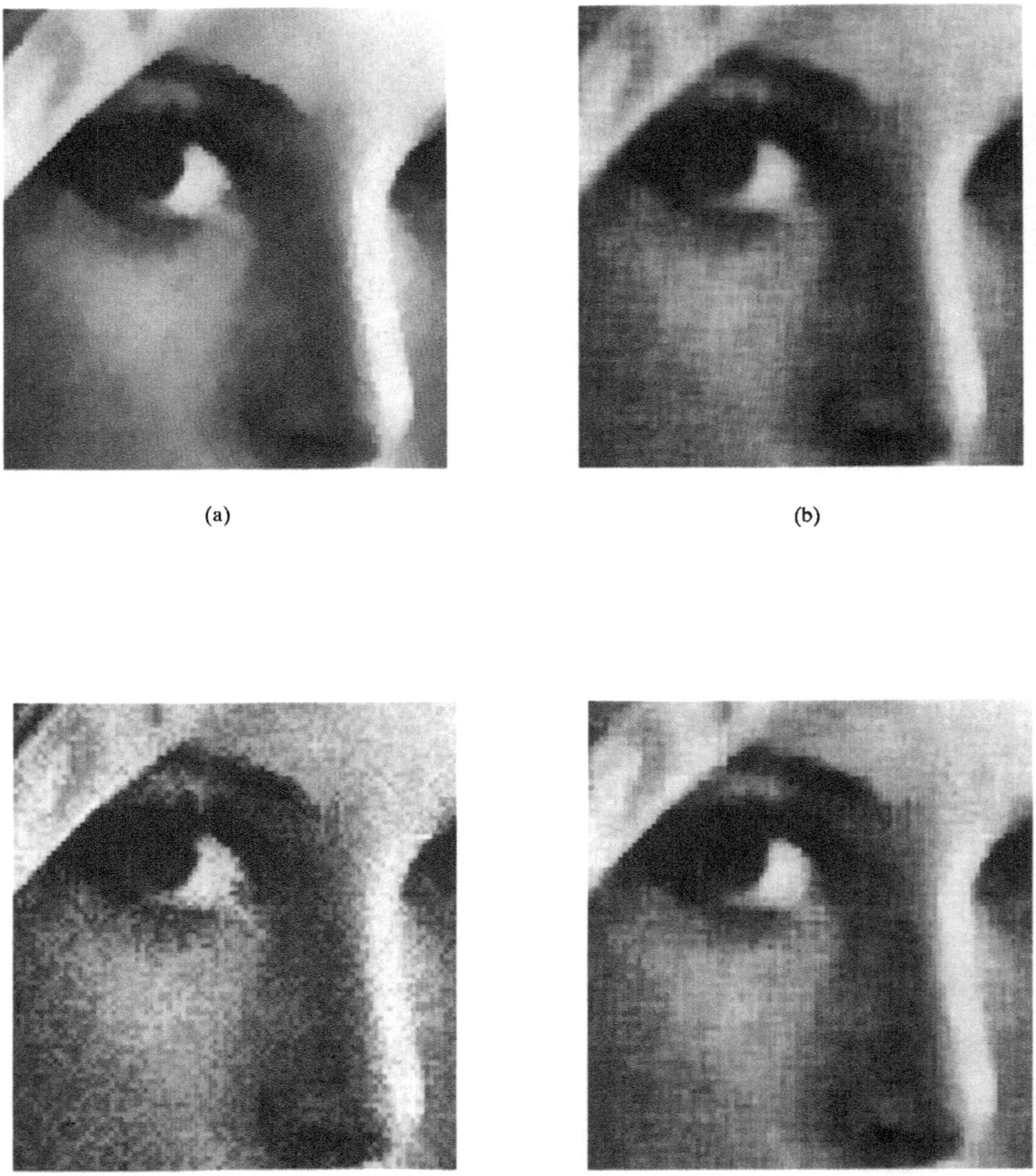

Fig. 7. Image estimates for (a) HMRF, $T = 5.0$, $\lambda = 0.5$, (b) α-trimmed mean, 7×7 cross, $\alpha = 3$, (c) multilevel median, 11×11, and (d) cross median, 7×7.

(a)

(b)

Fig. 8. (a) $\mathcal{I}_D$ for $D = 15$ and (b) reconstructed image

6 Conclusion

This paper proposes a new nonlinear estimation procedure for estimating image data that have been corrupted by Gaussian and impulsive noise. The method uses a MAP estimator with an image model based on the Huber–Markov random field. By using this model the designer has complete control over the degree to which discontinuities are retained and the degree to which the image is smoothed between discontinuities. Examples demonstrate the flexibility and the quality of results that can be obtained with this method. An impulse detector was also designed to indicate the locations of impulsive contamination. By using this detector impulsive noise can also be effectively removed from corrupted data. This makes this estimator design especially useful for environments that contain mixed-noise sources. Implementation of the estimator on both digital and analog computational devices was also discussed.

Currently we are studying several extensions that will make this approach useful in a greater variety of environments. First, when the entire image is not available at once, such as in filtering video data, it is necessary to form estimates based on partial information. Suboptimal estimates can be made by forming a window operator that estimates the value at a pixel based on some set of its neighbors. Depending on the degree of smoothing, suboptimal estimates that are very close to the optimal estimates can be obtained with windows as small as 5×5. We are also currently extending the image model so that it is effective for filtering color images (beyond the trivial extension of filtering each color plan separately). By including cross-correlation terms between color planes it should be possible to improve the color image estimate.

Acknowledgment

The authors would like to thank Ms. Danielle M. Walker for her assistance in simulating the many examples presented in this paper. This work was supported in part by the University of

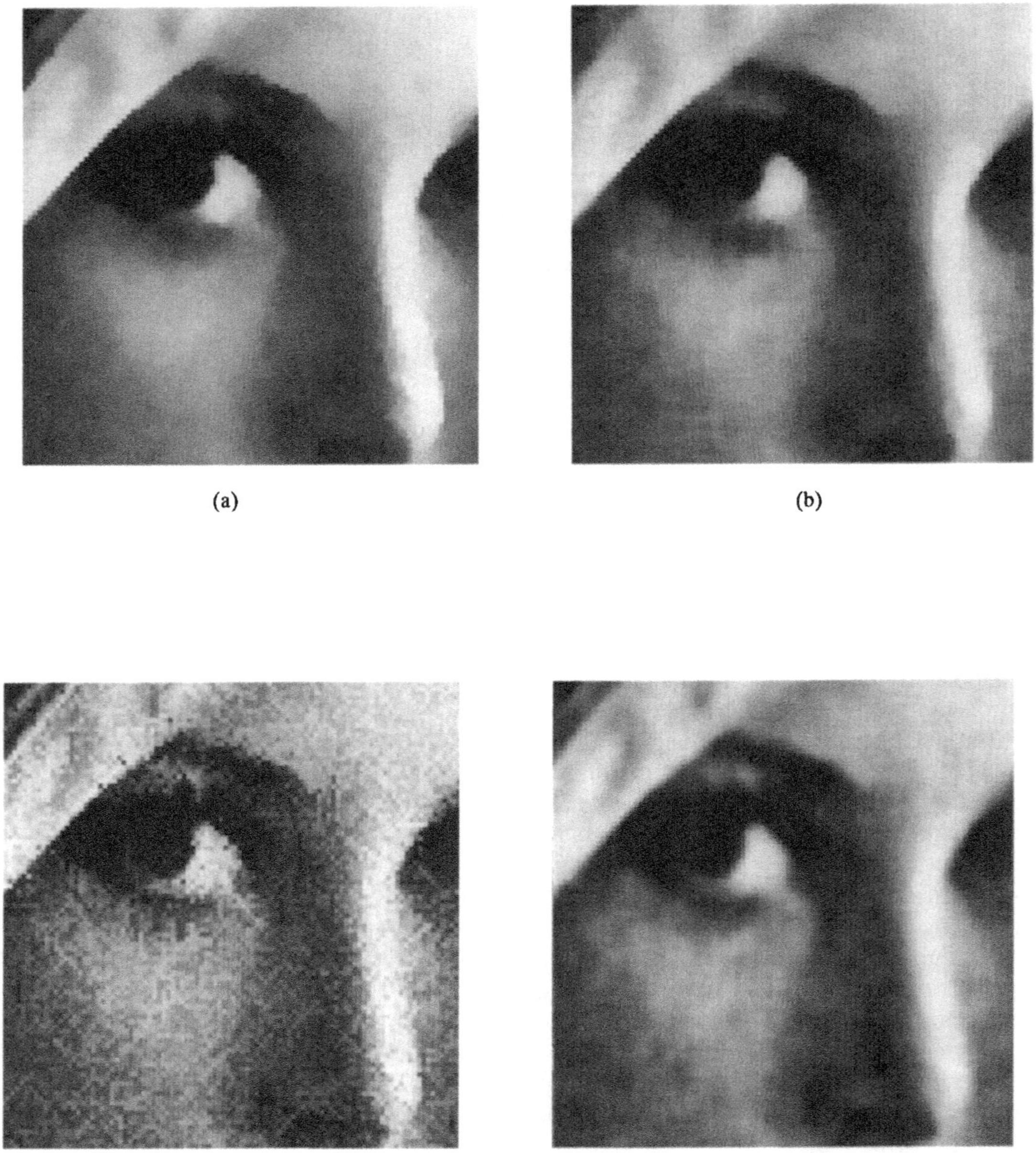

(a) (b) (c) (d)

Fig. 9. Image estimates for (a) multilevel/HMRF, $D = 15$, $T = 5.0$, $\lambda = 0.5$, (b) square median, 5×5, (c) multilevel median, 11×11, and (d) α-trimmed mean, $5\times$ square, $\alpha = 5$.

Table 2. Comparison of estimation methods for Gaussian contamination only.

Estimation Method	References	MSE	MAE
HMRF model ($\lambda = 1.0$, $T = 5.0$)	–	0.005327	0.061209
Mean (3×3 square)	–	0.006956	0.071079
Median (5×5 square)	1,22	0.008183	0.071100
α-Trimmed mean (5×5 cross, $6\alpha = 3$)	6	0.008275	0.078529
Morphological (4th-order open/close)	2,8,9	0.009299	0.082729

Table 3. Comparison of estimation methods for Gaussian and impulsive contamination.

Estimation Method	References	MSE	MAE
Multilevel median/HMRF model ($\lambda = 1.0, T = 5.0$)	–	0.009232	0.071335
Median (5×5 square)	1,22	0.011180	0.076306
α-Trimmed mean (5×5 square, $\alpha = 5$)	6	0.012390	0.078469
Morphological (4th-order open/close)	2,8,9	0.011375	0.089538

Notre Dame's Jesse H. Jones Faculty Research Fund.

References

1. G.R. Arce, N.C. Gallagher, and T.A. Nodes, "Median filter: theory for one- and two-dimensional filters," in *Advances in Computer Vision and Image Processing*, vol. 2, T.S. Huang, ed., JAI Press: Greenwich, CT, 1986, pp. 89–165.
2. R.L. Stevenson and G.R. Arce, "Morphological filters: statistics and further syntactic properties," *IEEE Trans. Circuits and Systems*, vol. CAS-34, pp. 1292–1304, 1987.
3. P. Maragos and R.W. Schafer, "Morphological filters—part I: their set-theoretic analysis and relations to linear shift-invariant filters," *IEEE Trans. Acoust., Speech, Signal Process.*, vol. ASSP-35, pp. 1153–1169, 1987.
4. P. Maragos and R.W. Schafer, "Morphological filters—part II: their relations to median, order-statistic, and stack filters," *IEEE Trans. Acoust., Speech, Signal Process.*, vol. ASSP-35, pp. 1170–1184, 1987.
5. A.C. Bovik, T.S. Huang, and D.C. Munson, "A generalization of median filtering using linear combinations of order statistics," *IEEE Trans. Acoust., Speech, Signal Process.*, vol. ASSP-31 pp. 1342–1349, 1983.
6. J.B. Bednar and T.L. Watt, "Alpha-trimmed means and their relationship to median filters," *IEEE Trans. Acoust., Speech, Signal Process.*, vol. ASSP-32, pp. 145–153, 1988.
7. A. Restrepo and A.C. Bovik, "Adaptive trimmed mean filters for image restoration," *IEEE Trans. Acoust., Speech, Signal Process.*, vol. ASSP-36, pp. 1326–1337, 1988.
8. J. Song and E.J. Delp, "A generalization of morphological filters using multiple structuring elements," in *Proc. 1989 IEEE Internat. Symp. on Circuits and Systems*, Portland, OR, May 1989, pp. 991–994.
9. J. Song, R.L. Stevenson, and E.J. Delp, "The use of mathematical morphology in image enhancement," in *Proc. 32nd Midwest Symp. on Circuits and Systems*, Champaign, IL, August 1989, pp. 89–165.
10. S. Haykin (ed.) *Array Signal Processing*, Prentice-Hall: Englewood Cliffs, NJ, 1985.
11. A.N. Tikhonov and A.V. Goncharsky (eds.), *Ill-Posed Problems in the Natural Sciences*, MIR Publishers, Moscow, 1987.
12. T. Poggio, V. Torre, and C. Kock, "Computational vision and regularization theory," *Nature*, vol. 317, pp. 314–319, 1985.
13. S. Geman and D. Geman, "Stochastic relaxation, Gibbs distributions, and the Bayesian restoration of images," *IEEE Trans. Patt. Anal. Mach. Intell.*, vol. PAMI-6, pp. 721–741, 1984.
14. S. Geman and D. McClure, "Statistical methods for tomographic image reconstruction," *Bull. Internat. Statist. Inst.*, vol. LII-4, pp. 5–21, 1987.
15. A. Blake and A. Zisserman, *Visual Reconstruction*, MIT Press: Cambridge, Massachusetts, 1987.
16. T. Hebert and R. Leahy, "A generalized EM algorithm for 3D Bayesian reconstruction from Poisson data using Gibbs priors," *IEEE Trans. Med. Imag.*, vol. 8, pp. 194–202, 1989.
17. K. Lange, "Convergence of EM image reconstruction algorithms with Gibbs priors," *IEEE Trans. Med. Imag.*, vol. 9, pp. 84–93, 1990.
18. S. Kirkpatrick, C.D. Gelatt, and M.P. Vecchi, "Optimization by simulated annealing," *Science*, vol. 220, pp. 671–680, 1983.
19. P.J. Huber, "Robust smoothing," in *Robustness in Statistics*, R.L. Launer and G.B. Wilkinson, eds., John Wiley: New York, 1981.
20. P.J. Huber, *Robust Statistics*, John Wiley: New York, 1981.
21. A. Rosenfeld and A.C. Kak, *Digital Picture Processing*, Academic Press: Orlando, FL, 1982.
22. N.C. Gallagher and G.L. Wise, "A theoretical analysis of the properties of median filters," *IEEE Trans. Acoust., Speech, Signal Process.*, vol. ASSP-29, pp. 1136–1141, 1981.
23. G.R. Arce and M.P. McLoughlin, "Theoretical analysis of the max/median filter," *IEEE Trans. Acoust., Speech,*

Signal Process., vol. ASSP-35, 1987, pp. 60–69.
24. G.R. Arce and R.E. Foster, "Detail-preserving ranked-order based filters for image processing," *IEEE Trans. Acoust., Speech, Signal Process.*, vol. 37, pp. 83–98, 1989.
25. J. Jou and A.C. Bovik, "Improved initial approximation and intensity-guided discontinuity detection in visible-surface reconstruction," *Comput. Vis. Graph., Image Process.*, vol. 4, pp. 292–325, 1989.
26. R.L. Stevenson and E.J. Delp, "Viewpoint invariant recovery of visual surfaces from sparse data," submitted to *IEEE Trans. Patt. Anal. Mach. Intell.*, vol. 14, No. 9, pp. 897–909, March 1992.
27. R.L. Stevenson, G.B. Adams, L.H. Jamieson, and E.J. Delp, "Three-dimensional surface reconstruction on the AT&T Pixel Machine," in *Proc. 24th Asilomar Conf. on Signals, Systems, and Computers*, Pacific Grove, CA, November 5–7, 1990, pp. 544–548.
28. C. Mead, *Analog VLSI and Neural Systems*, Addison-Wesley: Reading, MA, 1989.
29. J.G. Harris, "Analog models for early vision," Ph.D. thesis, California Institute of Technology, Pasadena, CA, 1991.
30. P. Penfield, R. Spence, and S. Duinker, *Tellegen's Theorem and Electrical Networks*, MIT Press: Cambridge, Massachusetts, 1970.
31. W.J. Karplus, "Analog circuits: solutions of field problems," McGraw-Hill: New York, 1958.
32. T. Poggio, "Early vision: from computational structure to algorithms and parallel hardware," in *Human and Machine Vision II*, A. Rosenfeld, ed., Academic Press: San Diego, CA, pp. 190–206, 1986.
33. J. Hutchinson, C. Koch, J. Luo, and C. Mead, "Computing motion using analog and binary resistive networks," *Computer*, vol. 21, No. 3, pp. 52-62, March 1988.
34. W.L. Hughes, *Nonlinear Electrical Networks*, Ronald Press: New York, 1960.

Robert L. Stevenson received a B.E.E. degree (summa cum laude) from the University of Delaware in 1986 and the Ph.D. in electrical engineering from Purdue University in 1990. While at Purdue he was supported by graduate fellowships from the National Science Foundation, duPont, and Phi Kappa Phi. He joined the faculty of the Department of Electrical Engineering at the University of Notre Dame in 1990, where he is currently an assistant professor. Dr. Stevenson is a member of Phi Kappa Phi, Tau Beta Pi, and Eta Kappa Nu. His research interests include multidimensional signal processing, electronic imaging, and computer vision.

Susan M. Schweizer is currently an undergraduate at the University of Notre Dame studying electrical engineering. Her expected graduation date is May 1993. Ms. Schweizer is the recipient of a Westinghouse Family Scholarship and a National Academy for Nuclear Training Scholarship. She is a member of Eta Kappa Nu and Tau Beta Pi.

Journal of Mathematical Imaging and Vision 2, 155–171 (1992).

Root-Signal Sets of Morphological Filters and their Use in Variable-Length BTC Image Coding

QIAOFEI WANG, MONCEF GABBOUJ, AND YRJÖ NEUVO
Signal Processing Laboratory, Tampere University of Technology, P.O. Box 553, SF-33101 Tampere, Finland

Abstract. The characterization of the root-signal set of a nonlinear operator has proved to be a crucial step in understanding the utility and usefulness of the operator. The set of root signals constitutes the passband of the nonlinear operator, and the complement of this set represents the stopband of the operator. Knowledge of these two sets for all operators determines which one must be used for any particular task. In this paper we investigate the root signals of the basic morphological filters, we study the properties of these signals, and we derive a system of equations to compute the number of binary-root signals for these morphological filters with structuring element of width k and signals of length n. The derivation is based on the state description for these root signals. Simple recursive equations are derived for counting the number of root signals of opening, closing, open-closing, and clos-opening. An application example in which these root signals are used in block truncation coding for image compression is discussed.

Key words.

1 Introduction

When linear filters do not produce satisfactory results, which occurs often in image-processing applications, the alternative is to pick a nonlinear filter. Which one to pick depends largely on the nature of the task to be performed. Several guidelines and lookup tables exist, but they provide the designer only with information on what filter class or classes are potential candidates. Choosing a particular filter among a whole class of filters is not usually a trivial task.

Optimality results based on certain criteria do exist for several classes of nonlinear filters; see, for instance, [1]–[6]. Another major criterion that is often used in certain classes of nonlinear filters is the root-signal set of an operator. This analysis tool, which has recently been developed for several classes of nonlinear filters [7]–[12], analyzes the deterministic behavior of nonlinear operators by characterizing their root-signal sets, which constitute their passbands. Knowledge of these signals for each operator determines which one must be used for a particular task.

Among the classes of nonlinear filters that have become increasingly popular is the class of morphological filters. Morphological filters are based on the theory of mathematical morphology developed by Matheron [13] and Serra [14]. The filters exploit the geometric rather than the analytic features of signals to achieve certain tasks, such as filtering. They were first defined in terms of the basic set operations and were then extended to gray-scale or function operations [15]–[17].

Fundamentally, mathematical morphology represents signals as sets, and a morphological operation consists of a set transformation that transforms one set into another. In practice, a signal is more customarily represented by a function defined on a finite or infinite region of support, although it can be alternatively represented by a set in Euclidean space. Filters are classified into set-processing filters and function-processing filters, according to whether a signal is represented by a set or a function, [15].

Morphological filters have been widely used in digital-signal processing for a number of years. They have found extensive applications in several areas, including biomedical image processing, shape recognition, edge detection, image restoration, and image enhancement.

The basic morphological operations are dilation, erosion, opening, and closing. Morphological opening and closing are important algorithms in signal processing. They eliminate specific signal structures smaller than some structuring element without distorting the global features of the signal.

Of particular significance is that morphological opening and closing possess the idempotency property, i.e., further iterations by the same filter on previous outputs do not cause further changes in the output produced by the first filter pass. In other words, one pass of opening or closing yields a root signal that is invariant to further passes of the same filter. Furthermore, morphological operations using iteratively applied openings and closings are also idempotent. Among these are open-closing (an opening followed by a closing by the same structuring element) and clos-opening (a closing followed by an opening by the same structuring element).

In many respects the closest nonlinear digital operations are rank-order operations, among which the median filter is the most popular. Several studies on the convergence behavior and root-signal sets of median filters have led to a deeper understanding of the median operation and to the application areas in which it excels [18], [19]. Of particular importance is the result found by Maragos and Schafer [17] that relates the roots of the median operator to those of open-closing and clos-opening. They showed that any root signal of the median filter is bounded from below by the output of the corresponding open-closing and from above by the output of the corresponding clos-opening.[1] Furthermore, they showed that the roots of the open-closing and the clos-opening are roots of the median filter. These results were generalized to stack filters in [11].

By defining a state model to describe the root-signal set of the median filter, Arce and Gallagher [20] characterized the root-signal set and computed the cardinality of this set for median filters with different window widths and for binary signals of arbitrary but finite lengths. (The multilevel case was studied by Fitch et al. [19]). The main feature of their state model is that each state in a median root produces other states in the state model as the signal length increases. Summing over all these states at a certain stage gives the total number of roots of the median filter for the specified window width and signal length. Any attempt to duplicate the above state model for morphological opening and closing will produce an incomplete set of states that cannot describe all possible states in opening and closing roots [21], [22]. This problem will be solved by incorporating the root number as a special state.

This paper is organized as follows. Section 2 presents some useful properties of morphological filters. Based on these properties, state models for opening and open-closing (or clos-opening) are developed in sections 3 and 4, respectively. These models lead to simple recursion relations that can be used to count the number of root signals for arbitrary-length structuring elements. Section 5 describes one application example in which root signals of morphological filters are used in a variable-length block truncation coding (BTC) image-coding scheme. Section 6 presents some conclusions.

2 Properties of One-Dimensional(1D) Root Signals

2.1 Definitions

In this paper morphological operations are considered as function-processing filters, although only binary operations are involved in the following. In a function-processing system it is convenient to define the signal length, which is a key parameter in the analysis of root signals. Definitions are given as follows [15]:

$$\begin{aligned}
\text{dilation: } & (f \oplus \check{K})(x) = \sup\{f(y) : y \in K_x\}, \\
\text{erosion: } & (f \ominus \check{K})(x) = \inf\{f(y) : y \in K_x\}, \\
\text{opening: } & f \circ K = (f \ominus \check{K}) \oplus K, \\
\text{closing: } & f \bullet K = (f \oplus \check{K}) \ominus K,
\end{aligned} \tag{1}$$

[1] Assume the window for the median filter is a convex symmetric (with respect to the origin) set W and that the window width is $|W| = 2k - 1$, where k is a positive integer. Then the corresponding structuring element for morphological filters is a convex set K with length $|K| = k$.

where $\check{K} = \{-z : z \in K\}$ denotes the symmetric set of K with respect to the origin and $K_x = \{z + x : z \in K\}$ denotes the translation of K by x. For sampled signals the structuring element K is a discrete and finite set that can be viewed as a moving window. Hence erosion (dilation) of a sampled function by a finite set K is equal to the moving local minimum (maximum).

2.2 *Threshold Decomposition Property and Positive Boolean Function Expressions*

Let $\underline{X} = [X_1, X_2, \ldots, X_L]^T$ (where ^{T} denotes transpose) be a multilevel, nonnegative signal vector. Without loss of generality, we require that $X_i \in \{0, 1, \ldots, m-1\}$. This signal vector can be decomposed into $m-1$ binary vectors $\underline{x}^t, t = 1, 2, \ldots, m-1$, by thresholding. This thresholding operation is T^t, so that

$$x_i^t = T^t(X_i) = \begin{cases} 1 & \text{if } X_i \geq t, \\ 0 & \text{otherwise.} \end{cases} \tag{2}$$

Note that

$$\underline{X} = \sum_{t=1}^{m-1} T^t(\underline{X}) = \sum_{t=1}^{m-1} \underline{x}^t. \tag{3}$$

Definition 1. An ordered set of sequences $\underline{X}_1, \underline{X}_2, \ldots, \underline{X}_k$ is said to obey the stacking property if

$$\underline{X}_1 \geq \underline{X}_2 \geq \cdots \geq \underline{X}_k. \tag{4}$$

Definition 2. A binary function g is said to possess the stacking property [1] if and only if

$$g(\underline{x}) \geq g(\underline{y}) \quad \text{whenever } \underline{x} \geq \underline{y}. \tag{5}$$

This is the same as the increasing property used to define morphological filters.

It has been shown that a necessary and sufficient condition for a binary function to possess the stacking property is that it is a positive Boolean function (PBF) [23].

Definition 3. A stack filter S_g based on the PBF $g(\cdot)$ is defined as follows:

$$S_g(\underline{X}) = \sum_{t=1}^{m-1} S_g(\underline{x}^t) = \sum_{t=1}^{m-1} g(\underline{x}^t). \tag{6}$$

We say that S_g obeys the threshold decomposition, which is a weak superposition property. Therefore a stack filter S_g is determined and hence represented by the PBF $g(\cdot)$.

The threshold decomposition property reduces the operation and the analysis of all stack filters to the analysis of the effects of these filters on binary signals. This simplifies the theoretical analysis of these highly nonlinear filters.

Since digital morphological filters with flat structuring elements are stack filters, we can find their PBF expressions. Let $g_o(\cdot), g_c(\cdot), g_{oc}(\cdot)$, and $g_{co}(\cdot)$ denote the PBF's of opening, closing, open-closing, and clos-opening by structuring element K, respectively. Then we have [24]

$$\begin{aligned} g_o(\underline{x}) &= \vee_{a\in\check{K}}(\wedge_{b\in K_a} x_b) \\ g_c(\underline{x}) &= \wedge_{a\in\check{K}}(\vee_{b\in K_a} x_b) \\ g_{oc}(\underline{x}) &= \wedge_{a\in\check{K}}(\vee_{b\in(K\oplus\check{K})_a}(\wedge_{c\in K_b} x_c)) \\ g_{co}(\underline{x}) &= \vee_{a\in\check{K}}(\wedge_{b\in(K\oplus\check{K})_a}(\vee_{c\in K_b} x_c)) \end{aligned} \tag{7}$$

It is important to note that $g_o(\cdot)$ is a dual function of $g_c(\cdot)$ and that $g_{oc}(\cdot)$ is a dual function of $g_{co}(\cdot)$.

Example. Consider the morphological opening by structuring element

$$K = \{-1, 0, 1\}. \tag{8}$$

Then the corresponding Boolean function is

$$\begin{aligned} g_o(\underline{x}) = &(x_{-2} \wedge x_{-1} \wedge x_0) \\ &\vee(x_{-1} \wedge x_0 \wedge x_1) \\ &\vee(x_0 \wedge x_1 \wedge x_2). \end{aligned} \tag{9}$$

2.3 *1D Root Structures*

In the 1D case the structuring element K is a line segment of length k. The root signal to be examined is a length-n binary signal taking on values in {0, 1}. To characterize the root sets of morphological filters we shall make use of the following 1D structures defined as follows: a *constant neighborhood* consists of at least k consecutive identically valued points. If the value is 1, it is called a *positive constant neighborhood*; if the value is 0, it is called a *negative constant neighborhood*. A *positive impulse* is a set of at most $k-1$ samples of value 1 surrounded by two negative constant neighborhoods. A *negative impulse* is a set of at most $k-1$ samples of value 0 surrounded by two positive constant neighborhoods. An *oscillation* is any sequence of samples that is not part of a constant neighborhood, a positive impulse, or a negative impulse.

The first major problem encountered in the filtering of finite support signals is the selection of an appending strategy. This was shown to have a major effect on the nature of the root signals of certain stack filters [11], [12]. In morphological filtering the commonly used appending strategy is called constant-value carry-on strategy. According to this strategy, constant values are appended to both ends of the finite-extent signal in order to compute the filter outputs at the boundaries of the signal. In 1D binary signals a sequence of $k-1$ 0's (respectively, 1's) is appended at both ends of a signal to be opened (respectively, closed) by a structuring element K with length k.

Stevenson and Arce [25] showed that constant neighborhoods are invariant to opening, closing, open-closing, and clos-opening. They further proved that negative (respectively, positive) impulses are invariant to opening (respectively, closing) and that oscillations are reduced to constant neighborhoods by the four types of morphological filters. This leads directly to the following properties of root signals of morphological filters.

Property 1. Binary-root signals of opening (respectively, closing) are composed of constant neighborhoods and negative (respectively, positive) impulses.

Furthermore, oscillations are reduced to negative and positive constant neighborhoods by opening and closing, respectively. Therefore a root signal of opening (respectively, closing) is composed solely of constant neighborhoods and negative (respectively, positive) impulses.

Since opening and closing are dual operators, their root signals are related by the following property.

Property 2. The complement of a binary root of opening (O) is a root of closing (C) and vice versa, i.e.,

$$\underline{x} = \mathrm{O}(\underline{x}) \Leftrightarrow \underline{1} - \underline{x} = \mathrm{C}(\underline{1} - \underline{x}), \qquad (10)$$

where $\underline{1} = [1, 1, \ldots, 1]^T$.

Property 2 has a major consequence concerning the number of root signals of opening and closing. In effect, the number of roots of opening is equal to the number of roots of closing for a given signal length. The proof is straightforward. Since the complement of a constant neighborhood is a constant neighborhood and the complement of a negative impulse is a positive impulse, the complement of a root of opening is therefore a root of closing, according to Property 1. Similarly, it is easy to show that the converse is true, that is, the complement of a root of opening is a root of closing. Therefore the cardinality of the root-signal set (for a given signal length) of opening is equal to that of closing. Let $R(f)$ denote the roots of f. We have

$$R(\mathrm{O}) = R(\mathrm{C}). \qquad (11)$$

The root-signal sets of open-closing and clos-opening behave in a similar manner, as described by the following property:

Property 3. A binary root of open-closing (OC) is a root of clos-opening (CO) and vice versa, i.e.,

$$\underline{x} = \mathrm{OC}(\underline{x}) \Leftrightarrow \underline{x} = \mathrm{CO}(\underline{x}). \qquad (12)$$

Hence the cardinality of the root-signal set of open-closing is equal to that of clos-opening for a given signal length, i.e.,

$$R(\mathrm{OC}) = R(\mathrm{CO}). \qquad (13)$$

Impulses and oscillations are shown to reduce the constant neighborhoods by open-closing and clos-opening, whereas positive and negative constant neighborhoods are invariant to these operators [25]. Therefore binary roots of both open-closing and clos-opening are composed of positive and negative neighborhoods only. Consequently, the two morphological operators have the same sets of binary-root signals. However, the same signal will not in general reduce to the same root by open-closing and clos-opening. An example is shown in figure 1.

input	1 1 0 0 0 1 1 1 0 0 1 1 1 1 0 0 0 1 0 0 1 1 1 0 1 1 0
opening	0 0 0 0 0 1 1 1 0 0 1 1 1 1 0 0 0 0 0 0 1 1 1 0 0 0 0
closing	1 1 0 0 0 1 1 1 1 1 1 1 1 1 0 0 0 1 1 1 1 1 1 1 1 1 1
open-closing	0 0 0 0 0 1 1 1 1 1 1 1 1 1 0 0 0 0 0 0 1 1 1 0 0 0 0
clos-opening	0 0 0 0 0 1 1 1 1 1 1 1 1 1 0 0 0 1 1 1 1 1 1 1 1 1 1

Fig. 1. Signal filtered by four different morphological filters by a length-3 structuring element.

In Properties 1–3 structuring element of the same size is used for all morphological filters involved. Filtering by a smaller structuring element leads to the following property of root signals.

Property 4. A binary root of opening, closing, open-closing, and clos-opening by a particular structuring element is also a root of the same opening, closing, open-closing, and clos-opening by a smaller structuring element, respectively, i.e.,

$$\underline{x} = f(\underline{x}; k_1) \Rightarrow \underline{x} = f(\underline{x}; k_2) \quad \text{provided that } k_2 \leq k_1, \tag{14}$$

where f = O, C, OC, CO.

The proof of Property 4 follows directly from the definitions of constant neighborhoods, impulses, and oscillations.

3 State Model for Opening

3.1 Opening by a Length-2 Structuring Element

Based on the above analysis, a state model for the root-signal set of morphological filters will be developed. To obtain a clearer insight into the proposed state model, we will first consider morphological filtering by a length-2 structuring element. Generalizations to length-3 and larger structuring elements will be carried out in subsection 3.2.

Consider the opening operation by a length-2 structuring element. The results for the closing operation follow immediately; see Property 2 above.

The idea of forming and counting root signals of morphological filters follows basically the same methodology used to count the roots of median filters [20]. New bits are appended, one at a time, to the finite-length signal as long as the overall signal is still a root signal. Since this subsection deals with length-2 structuring elements, when a new bit is added to a root signal, the overall signal is a root signal if the last two bits of the signal fall in one of the following states: in state A (00) the next digit must be 0, in state B (10) the next digit must be 0, and in state C (11) the next digit can be either 0 or 1.

Notice that by extending a root signal by one bit, each state will generate another state. The state diagram in figure 2(a) shows the allowable paths. Forbidden paths lead to the "sink" state. State A generates another state A, state B generates a state A, and state C generates either a state C or a state B. What is missing from these states is state 01. This pattern is, of course, not allowed at the end of a root signal of an opening, but it can indeed appear inside of an opening root signal, provided that the 1 in 01 is followed by another 1, e.g., as in 11011. The set of state models A, B, C given above is therefore incomplete. In solving this problem a closer look at state 01 shows that whenever the 01 pattern is present in a root signal, the next bit must always be a 1 for the overall signal to remain a root signal. But this corresponds to state C, that is, state C is not generated by just A, B, and C. The number of root signals ending with 011 is then equal to the number of C states generated by 01; see figure 3(a).

Let $A(n)$, $B(n)$, and $C(n)$ denote the number of states A, B, and C at the nth stage, respectively. From the above discussion it follows that

$$\begin{aligned} A(n+1) &= A(n) + B(n), \\ B(n+1) &= C(n), \\ C(n+1) &= C(n) + R(n-2), \end{aligned} \tag{15}$$

where $R(n)$ is the total number of root signals for a given signal length n. In the last recursive equation $R(n-2)$ can be viewed as a special state that is used to complete the state generation. The total number of states is then equal to the

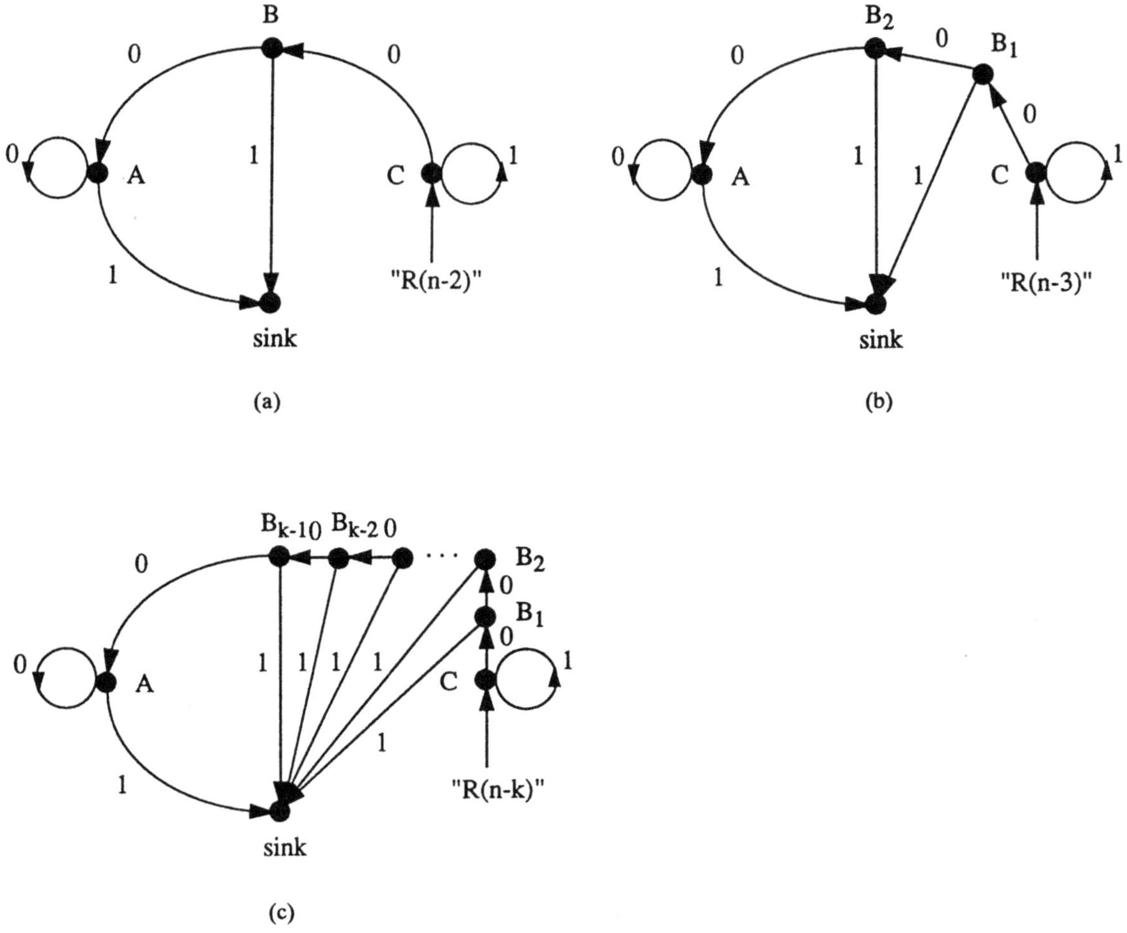

Fig. 2. (a) State diagram for $k = 2$; (b) state diagram for $k = 3$; (c) state diagram for general structuring element.

number of root signals.

$$R(n+1) = A(n+1) + B(n+1) + C(n+1). \quad (16)$$

Substituting (15) into (16) produces the following third-order recursive relation:

$$R(n+1) = A(n) + B(n) + 2C(n) + R(n-2). \quad (17)$$

With the initial conditions $A(2) = C(2) = R(0) = R(1) = 1, B(2) = 0$, and $R(2) = 2$, the recursive relation given by (15) and (17) can be used to count the number of root signals of any length. Relation (17) can further be manipulated as follows:

$$\begin{aligned} R(n+1) &= R(n) + C(n) + R(n-2) \\ &= R(n) + C(n-1) \\ &\quad + R(n-3) + R(n-2) \\ &= R(n) + C(2) + R(0) + R(1) \\ &\quad + \cdots + R(n-3) + R(n-2) \\ &= R(n) + \sum_{i=0}^{n-2} R(i) + 1. \end{aligned} \quad (18)$$

This difference equation can be further simplified. By (18)

$$R(n) = R(n-1) + \sum_{i=0}^{n-3} R(i) + 1. \quad (19)$$

Compute the difference $R(n+1) - R(n)$ by (18) and (19) as

$$\begin{aligned} R(n+1) - R(n) &= R(n) - R(n-1) \\ &\quad + \sum_{i=0}^{n-2} R(i) - \sum_{i=0}^{n-3} R(i) \\ &= R(n) - R(n-1) \\ &\quad + R(n-2). \end{aligned} \quad (20)$$

Adding $R(n)$ to both sides of (20) produces the following simple third-order recursive relation:

$$R(n+1) = 2R(n) - R(n-1) + R(n-2). \quad (21)$$

3.2 Opening by an Arbitrary Fixed-Length Structuring Element

The results of subsection 3.1 will be extended to arbitrary but finite-length structuring elements. The goal is again to set up a state model for the root-signal set of morphological openings.

By increasing the length of the structuring element, the system memory is increased because the minimum length of a constant neighborhood increases accordingly. This increases the number of the intermediate or transitional B states in the system. However, as will be demonstrated later, the final recursive relation involves only the total number of root signals.

The generalization of the state model is better understood when structuring elements of length 3 are considered first. When the length of the structuring element is increased to 3, the minimum length of the constant neighborhood is also increased to 3. The states that are allowed to appear at the end of a root signal are given as follows: in state A (000) the next digit must be 0, in state B_1 (110) the next digit must be 0, in state B_2 (100) the next digit must be 0, and in state C (111) the next digit can be either 0 or 1.

Again, this is an incomplete state model since state C can be generated by other states not yet included in the above state model. The problem is solved by incorporating the total number of root signals as an extra state; see figure 3(b). The following recursive equations are then obtained for the state model:

$$\begin{aligned} A(n+1) &= A(n) + B_2(n) \\ B_1(n+1) &= C(n) \\ B_2(n+1) &= B_1(n) \\ C(n+1) &= C(n) + R(n-3). \end{aligned} \quad (22)$$

The state diagram is shown in figure 2(b). Adding up all these states results in the total number of root signals,

$$\begin{aligned} R(n+1) &= A(n+1) + B_1(n+1) \\ &\quad + B_2(n) + C(n+1) \\ &= A(n) + B_1(n) + B_2(n) \\ &\quad + 2C(n) + R(n-3) \\ &= R(n) + C(n) + R(n-3) \end{aligned}$$

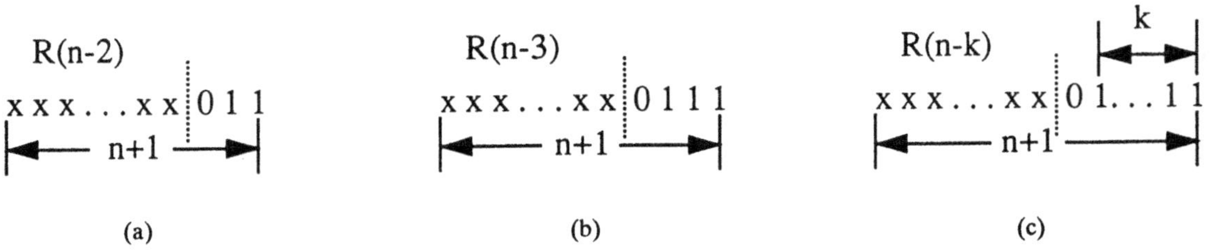

Fig. 3. A C state preceded by 0: (a) $k = 2$; (b) $k = 3$; (c) k is an arbitrary integer.

$$\begin{aligned}&= R(n) + C(n-1)\\&\quad +R(n-4) + R(n-3)\\&= R(n) + C(3)\\&\quad +R(0) + R(1) + \cdots\\&\quad +R(n-4) + R(n-3)\\&= R(n) + \sum_{i=0}^{n-3} R(i) + 1.\end{aligned} \tag{23}$$

From the above equation compute the difference,

$$\begin{aligned}R(n+1) - R(n) &= R(n) - R(n-1)\\&\quad + \sum_{i=0}^{n-3} R(i) - \sum_{i=0}^{n-4} R(i)\\&= R(n) - R(n-1)\\&\quad +R(n-3).\end{aligned} \tag{24}$$

Adding $R(n)$ to both sides of (24) yields

$$\begin{aligned}R(n+1) &= 2R(n)\\&\quad -R(n-1) + R(n-3).\end{aligned} \tag{25}$$

Extension of the above model to arbitrary-length structuring elements is now immediate. States A and C remain the same except for an appropriate increase in their number of bits, whereas the number of transitional states B increases. State C is again generated by using the states in the state model and the total number of root signals as an extra state; see figure 3(c). The new generalized state model is composed of the following states, where the B_i's are the transitional states:

$$\begin{aligned}A(n+1) &= A(n) + B_{k-1}(n)\\B_1(n+1) &= C(n)\\B_2(n+1) &= B_1(n)\\B_3(n+1) &= B_2(n)\\&\vdots\\B_{k-1}(n+1) &= B_{k-2}(n)\\C(n+1) &= C(n) + R(n-k).\end{aligned} \tag{26}$$

Figure 2(c) shows the state diagram for arbitrary-length structuring elements. As before, we have

$$\begin{aligned}R(n+1) &= A(n+1)\\&\quad +B_1(n+1) + B_2(n)\\&\quad + \cdots + B_{k-1}(n+1)\\&\quad +C(n+1)\\&= A(n) + B_1(n) + B_2(n)\\&\quad + \cdots + B_{k-1}(n)\\&\quad +2C(n) + R(n-k)\\&= R(n) + C(n)\\&\quad +R(n-k)\\&= R(n) + C(n-1)\\&\quad +R(n-k-1)\\&\quad +R(n-k)\\&= R(n) + C(k)\\&\quad +R(0) + R(1)\\&\quad + \cdots + R(n-k-1)\\&\quad +R(n-k)\end{aligned}$$

$$= R(n) + \sum_{i=0}^{n-k} R(i) + 1 \qquad (27)$$

and the difference,

$$\begin{aligned} R(n+1) - R(n) &= R(n) - R(n-1) \\ &\quad + \sum_{i=0}^{n-k} R(i) - \sum_{i=0}^{n-1-k} R(i) \\ &= R(n) - R(n-1) \\ &\quad + R(n-k). \end{aligned} \qquad (28)$$

Adding $R(n)$ to both sides of (28), we have

$$\begin{aligned} R(n+1) &= 2R(n) - R(n-1) \\ &\quad + R(n-k). \end{aligned} \qquad (29)$$

This is a $(k+1)$st-order difference equation for counting the number of root signals for morphological opening by a length-k structuring element. This relation reduces to (21) and (25) when $k = 2$ and $k = 3$, respectively. Table 1 summarizes the total number of root signals for morphological opening and closing with structuring elements of length up to 5 and signal lengths up to 12.

4 State Model for Open-Closing and Clos-Opening

As was stated earlier, a binary root of open-closing and clos-opening is composed solely of positive and negative constant neighborhoods. Therefore only two states are needed to describe it, i.e., state A and state C, where every digit is 0 and 1, respectively. State A generates another state A; and state C generates another state C. What is lacking from this state model is the transition from a state A to a state C and vice versa, e.g., from 0 to 1 and from 1 to 0. These transitions represent edges in 1D signals that are preserved under both morphological filters. Fortunately, these transitions can be accounted for without introducing any new states. Note that those A states not generated by an A state in the previous stage are preceded by a 1 that must belong to a C state. Similarly, those C states not generated by a C state in the previous stage are preceded by a 0 that must belong to an A state; this is illustrated in figure 4. From the above analysis we have

$$\begin{aligned} A(n+1) &= A(n) + C(n+1-k), \\ C(n+1) &= C(n) + A(n+1-k). \end{aligned} \qquad (30)$$

Thus

$$\begin{aligned} R(n+1) &= A(n+1) + C(n+1) \\ &= A(n) + C(n) \\ &\quad + A(n+1-k) \\ &\quad + C(n+1-k) \\ &= R(n) + R(n+1-k). \end{aligned} \qquad (31)$$

The state diagram is shown in figure 5. Table 2 summarizes the total number of root signals for open-closing and clos-opening with structuring elements of lengths up to 5 and signal lengths up to 12.

The results obtained so far have been extended to multiple-level root signals [26]. In section 5 we will show how to use these results in a BTC image-coding application.

5 Variable-Length BTC Image Coding

BTC [27] is an efficient image-coding scheme. Its performance and simplicity make it attractive in real-time image transmission. The principle of the BTC algorithm is to use a nonparameter one-bit quantizer to preserve the local statistics of the image, e.g., the first- and second-order moments of an image.

For a more detailed description of the BTC image-coding scheme, consider an image block of size $n \times n$ (usually $n = 4$). Set $m = n^2$, and denote by $X_i, i = 1, 2, \ldots, m$, the values of the pixels in the block. Then the local mean and variance can be expressed as

$$\overline{X} = \frac{1}{m} \sum_{i=1}^{m} X_i,$$

Table 1. Number of roots for opening and closing.

Structuring Element Length k	Signal Length n										
	2	3	4	5	6	7	8	9	10	11	12
2	2	4	7	12	21	37	65	114	200	351	616
3	1	2	4	7	11	17	27	44	72	117	189
4	1	1	2	4	7	11	16	23	34	52	81
5	1	1	1	2	4	7	11	16	22	30	42

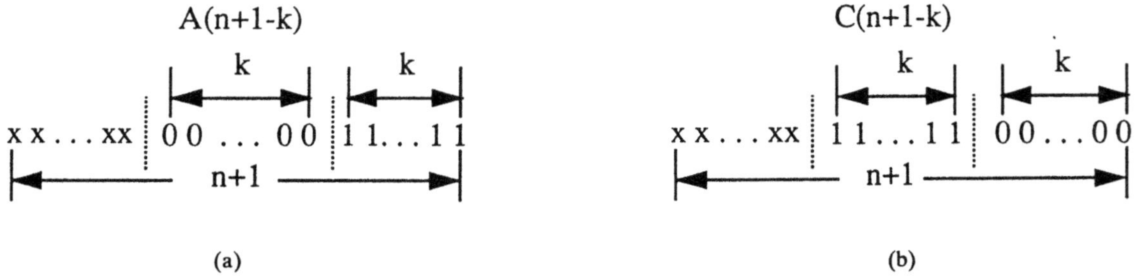

Fig. 4. Transition in a root of open-closing and clos-opening: (a) a C state preceded by 0; (b) an A state preceded by 1.

Table 2. Number of roots for open-closing and clos-opening.

Structuring Element Length k	Signal Length n										
	2	3	4	5	6	7	8	9	10	11	12
2	2	2	4	6	10	16	26	42	68	110	178
3	0	2	2	2	4	6	8	12	18	26	38
4	0	0	2	2	2	2	4	6	8	10	14
5	0	0	0	2	2	2	2	2	4	6	8

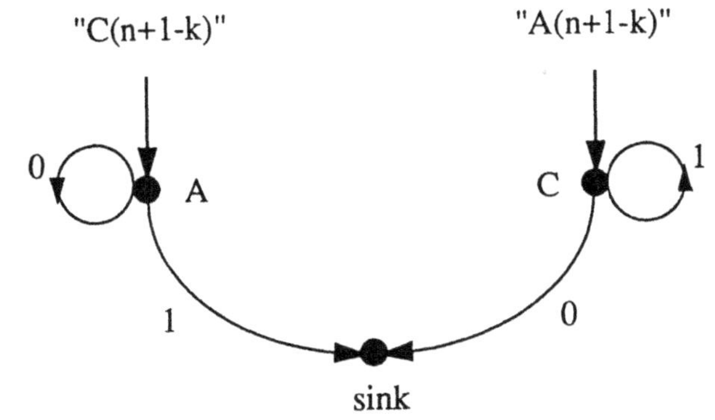

Fig. 5. State diagram for open-closing and clos-opening.

$$\sigma^2 = \frac{1}{m}\sum_{i=1}^{m} X_i^2 - \overline{X}^2. \quad (32)$$

When a one-bit quantizer is used, a threshold T and two output levels S_0 and S_1 must be computed so that

$$\begin{aligned} &\text{if } X_i < T, \quad \text{output} = S_0, \\ &\text{if } X_i \geq T, \quad \text{output} = S_1 \end{aligned} \quad (33)$$

for $i = 1, 2, \ldots, m$. Set $T = \overline{X}$, and denote by q the number of X_i's greater than or equal to T. Then, to preserve the sample mean and sample variance, set S_0 and S_1 to

$$\begin{aligned} S_0 &= \overline{X} - \sigma\sqrt{\frac{q}{m-q}}, \\ S_1 &= \overline{X} + \sigma\sqrt{\frac{m-q}{q}}. \end{aligned} \quad (34)$$

It is easy to see that the block is now described by $\overline{X}, \sigma$, and an $n \times n$ bit plane consisting of zeros and ones corresponding to pixels with values S_0 and S_1, respectively. For coding the bit plane, n^2 bits are required. Therefore, when $n = 4$, 2 bits/pixel will be required to code an image, assuming that 16 bits are used to code $\overline{X}$ and σ (8 bits for each).

To further compress the number of bits per pixel, Arce and Gallagher [28] proposed an algorithm that uses the roots of a 1D-window-width three-median filter to perform BTC. They showed that the bit plane has a high probability of belonging to the root-signal set of the median filter and can be reduced to a root signal through a single filter pass. Applying this 1D median filter to each row of the bit plane, one can use only 12 bits to code it, reducing the bit rate in this way from 2 to $28/16 = 1.75$ bits/pixel.

This success is partially due to the correlation among the root signals in the bit plane. However, 1D root signals exploit the 1D correlation only. Therefore 2D root signals can be expected to improve the results. Recently, a new scheme that uses root signals of the cross-median and separable-median filters was proposed for compressing the bit rate [29]. It has been shown that 13 and 11 bits are required to code binary-root signals (of 4×4 size) of the 3×3 cross-median and separable-median filters, respectively. The number of bits per pixel achieved this way thus become $29/16 = 1.81$ and $27/16 = 1.69$, respectively.

The standard BTC uses 16 bits to code the 4×4 bit plane while using 16 bits to code $\overline{X}$ and σ. In this section we develop a coding scheme that keeps the same bit rate for the local mean and variance but reduces the bits for coding the bit plane.

The roots of morphological filters are used to compress the bits required for the bit plane. The root properties of morphological filters allow us to develop a variable-length BTC image-coding scheme. Compared to the standard BTC coding method, the bits per pixel needed are reduced according to the local statistics of the image.

From table 1 we find that for binary signals of length 4, the number of roots of opening by structuring elements of lengths 4, 3, and 2 are 2, 4, and 7, respectively. By Property 4 a root of opening by a particular structuring element is also a root of opening by a smaller structuring element. Therefore the root sets by structuring elements of lengths 4, 3, and 2 are inclusive; they are shown in figure 6(a). In our coding scheme the bit plane is subject to morphological opening and closing row by row. The structuring elements are chosen from 2 to 4. First, the largest structuring element is applied, and the possible outputs are 0000 and 1111. Only one bit is required to code them. If the output is 1111, the outputs, when smaller structuring elements are applied, must also be 1111. Therefore

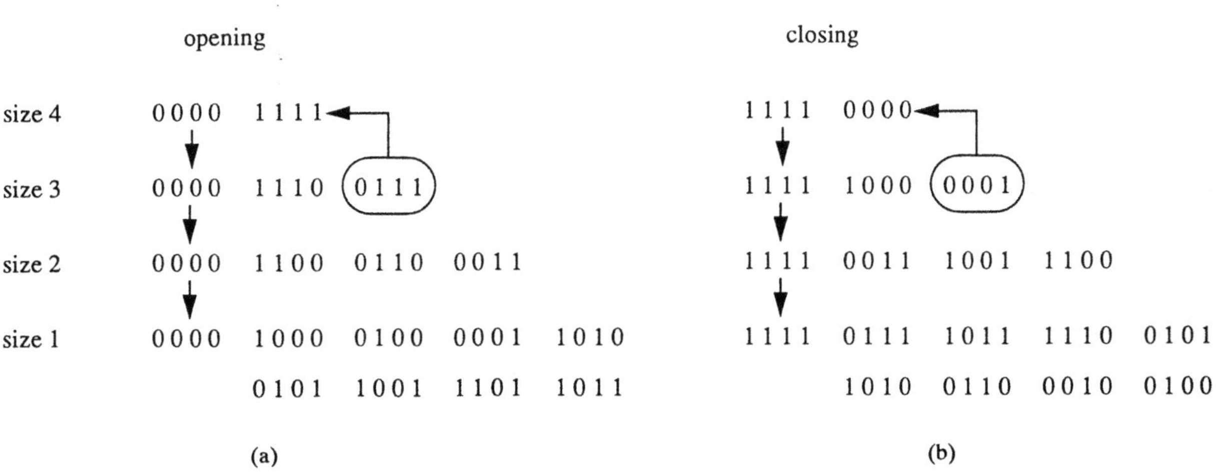

Fig. 6. Roots of opening and closing by structuring elements of lengths 1, 2, 3, and 4: (a) roots of opening; (b) roots of closing.

(a)	(b)
1 1 ---> 1 1 1 1	1 1 ---> 0 0 0 0
1 0 1 ---> 1 1 1 0	1 0 1 ---> 1 0 0 0
1 0 0 1 ---> 0 0 1 1	1 0 0 1 ---> 0 0 1 1
1 0 0 0 1 ---> 0 1 1 0	1 0 0 0 1 ---> 1 0 0 1
1 0 0 0 0 ---> 0 0 1 1	1 0 0 0 0 ---> 1 1 0 0
0 ---> 0 0 0 0	0 ---> 1 1 1 1

Fig. 7. Root coding: (a) opening; (b) closing.

we stop and move on to the next row or next block. If the output is 0000, then we continue by applying a length-3 structuring element. The possible roots are 0000, 1100, 0111, and 1111. Obviously, the last root will never appear at the output because of the first-stage operation. Therefore the states (roots) that are not present in the first stage are only the first three (we call them independent states). If we discard one of these three states, only one bit is required. Using a length-2 structuring element, we have four independent states that are not present in the previous stages. For completeness we also give the independent states of opening by a length-1 structuring element, which is actually the trivial identical filter. Note that all 16 possible states of four bits are included in figure 6(a). However, the states in the last row are unlikely to be present in the bit plane, and the opening by a structuring element with length greater than one will take them to the upper levels. We shall discard them in our coding scheme.

The variable-length codes are designed in such a way that self-punctuation is possible. Figure 7(a) depicts the coding scheme. Since 0000 appears with very high probability after openings, we add one bit at the beginning of a code to indicate its presence. After the first bit the presence of 1 implies the end of the code for the current row. In this way we avoid transmitting side information for punctuation.

A similar coding scheme can be established by using the roots of closing shown in figure 6(b). The codes are designed as shown in figure 7(b). However, it is difficult to say whether opening or closing produces better results. According to Property 1, opening preserves negative structures, including negative constant neighborhoods and negative impulses, whereas closing preserves positive structures.

One way to further improve this proposed scheme would be to exploit the 2D correlation that exists in the bit plane. As was done in [26], an adaptive algorithm that switches between opening and closing could be used to improve the picture quality while keeping the same bit rate. However, a much simpler method is proposed. For each bit plane, we use opening or closing, depending on which one produces a better result (an output that is closer to the original input). Only one extra bit is added to denote whether an opening or a closing should be used for a block.

With minor adjustment this coding scheme can be designed for progressive or multiresolution image transmission in which a coarse rendition of the image is sent first to give the receiver an early impression of the image contents, and then subsequent transmission provides image details of progressively finer resolution.

Image "Lena" was used to test the feasibility of the proposed coding scheme. The results are also compared with those of the median-based BTC coding scheme. Figure 8 shows the results of the different BTC coding schemes. The mean absolute error (MAE), mean-squared error (MSE), and signal-to-noise ratio (SNR) were computed for each method.

6 Conclusions

In this paper the root-signal sets of the basic morphological filters have been characterized. Several properties of these root signals were studied, based on which state models for the root-signal sets were developed. These models have led to simple recursive relations that can be used to compute the total number of root signals for arbitrary finite-length structuring elements.

Among several potential applications, these root signals have been used in a BTC image-coding scheme. The results were satisfactory and comparable with previous results obtained by a median-based BTC image-coding scheme. The aim was not to improve performance so much as to show feasibility and possible use of the results obtained in this paper. Other potential applications of these results exist in the areas of image restoration and image enhancement where image features are the prime criteria for selecting a nonlinear filter. The theoretical background has already been developed for stack filters [8], [11], and the results should be useful for morphological filters.

(a)

(b)

(c)

(d)

Fig. 8. Examples of BTC image coding: (a) the original "Lena" of 256×256 size at 8.0 bits/pel; (b) the BTC-coded "Lena" at 2.0 bits/pel (MAE = 3.96, MSE = 44.76, and SNR = 31.62 dB); (c) the 1D- median-root-coded "Lena" at 1.75 bits/pel (MAE = 4.42, MSE = 55.21, and SNR = 30.71 dB); (d) the opening- and closing-coded "Lena" at 1.72 bits/pel (MAE = 4.55, MSE = 60.37, and SNR = 30.32 dB).

References

1. E.J. Coyle and J.H. Lin, "Stack filters and the mean absolute error criterion," *IEEE Trans. Acoust., Speech, Signal Process.*, vol. ASSP-36, pp. 1244–1254, 1988.
2. E.J. Coyle, J.H. Lin, and M. Gabbouj, "Optimal stack filtering and the estimation and structural approaches to image processing," *IEEE Trans. Acoust., Speech, Signal Process.*, vol. ASSP-37, pp. 2037–2066, 1989.
3. L. Yin, J. Astola, and Y. Neuvo, "Optimal weighted order statistic filter under the mean absolute error criterion," in *Proc. ICASSP 91, Int. Conf. on Acoustics, Speech, and Signal Processing*, Toronto, May 1991, pp. 2539–2532.
4. L. Yin, J. Astola, and Y. Neuvo, "Adaptive stack filtering with application to image processing," *IEEE Trans. Signal Process.*, to be published.
5. E.R. Dougherty, "Minimal search for the optimal mean-square digital gray-scale morphological filter," *Visual Communications and Image Processing '90: Fifth in a Series*, M. Kunt, ed., *Proc. Soc. Photo-Opt. Instrum. Eng.*, vol. 1360, pp. 214–226, 1990.
6. J. Nieweglowski, L. Yin, M. Gabbouj, and Y. Neuvo, "Optimal weighted order statistic filters under structural constraints," in *Proc. 1992 IEEE Int. Symp. on Circuits and Systems*, San Diego, CA, May 1992, pp. 2621–2624.
7. Q. Wang and Y. Neuvo, "On two-dimensional root structures of separable and cross median filters," in *Proc. 1991 IEEE Int. Symp. on Circuits and Systems*, Singapore, June 1991, pp. 104–107.
8. P.T. Yu and E.J. Coyle, "Convergence behavior and N-roots of stack filters," *IEEE Trans. Acoust., Speech, Signal Process.*, vol. ASSP-38, pp. 1529–1544, 1990.
9. M. Gabbouj and E.J. Coyle, "Minimum mean absolute error stack filtering with structural constraints and goals," *IEEE Trans. Acoust., Speech, Signal Process.*, vol. ASSP-38, pp. 995–968, 1990.
10. P.T. Yu and E.J. Coyle, "On the existence and design of the best stack filter based associative memory," *IEEE Trans. Circuits and Systems II*, vol. 39, pp. 171–184, 1992.
11. M. Gabbouj, P.T. Yu, and E.J. Coyle, "Convergence behavior and root signal sets of stack filters," *Circuits, Systems, and Signal Processing*, Special Issue on Median and Morphological Filtering, 1991.
12. P.T. Yu, W.L. Wang, and S.S. Hung, "Root properties of median filters under three appending strategies," in *Proc. 1991 IEEE Int. Symp. on Circuits and Systems*, Singapore, June 1991.
13. G. Matheron, *Random Sets and Integral Geometry*, Wiley: New York, 1975.
14. J. Serra, *Image Analysis and Mathematical Morphology*, Academic Press: New York, 1982.
15. P.A. Maragos and R.W. Schafer, "Morphological filters—part I: their set-theoretic analysis and relations to linear shift-invariant filters," *IEEE Trans. Acoust., Speech, Signal Process.*, vol. ASSP-35, pp. 1153–1169, 1987.
16. R.M. Haralick, S.R. Sternberg, and X. Zhuang, "Image analysis using mathematical morphology," *IEEE Trans. Patt. Anal. Mach. Intell.*, vol. PAMI-9, pp. 532–550, 1987.
17. P.A. Maragos and R.W. Schafer, "Morphological filters—part II: their relations to median, order-statistic, and stack filters," *IEEE Trans. Acoust., Speech, Signal Process.*, vol. ASSP-35, pp. 1170–1184, 1987.
18. S.G. Tyan, "Median filtering: deterministic properties," in *Topics in Applied Physics, Two-Dimensional Digital Signal Processing II*, T.S. Huang, ed., Springer: Berlin, 1981, pp. 197–217.
19. J.P. Fitch, E.J. Coyle, and N.C. Gallagher, Jr., "Root properties and convergence rates of median filters," *IEEE Trans. Acoust., Speech, Signal Process.*, vol. ASSP-33, pp. 230–240, 1985.
20. G.R. Arce and N.C. Gallagher, "State description for the root-signal set of median filters," *IEEE Trans. Acoust., Speech, Signal Process.*, vol. ASSP-30, pp. 894–902, 1982.
21. Q. Wang, M. Gabbouj, and Y. Neuvo, "State description for the root signal sets of morphological filters," in *Proc. 1992 IEEE Int. Symp. on Circuits and Systems*, San Diego, CA, May 1992, pp. 113–116.
22. Q. Wang, M. Gabbouj, and Y. Neuvo, "Root signal sets of morphological filters," *Electron. Lett.*, vol. 28, pp. 952–953, 1992.
23. E.N. Gilbert, "Lattice-theoretic properties of frontal switching functions," *J. Math. Phys.*, vol. 33, April 1954.
24. L. Koskinen, J. Astola, and Y. Neuvo, "Morphological filtering of noisy images," in *Visual Communications and Image Processing '90: Fifth in a Series*, M. Kunt, ed., *Proc. Soc. Photo-Opt. Instrum. Eng.*, vol. 1360, pp. 155–165, 1990.
25. R.L. Stevenson and G.R. Arce, "Morphological filters: statistics and further syntactic properties," *IEEE Trans. Circuits and Systems*, vol. CAS-34, pp. 1292–1305, 1987.
26. Q. Wang, M. Gabbouj, and Y. Neuvo, "Root properties of morphological filters," submitted to *IEEE Trans. Signal Process.*
27. E.J. Delp and O.R. Mitchell, "Image compression using block truncation coding," *IEEE Trans. Commun.*, vol. COM-27, pp. 1335–1342, 1979.
28. G.R. Arce and N.C. Gallagher, "BTC image coding using median filter roots," *IEEE Trans. Commun.*, vol. COM-31, pp. 784–793, 1983.
29. B. Zeng, Q. Wang, and Y. Neuvo, "BTC image coding using two-dimensional median filter roots," in *Proc. 1991 IEEE Int. Symp. on Circuits and Systems*, Singapore, June 1991, pp. 400–403.

Qiaofei Wang received the B.S. and M.S. Degrees from the University of Electronic Science and Technology of China, Chengdu, Sichuan, China, in 1983 and 1986, respectively. Since 1990 he has been with the Signal Processing Laboratory, Tampere University of Technology, Finland, where he works as a research assistant and studies toward the doctoral degree. His research interests include nonlinear signal and image processing, and mathematical morphology and their applications to biomedical data.

Moncef Gabbouj received the B.S. degree in electrical engineering from Oklahoma State University in 1985, and the M.S. and Ph.D. degrees in electrical engineering from Purdue University in 1986 and 1989, respectively. Since 1990 he has been with the Research Institute of Information Technology, Tampere, Finland, where he is currently a senior research scientist. He also holds a teaching position in the Signal Processing Laboratory at Tampere University of Technology. His research interests include nonlinear signal and image processing, mathematical morphology, neural networks, and artificial intelligence.

Dr. Gabbouj was a corecipient of the Myril B. Reed Best Paper Award from the 32nd Midwest Symposium on Circuits and Systems. He is currently the Secretary of the Technical Committee on Digital Signal Processing of the IEEE Circuits and Systems Society and the Director of the International University Program in Digital Signal Processing at Tampere University of Technology.

Yrjö Neuvo received the Diploma Engineer and Licentiate of Technology degrees from Helsinki University of Technology in 1968 and 1971, respectively, and the Ph.D. degree in electrical engineering from Cornell University in 1974. He held various research and teaching positions at Helsinki University of Technology, the Academy of Finland, and Cornell University from 1968 to 1976. Since 1976 he has been a professor of electrical engineering at the Tampere University of Technology, Finland. During the academic year 1981–1982 he was with the University of California, Santa Barbara as a visiting professor. Since 1983 *he has held the position of National Research Professor at the Academy of Finland. His main research interests are digital signal and image-processing algorithms and their implementations. Currently, median-type nonlinear filters and their applications in video signal processing are his favorite research topics.

Dr. Neuvo is a Fellow of IEEE. He is member of EURASIP, Phi Kappa Phi, and the Finnish Academy of Technical Sciences, a Foreign Member of the Royal Swedish Academy of Technical Sciences, and a member of Academiae Europae. He is a member of the Editorial Board of *Signal Processing*, associate editor of *Circuits, Systems and Signal Processing*, and associate editor of *IEEE Transactions on Circuits and Systems for Video Technology*. He serves as a member of the Signal Processing Committee of the IEEE Circuits and Systems Society. He is a member of the Committee for Development of Science and Technology in Europe and a board member of Nordisk Industrifond. He was president of the Society of Electronics Engineers in Finland from 1978 through 1980. He was general chairman of the 1988 IEEE International Symposium on Circuits and Systems held in Helsinki, Finland.

Journal of Mathematical Imaging and Vision, 2, 173–183 (1992).

Unification of Nonlinear Filtering in the Context of Binary Logical Calculus, Part I: Binary Filters

EDWARD R. DOUGHERTY
Center for Imaging Science, Rochester Institute of Technology, Rochester, NY 14623-0887

ROBERT M. HARALICK
Department of Electrical Engineering, University of Washington, Seattle, WA 98105

Abstract. The mathematical structure of nonlinear filtering is expressed in the context of binary logic. This first part of a two-part study concerns the binary setting. Operator properties, such as antiextensivity and idempotence, are expressed in finite logical expressions, as are the Matheron representation for morphological filters and its extension to translation-invariant operators, thereby giving simplicity to both operational properties and representations and also exposing the manner in which logic methods can be used for filter design and analysis. The second part of the study treats gray-scale filters.

Key words. nonlinear filter, morphological filter, image algebra, representation, cellular logic

1 Introduction

The present two-part paper seeks to address a basic question regarding mathematical imaging: What is an appropriate algebraic framework for image processing? The goal is not to seek the most abstract setting or the most complete, nor is it to present new mathematics or a new representation theory. In fact, in it we step back from some existing algebraic frameworks to reformulate some basic filter theory in a practical setting, that of binary logical calculus. The word *nonlinear* in the title reflects the fact that the filters considered herein are nonlinear, but given the limitation to practical cellular-logic implementation, it is in a sense redundant since there is no vector-space structure within which to ground linearity. Nevertheless, since we are concerned with algebraic structures classically identified with nonlinear image processing, the title appears appropriate.

Loosely, an *image algebra* is a collection of objects and operations between these objects that form an algebraic structure in which to formulate image-processing algorithms. Sternberg [1] has used the term *image algebra* to refer to morphological (Minkowski) algebra, and it is in this context that Crimmons and Brown [2] use the same term in relation to automatic shape recognition. More recently, *image algebra* has been used to describe two algebraic structures containing more structure than morphological algebra. One image algebra, developed by Ritter and colleagues [3], [4], is heterogeneous in that it contains many sorts of entities, the essential two being images and templates. The other, developed by Dougherty and Giardina [5], possesses a heterogeneous form and, as further developed by Dougherty [6], [7], a homogeneous form. Both image algebras serve to represent image-processing operations, both linear and nonlinear. In particular, they serve as a framework for linear operations because each contains the linear algebra of matrices as a subalgebra, and they serve as a framework for nonlinear operations because each possesses the necessary lattice structure. Regarding the necessity of structure within image algebra, Dougherty and Giardina [8] take special note of the induced nature of subalgebras. But what subalgebras need to be induced? Indeed, what subalgebras should be induced, given the computational nature of image processing?

In point of fact, digital images do not form a

vector space relative to induced image addition and scalar multiplication. Strictly speaking, because the gray range is discrete and finite, image addition is not even closed. Even if we ignore the finitude of the gray range, we are still confronted by discreteness, so that the whole notion of linearity, including linear operators, cannot be subsumed within any image algebra that remains faithful to digital processing. This does not mean that richer mathematical structures cannot be of use, only that one should not see these as fully exhausting the algebraic question. It should be kept in mind that digital processing involves logic gates and bounded finite 0–1 representations. If we are to stick more closely to actual processing when we propose mathematical representations, then we need to stay within the confines of digital logic (or discrete set theory).

The set-theoretic properties of binary image filtering have been laid down by Matheron [9]. These include monotonicity, extensivity (antiextensivity), and idempotence. These are logical (or set-theoretic) concepts, and they play dominant roles in binary filtering. Matheron also carefully examines the role of translation invariance. This latter concept involves the translational structure within which image processing takes place and is related to image stationarity. If one reads Matheron closely, it is clear that the basic morphological operations of erosion, dilation, opening, and closing do not appear by chance in image processing. Indeed, he recognizes that any translation-invariant, monotonically increasing operator must be formed from a union of erosions and that any translation-invariant, increasing, antiextensive, and idempotent operator must be formed from a union of openings. Hence, by the very nature of digital image processing, mathematical morphology must play a key algebraic role, and therefore it is not surprising that Minkowski (morphological) algebra is central to the image algebras of both Ritter et al. and Dougherty and Giardina.

The algebraic framework for binary images established by Matheron [9] is extended to gray-scale images by means of lattice theory by Serra [10], [11] and Matheron [12]. They recognize that a complete lattice is the appropriate framework for the algebraic properties central to morphological processing and that the basic Matheron propositions thereby apply. Perhaps more importantly, the abstract lattice setting provides a framework for image processing that is more directly related to its logical (computable) nature than are richer algebras (containing vector-space subalgebras). More recent papers have further substantiated the proposition that a complete lattice provides the setting for the "algebraic basis of mathematical morphology" (Heijmans and Ronse [13], [14], Heijmans [15], Ronse [16]).

The central role of binary mathematical morphology arises from the set-theoretic aspects of binary processing; the key role of cellular logic arises from the manner in which image operators must be implemented on a digital computer. Cellular logic, and relevant related architectures, impress themselves on the algebraic analysis of image processing because processing is digital. It might be tempting to separate the computational and the abstract–mathematical problems, treating the former as architectural and the latter as algebraic; in fact, however, they are interrelated. Here is where we step back from the abstract set-theoretic analysis of Matheron and the subsequent lattice-theoretic approaches.

In part I of the present study we begin with cellular logic and explain the manner in which algebraic binary-filter theory emerges therefrom (see part II for gray-scale analysis). Such an approach naturally places those concepts typically considered to be morphological directly into the framework of cellular logic, which, of course, explains (in hindsight) the major role of cellular logic in the implementation of morphological processing. In particular, the minimal Matheron representation of increasing, translation-invariant binary-image operators as unions of erosions reduces to the well-known proposition that every finite positive Boolean expression possesses a minimal sum-of-products form. In addition, the extension of the Matheron representation by Banon and Berrera [17] to binary-image operators that are merely translation invariant, specifically, that these can be represented by unions of hit-or-miss operators, is seen to have a straightforward interpretation in cellular logic.

One might ask whether there is anything to be gained by the exercise of stepping back from the more general lattice framework, other than perhaps some readjustment of thinking. In fact, as will become evident, there is much more to be gained. By recognizing the practical Boolean nature of morphology and by formulating filter theory in the language of logic design, we see that standard computing tools, such as Karnaugh maps and Quine–McClusky reduction, can be applied to the construction of morphological operators. In operator design we are confronted by the need both to compose operator representations and, on the other hand, to decompose operators into constituent parts satisfying different algebraic constraints. Having a clear appreciation of the discrete logical character of nonlinear operators permits application of existing automatic routines.

2 Cellular-Logic Filters

We consider translation-invariant, moving-window operators on the space $\mathbf{S_B}$ of binary signals defined on Z, the set of integers. We assume that the window $W\langle m\rangle$ is centered at m and is of length $2M + 1$. If Ψ is an operator of the specified type and $x = \{x[m]\}$ is a binary signal in $\mathbf{S_B}$, then

$$\Psi(x)[m] = \Psi(x[m-M], x[m-M+1], \ldots, x[m+M]), \tag{1}$$

where we do not distinguish between the operator and the function rule defining the operator, calling them both Ψ. Relative to computer architecture, window logic is manifested as cellular logic, and therefore the filter Ψ will be called a *cellular-logic filter.* The choice of Z as the domain space for $\mathbf{S_B}$ is for convenience. Insofar as the subsequent logical analysis is concerned, the operative functional expression is (1), which depends only on denumerability (discreteness) of the domain and the finiteness of the window. In a setting different from Z (say, binary images defined on $Z \times Z$) the window can be of any shape and the ordering $x[m-M], x[m-M+1], \ldots, x[m+M]$ merely represents some given listing of the way in which the window is to be scanned. In particular, the assumption that the window is centered in (1) serves only the purpose of notational convenience.

Since Ψ is translation invariant, much of its analysis can be accomplished by considering the single output value $\Psi(x)[0]$, whose value depends on the window $W\langle 0\rangle = \{-M, -M+1, \ldots, M\}$ centered at the origin. $\Psi(x)[0]$ can be considered to be a binary functional on the set $\mathbf{M}$ of $\{0,1\}$-valued functions defined on $W\langle 0\rangle$. Every element of $\mathbf{M}$ can be represented by a string of ones and zeros, $(x[-M], x[-M+1], \ldots, x[M])$. From a set-theoretic perspective, every element of $\mathbf{M}$ is a subset of $W\langle 0\rangle$, where j lies in the subset if and only if $x[j] = 1$. Union and intersection in $W\langle 0\rangle$ are replaced by the logical maximum and minimum operations

$$x \vee y = (x[-M] \vee y[-M], \ldots, x[M] \vee y[M]), \tag{2}$$

$$x \wedge y = (x[-M] \wedge y[-M], \ldots, x[M] \wedge y[M]) \tag{3}$$

in $\mathbf{M}$. Moreover, the order relation $x \leq y$ if and only if $x[i] \leq y[i]$ for $i = -M, \ldots, M$ corresponds to the subset relation in $W\langle 0\rangle$.

As a binary functional, $\Psi(x)[0]$ can be written in logical format as a maximum of minima or, by using logical notation, as a canonical sum of products

$$\Psi(x)[0] = \sum x[-M]^{p[-M]} x[-M+1]^{p[-M+1]} \cdots x[M]^{p[M]}, \tag{4}$$

where, for $j = -M, \ldots, M$, $p[j]$ is -1, 0, or 1 and where $x[j]^{-1}$ is the negation of $x[j]$ (also written $x[j]'$) and $x[j]^0$ means that the logical variable $x[j]$ does not appear in the product. In other words, $\Psi(x)[0]$ is a Boolean expression over $2M + 1$ binary variables. As is well known, there are many expressions equivalent to (4), and, in fact, there are methods, such as

Karnaugh maps and the Quine–McClusky procedure, for minimizing the number of logic gates forming canonical sum-of-product expressions.

Owing to translation invariance the logical expression (4) applies to $\Psi(x)[m]$ for any m; the $p[j]$ remain the same:

$$\Psi(x)[m] = \sum x[m - M]^{p[-M]} \cdots x[m + j]^{p[j]} \cdots x[m + M]^{p[M]}. \tag{5}$$

The variables $x[j]$ lie in the translated window $W\langle m\rangle = W\langle 0\rangle + m$ centered at m.

Another way of looking at the expansion (4) [and therefore at the expansion (5)] is to proceed in the following manner: (i) group the variables with +1 exponents in each product, and let $W_i\langle 0\rangle$ denote their product; (ii) group the variables with −1 exponents in each product, and let $W_i\langle 0\rangle'$ denote their product; (iii) ignore all variables with 0 exponent. Then, omitting null products, unless $\Psi(x)[0]$ is the zero function, expression (4) takes the form

$$\Psi(x)[0] = \sum W_i\langle 0\rangle W_i\langle 0\rangle', \tag{6}$$

where it is possible for $W_i\langle 0\rangle$ or $W_i\langle 0\rangle'$ to be null, in which case it is denoted by 1.

Geometrically, $W_i\langle 0\rangle$ can be interpreted as a subwindow of $W\langle 0\rangle$ corresponding to positive Boolean variables (exponent +1) and $W_i\langle 0\rangle'$ can be interpreted as a subwindow corresponding to negative Boolean variables (exponent −1). We will subsequently make use of this convention by considering translates $W_i\langle m\rangle$ and $W_i\langle m\rangle'$. For instance, if $W\langle 0\rangle$ is the five-point window and

$$\Psi(x)[0] = x[-1]x[0]x[1] + x[0]x[1] + x[-1]x[1]'x[2]', \tag{7}$$

then $W_1\langle 0\rangle = x[-1]x[0]x[1]$, $W_1\langle 0\rangle' = 1$, $W_2\langle 0\rangle = x[0]x[1]$, $W_2\langle 0\rangle' = 1$, $W_3\langle 0\rangle = x[-1]$, and $W_3\langle 0\rangle' = x[1]'x[2]'$. Among other things, the subwindow notation facilitates writing outputs at points other than the origin. Here, for instance, $\Psi(x)[m]$ is written simply as

$$\Psi(x)[m] = W_1\langle m\rangle + W_2\langle m\rangle + W_3\langle m\rangle W_3\langle m\rangle', \tag{8}$$

where $W_i\langle m\rangle$ and $W_i\langle m\rangle'$ refer to the translated subwindows $W_i\langle 0\rangle + m$ and $W_i\langle 0\rangle' + m$, respectively.

3 Increasing Filters

A cellular-logic filter Ψ is *monotonically increasing* if $x \le y$ implies $\Psi(x) \le \Psi(y)$. Owing to translation invariance, Ψ is increasing if and only if $(x[-M], \ldots, x[M]) \le (y[-M], \ldots, y[M])$ implies $\Psi(x)[0] \le \Psi(y)[0]$. Ψ is increasing if and only if it can be expressed as a minimal sum of products for which there exists no negation in the expansion; i.e., $W_i\langle 0\rangle' = 1$ for all i. (In logical terminology, Ψ is a positive Boolean function.) Hence an increasing filter Ψ has a canonical representation

$$\Psi(x)[0] = W_1\langle 0\rangle + W_2\langle 0\rangle + \cdots + W_p\langle 0\rangle \tag{9}$$

that possesses a minimal number of product terms. The minimal expression is unique and can be obtained from any other sum-of-products expression. In general, any number of products can be adjoined by summation to the minimal expression without changing the filter so long as each is formed from an existing product by adjoining positive factors. In the minimal expression (9) the factors of $W_i\langle 0\rangle$ do not form a subset of the factors of $W_j\langle 0\rangle$, for $j \ne i$.

There is a natural ordering on binary operators. Suppose Ψ_1 and Ψ_2 are two operators. We write $\Psi_1 \le \Psi_2$ if and only if $\Psi_1(x) \le \Psi_2(x)$ for any signal x. Now suppose Ψ_1 and Ψ_2 are increasing and in minimal sum-of-product form. Then $\Psi_1 \le \Psi_2$ if and only if for any product of Ψ_1 there exists a product of Ψ_2 whose factors form a subset of the factors of the given product for Ψ_1.

An increasing cellular-logic filter Ψ is said to be *antiextensive* [*extensive*] if $\Psi(x) \le x$ [$\Psi(x) \ge x$] for all x. Relative to a sum-of-products expression for $\Psi(x)[0]$, Ψ is antiextensive if and only if each product term of $\Psi(x)[0]$ contains $x[0]$. Ψ is extensive if and only if it possesses the singleton product term $x[0]$ in its minimal sum-of-products representation.

4 Iteration

Of great concern in filtering is iteration: given filters Ψ and Φ, what can be said about the product $\Phi\Psi$? For the moment, we consider arbitrary filters Ψ and Φ, not necessarily increasing, and we examine the sum-of-products representation for $\Phi\Psi$. The cumbersome part of the problem is this: when Φ operates on $\Psi(x)$, each variable $y[m]$ in $y = \Psi(x)$ is expressed as a sum-of-products of the original x variables lying in the window $W\langle m\rangle$ about $x[m]$. Thus the expression for $\Phi\Psi(x)[0]$ potentially includes the variables $x[-2M]$, $x[-2M+1], \ldots, x[2M]$. The expression for $\Phi\Psi(x)[0]$ results from putting the expressions for $\Psi(x)[-M]$, $\Psi(x)[-M+1], \ldots, \Psi(x)[M]$ into the expression for $\Phi(x)[0]$ in place of $x[-M]$, $x[-M+1], \ldots, x[M]$, respectively. A key point is that once this has been done, reduction can be done to achieve a minimal-gate representation, and this can be accomplished automatically by some procedure such as the Quine–McClusky algorithm.

If Ψ and Φ happen to be increasing, the same reasoning applies; however, here reduction is much simpler. We need only expand the terms within the minimal sum-of-products representation for Φ when we replace the variables $x[j]$ by $\Psi(x)[j]$ and eliminate redundant products. This can always be done automatically.

As an illustration, consider the three-point window about the origin and let

$$\Psi(x)[0] = x[-1]x[0] + x[0]x[1]. \tag{10}$$

Then

$$\begin{aligned}\Psi\Psi(x)[0] &= (x[-2]x[-1] + x[-1]x[0])\\ &\quad (x[-1]x[0] + x[0]x[1])\\ &\quad +(x[-1]x[0] + x[0]x[1])\\ &\quad (x[0]x[1] + x[1]x[2])\\ &= \Psi(x)[0](x[-2]x[-1] + x[-1]x[0]\\ &\quad +x[0]x[1] + x[1]x[2])\\ &= \Psi(x)[0](x[-2]x[-1]\\ &\quad +\Psi(x)[0] + x[1]x[2])\\ &= \Psi(x)[0],\end{aligned} \tag{11}$$

the last equality following from the fact that for any logical expression ab, where $a \leq b$, $ab = a$. For this particular example we obtain the very special relation $\Psi\Psi = \Psi$.

5 Idempotence

A filter Ψ is said to be *idempotent* if $\Psi\Psi = \Psi$. For increasing filters idempotence can be characterized in terms of sum-of-products expressions. Consider the minimal sum-of-products expression for an increasing filter Ψ. Some product terms of $\Psi(x)[0]$ contain $x[0]$, and some do not. Thus we can express $\Psi(x)[0]$ as

$$\begin{aligned}\Psi(x)[0] = x[0]&\sum f_i(x[-M], \ldots, x[-1],\\ &x[1], \ldots, x[M])\\ &+\sum g_j(x[-M], \ldots, x[-1],\\ &x[1], \ldots, x[M]),\end{aligned} \tag{12}$$

where f_i and g_j are products of the variables in the centered window $W\langle 0\rangle$, excluding the variable $x[0]$. If $x[0]$ happens to be a product term of $\Psi(x)[0]$, then one of the f_i is 1 and without loss of generality we assume $f_1 = 1$. If the second sum is empty, then Ψ is antiextensive; otherwise, it is not. If $f_1 = 1$, then Ψ is extensive; otherwise, it is not. We write the decomposition (12) as

$$\Psi(x)[0] = x[0]\Psi_0(x)[0] + \Psi_1(x)[0]. \tag{13}$$

Operating a second time by Ψ yields

$$\begin{aligned}\Psi\Psi(x)[0] = \Psi(x)[0]&\sum f_i(\Psi(x)[-M],\\ &\ldots, \Psi(x)[-1], \Psi(x)[1],\\ &\ldots, \Psi(x)[M]).\\ &+\sum g_j(\Psi(x)[-M],\\ &\ldots, \Psi(x)[-1], \Psi(x)[1],\\ &\ldots, \Psi(x)[M]).\end{aligned} \tag{14}$$

In terms of the decomposition (13) idempotence takes the form

$$\begin{aligned}\Psi(x)[0] = \Psi(x)[0]&\Psi_0(\Psi(x))[0]\\ &+\Psi_1(\Psi(x))[0],\end{aligned} \tag{15}$$

which is a logical identity of the form $a = ab + c$. A necessary condition for the identity is $c \leq a$.

Two sufficient conditions are $c = a$ and $b \geq a \geq c$.

A key subcase concerning idempotence for an increasing filter Ψ is when the operator is antiextensive. In such a situation Ψ_1 is null, so that equation (15) is of the logical form $a = ab$, and hence a necessary and sufficient condition for idempotence is

$$\Psi(x)[0] \leq \Psi_0(\Psi(x))[0]. \quad (16)$$

This is precisely what happened for the antiextensive filter of equation (10). For it,

$$\Psi_0(x)[0] = x[-1] + x[1], \quad (17)$$

$$\Psi_0(\Psi(x))[0] = x[-2]x[-1] + x[-1]x[0] + x[0]x[1] + x[1]x[2]. \quad (18)$$

The filter of equation (10) belongs to the important subclass of all increasing antiextensive, idempotent cellular-logic filters.

The filters in this special class are called τ-*openings*, and within this class are the *openings*, which, in the context of a fixed window $W\langle 0\rangle$, will be called $W\langle 0\rangle$-*openings*. A $W\langle 0\rangle$-opening is defined by specifying a primitive product whose first factor is $x[0]$. To wit, let

$$h_0 = x[0]x[j_1]x[j_2]\cdots x[j_r], \quad (19)$$

where $0 < j_1 < \cdots < j_r \leq M$, be the primitive product. For $k = 1, 2, \ldots, r$ let

$$h_k = x[-j_k]x[j_1 - j_k]\cdots x[j_r - j_k]. \quad (20)$$

Define the $W\langle 0\rangle$-opening $\Psi(x)$ by

$$\Psi(x)[0] = h_0 + h_1 + \cdots + h_r. \quad (21)$$

Since $x[0]$ appears in every product, Ψ is antiextensive.

By using strictly logical calculus it can be shown that every $W\langle 0\rangle$-opening is idempotent. We consider $r = 2$, the proof for general r being similar but tedious. For $r = 2$ express h_0, h_1, and h_2, as $h_0 = x[0]x[p]x[q]$, $h_1 = x[-p]x[0]x[q-p]$, and $h_2 = x[-q]x[p-q]x[0]$, so that

$$\Psi(x)[0] = x[0](x[p]x[q] + x[-p]x[q-p] + x[-q]x[p-q]), \quad (22)$$

the sum being $\Psi_0(x)[0]$. Moreover,

$$\Psi_0(\Psi(x))[0] = \Psi(x)[p]\Psi(x)[q] + \Psi(x)[-p]\Psi(x)[q-p] + \Psi(x)[-q]\Psi(x)[p-q]. \quad (23)$$

There are three summands, each possessing two factors, forming $\Psi_0(\Psi(x))[0]$. The factors of the first summand are

$$\Psi(x)[p] = x[p](x[2p]x[p+q] + x[0]x[q] + x[p-q]x[2p-q]),$$

$$\Psi(x)[q] = x[q](x[p+q]x[2q] + x[q-p]x[2q-p] + x[0]x[p]). \quad (24)$$

Since each factor contains the summand $x[0]x[p]x[q] = h_0$, the product $\Psi(x)[p]\Psi(x)[q]$ also contains the summand h_0. The factors of the second summand forming $\Psi_0(\Psi(x))[0]$ are

$$\Psi(x)[-p] = x[-p](x[0]x[q-p] + x[-2p]x[q-2p] + x[-p-q]x[-q]),$$

$$\Psi(x)[q-p] = x[q-p](x[q]x[2q-p] + x[q-2p]x[2q-2p] + x[-p]x[0]). \quad (25)$$

Each factor contains the summand $x[-p]x[0]x[q-p] = h_1$. Hence $\Psi(x)[-p]\psi(x)[q-p]$ also contains the summand h_1. Finally, a similar computation shows that the final summand forming $\Psi_0(\Psi(x))[0]$ contains h_2, and therefore $\Psi_0(\Psi(x))[0] \geq \Psi(x)[0]$ and Ψ is idempotent.

Although every opening is a τ-opening, not every increasing, antiextensive, idempotent cellular-logic filter is an opening. For instance,

$$\Psi(x)[0] = x[0](x[-2] + x[-1] + x[1] + x[2]) \quad (26)$$

is increasing, antiextensive, and idempotent but is not an opening.

It is, however, a sum (union) of openings since

$$\Psi(x)[0] = (x[-1]x[0] + x[0]x[1]) + (x[-2]x[0] + x[0]x[2]) \quad (27)$$

and both summands are openings. The expression of τ-openings as unions of openings is a question that, starting with Matheron [9], has been long addressed in mathematical morphology.

6 Monotonic Cellular Logic and Binary Mathematical Morphology

The advantages of implementing binary mathematical morphology in cellular-logic architectures have long been recognized. The success of the cellular approach is based on the fact that binary morphological operations are actually reformulations of Boolean expressions, so that binary Minkowski (morphological) algebra is equivalent to cellular-logic algebra, which is itself simply Boolean algebra with translations. We examine this equivalence.

Suppose $\Psi(x)[0]$ is defined by a single product

$$\Psi(x)[0] = x[j_1]x[j_2]\cdots x[j_r], \tag{28}$$

where $-M \leq j_1 < j_2 < \cdots < j_r \leq M$. Let $A_\Psi\langle 0\rangle = \{j_1, j_2, \ldots, j_r\}$ be the subset of $W\langle 0\rangle$ associated with the product $\Psi(x)[0]$. Then $\Psi(x)[0] = 1$ if and only if $A_\Psi\langle 0\rangle$ is a subset of the set corresponding to x, this latter set to be denoted by $\langle x\rangle$. In general, $\Psi(x)[m] = 1$ if an only if $A_\Psi\langle m\rangle$ is a subset of $\langle x\rangle$. Since $A_\Psi\langle m\rangle = A_\Psi\langle 0\rangle + m$, this equivalence can be expressed in morphological terms: if we let $\Psi^\wedge$ denote the set mapping corresponding to the logical mapping Ψ, then

$$\Psi^\wedge(\langle x\rangle) = \langle x\rangle \ominus A_\Psi\langle 0\rangle, \tag{29}$$

where $\ominus$ denotes erosion. Because the collection of 0–1 signals is isomorphic to the collection of integer subsets, Ψ and $\Psi^\wedge$ are actually the same operator, so that (29) states that every single-product increasing logical binary operator defined over the window $W\langle 0\rangle$ is equivalent to an erosion whose structuring element lies in $W\langle 0\rangle$.

More generally, a cellular-logic operator Ψ is defined by a sum of products possessing no negations if and only if Ψ is monotonically increasing. Since the logical operation + is equivalent to union, Ψ is a positive Boolean expression if and only if it is equivalent to a union of erosions, the structuring elements in the erosion expansion corresponding to products in the logical expansion. In sum, we have four equivalent conditions: (i) Ψ can be expressed as a sum of products possessing no negations; (ii) Ψ is monotonically increasing as a logical operator; (iii) $\Psi^\wedge$ is monotonically increasing as a set operator; (iv) $\Psi^\wedge$ is a union of erosions.

Define the *kernel* of an increasing logical filter Ψ to be the collection Ker$[\Psi]$ of all signals z for which $\Psi(z)[0] = 1$. Then $z \in$ Ker$[\Psi]$ if and only if there is a product $x[j_1]\cdots x[j_r]$ in the sum-of-products expansion defining Ψ such that $z[j_1] = \cdots = z[j_r] = 1$, which is equivalent to saying that $A = \{j_1, \ldots, j_r\}$ is a subset of $\langle z\rangle$, which in turn means that 0 lies in $\langle z\rangle \ominus A$. Since A is one of the structuring elements forming the union of erosions comprising $\Psi^\wedge, 0 \in \Psi^\wedge(\langle x\rangle)$. By definition, a set lies in the kernel of a set mapping if and only if the filtered version of the set contains the origin. Hence the kernel of Ψ as a logical operator is equivalent to the kernel of $\Psi^\wedge$ as a morphological filter.

If a set operator $\Psi^\wedge$ is increasing and translation invariant, the Matheron representation [9] states that $\Psi^\wedge$ is expressed as the union of erosions by kernel elements, namely,

$$\Psi^\wedge(S) = \cup\{S \ominus A : A \in \text{Ker}[\Psi^\wedge]\}. \tag{30}$$

It was noticed by Maragos and Schafer [18] and by Dougherty and Giardina [19], [20] that the kernel expression is redundant. Bas$[\Psi^\wedge]$ is called the *basis* for $\Psi^\wedge$ if (a) every element in the kernel possesses a subset in Bas$[\Psi^\wedge]$ and (b) no two elements in Bas$[\Psi^\wedge]$ are properly related by the subset relation. Bases are unique. If there exists a basis for $\Psi^\wedge$, then the kernel expansion of equation (30) can be replaced by an expansion over the basis of the filter. The defining conditions of a basis mean there is no redundancy in the Matheron representation.

A monotonically increasing cellular-logic operator Ψ possesses a minimal sum-of-products representation. In that minimal form no product is a proper subproduct of another product. But this says that no structuring element is a proper subset of another structuring element in the erosion expansion representing $\Psi^\wedge$, which is then precisely the basis form of the Matheron representation for $\Psi^\wedge$. Thus in the discrete-window context the Matheron basis representation of a translation-invariant, increasing set mapping is actually a restatement of the fact that every increasing logical operator over a finite set of variables has a minimal sum-of-products representation, the minimizing products being the filter basis.

As an illustration, consider the logical operator defined by

$$\Psi(x)[0] = x[-1]x[0](x[1] + x[2]) \\ +x[-1](x[-2] + x[0]). \qquad (31)$$

Logical calculus yields

$$\Psi(x) = x[-1]x[0]x[1] \\ +x[-1]x[0]x[2] \\ +x[-2]x[-1] + x[-1]x[0]. \qquad (32)$$

Reduction yields the minimal sum-of-products representation

$$\Psi(x) = x[-1]x[0] + x[-2]x[-1]. \qquad (33)$$

Direct translation of equation (32) yields an erosion representation of the set operator corresponding to Ψ, namely,

$$\Psi^\wedge(\langle x\rangle) = (\langle x\rangle \ominus \{-1,0,1\}) \\ \cup(\langle x\rangle \ominus \{-1,0,2\}) \\ \cup(\langle x\rangle \ominus \{-2,-1\}) \\ \cup(\langle x\rangle \ominus \{-1,0\}). \qquad (34)$$

Since $\{-1,0\}$ is a subset of both $\{-1,0,1\}$ and $\{-1,0,2\}$, the Matheron basis representation

$$\Psi^\wedge(\langle x\rangle) = (\langle x\rangle \ominus \{-1,0\}) \\ \cup(\langle x\rangle \ominus \{-2,-1\}) \qquad (35)$$

is obtained, and this representation corresponds to the minimal sum-of-products representation of equation (33).

In the context of the Matheron representation we see the morphological interpretation of openings. As defined by Matheron, an operator that is translation-invariant, increasing, antiextensive, and idempotent is called a τ-opening. The most basic τ-opening is the elementary opening defined as erosion followed by dilation with the same structuring element: for signal $\langle x\rangle$ and structuring element A the opening of $\langle x\rangle$ by A is defined by $\langle x\rangle \circ A = (\langle x\rangle \ominus A) \oplus A$. The morphological basis of the opening $\langle x\rangle \circ A$ consists of all translates of A that contain the origin. Consequently, if A is finite, $A = \{j_0, j_1, \ldots, j_r\}$, then

$$\langle x\rangle \circ A = \cup\{\langle x\rangle \ominus (A - j_k) \\ : k = 0,1,\ldots,r\}. \qquad (36)$$

Letting $h_0 = x[0]x[j_1 - j_0]\cdots x[j_r - j_0]$, we see that $\langle x\rangle \circ A$ is equivalent to $\Psi(x)$, where $\Psi(x)[0]$ is defined by equation (21). Hence, a cellular-logic opening (as we have defined it) is equivalent to a morphological opening. Regarding τ-openings, Matheron [9] has shown that an operator is a τ-opening if and only if it can be represented as a union of openings, and this is precisely the import of equation (27), which expresses the τ-opening Ψ defined in equation (26) as a sum (union) of openings.

A key advantage of the logical formulation of mathematical morphology is the ability to check properties and relationships automatically. For instance, since idempotence for binary morphological operators is equivalent to idempotence for logical operators and since the latter characterization is machine checkable, we ipso facto have machine algorithms to check the morphological property. A second important example concerns the Matheron representation. Given the Matheron representations of several filters, the Matheron representation of an iteration can be found by the same algorithm that reduces an iteration of sum-of-product expansions to a single minimal sum of products.

7 Cellular Logic and Hit-or-Miss Transformations

Positive Boolean expansions are equivalent to the Matheron representation; what about the general sum-of-products expression (4)? Let us again begin with a single product

$$\Psi(x)[0] = x[j_1]x[j_2] \\ \cdots x[j_r]x[i_1]'x[i_2]' \\ \cdots x[i_s]', \qquad (37)$$

where $-M \le j_1 < \cdots < j_r \le M, -M \le i_1 < \cdots < i_s \le M$ and there does not exist a pair of indices j_a and i_b such that $j_a = i_b$. If we let $A = \{j_1, \ldots, j_r\}$ and $B = \{i_1, \ldots, i_s\}$, then $\Psi(x)[m] = 1$ if and only if $A + m$ is a subset of $\langle x \rangle$ and $B+m$ is a subset of $\langle x \rangle^c$, the complement of $\langle x \rangle$. This means that m lies in both $\langle x \rangle \ominus A$ and $\langle x \rangle^c \ominus B$. But the intersection of these two erosions is the hit-or-miss transform (Serra [21]) generated by the structuring pair (A, B):

$$\langle x \rangle \circledast (A,B) = (\langle x \rangle \ominus A) \cap (\langle x \rangle \ominus B), \quad (38)$$

and Ψ is equivalent to the hit-or-miss operator.

If we now consider the most general form of the sum-of-products Boolean expression in equation (4), we see that every translation-invariant, moving-window binary logical function is equivalent to a union of hit-or-miss operators with structuring elements in the window. Thus a general Boolean operator Ψ possesses a morphological equivalent $\Psi^\wedge$. In the discrete, moving window case, minimal-gate expressions can be found by considering the operator as a sum of products and applying some reduction algorithm.

As an illustration of how to employ the logic–morphology isomorphism, consider a four-point image window with the origin in the lower-left corner, so that $W\langle 0 \rangle = \{(0,0), (0,1), (1,0), (1,1)\}$ in the Cartesian grid. If we let x, y, z, and w denote the left–right, top-down raster scan of the four-point square, then every moving-window operator can be defined by a truth table consisting of strings of the form $xyzw$, where the operator Ψ takes the form $xyzw \to \Psi(xyzw)$. Suppose that we wish to find the minimal morphological implementation of Ψ, where Ψ is defined by the truth table output (in the usual order): 0, 0, 0, 0, 1, 1, 1, 1, 0, 0, 0, 0, 0, 0, 1, 1. A Karnaugh map reduction yields $\Psi(xyzw) = yx' + yz$. Its morphological equivalent is given by

$$\Psi^\wedge(S) = [S \circledast (A,B)] \cup [S \ominus C], \quad (39)$$

where $A = \{(1,1)\}$, $B = \{(0,1)\}$, and $C = \{(0,0),(1,1)\}$.

Just as the representation of a monotonically increasing cellular-logic operator as a minimal sum of products constitutes a finite logical realization of the Matheron erosion representation the interpretation of a general cellular-logic operator as a union of hit-or-miss transforms constitutes a finite logical realization of a different morphological representation theorem, namely, Banon and Berrera's extension [17] of the Matheron representation to translation-invariant set mappings that are not necessarily increasing. If $\Psi^\wedge$ is a translation-invariant set operator, then

$$\Psi^\wedge(S) = \cup\{S \otimes (A,B) : [A,B] < \text{Ker}[\Psi^\wedge]\}, \quad (40)$$

where $[A,B] = \{T : A < T < B\}$, a subset of the power set, is called the *closed interval* with *extremities* A and B (Birkhoff [22]) and $S \otimes (A,B)$ is the hit-or-miss operator applied to S with the structuring pair (A, B^c). In certain circumstances, the representation (40) can be reduced. A closed interval in Ker[$\Psi^\wedge$] is said to be *maximal* if no other closed interval contained in Ker[$\Psi^\wedge$] properly contains it. The set $\mathbf{B}[\Psi^\wedge]$ of all maximal closed intervals in Ker[$\Psi^\wedge$] is called the *basis* of $\Psi^\wedge$, and $\mathbf{B}[\Psi^\wedge]$ is said to *satisfy the representation condition* for $\Psi^\wedge$ if and only if for any closed interval in Ker[$\Psi^\wedge$] there exists a closed interval in $\mathbf{B}[\Psi^\wedge]$ containing it. If the basis $\mathbf{B}[\Psi^\wedge]$ satisfies the representation condition, then the expansion (40) reduces to

$$\Psi^\wedge(S) = \cup\{S \otimes (A,B) : [A,B] \in \mathbf{B}[\Psi^\wedge]\}. \quad (41)$$

As with the Matheron representation for increasing cellular-logic operators, the representation (41) is related to a general cellular-logic operator Ψ by recognizing, as we have in equation (39), that Ψ corresponds to a set mapping $\Psi^\wedge$. To illustrate the relationship, we consider two examples over a three-point window.

First suppose

$$\Psi(x)[0] = x[-1]x[0]' + x[0]x[1]. \quad (42)$$

To lie in the kernel of Ψ, a signal z must be defined in one of the following four ways over $\{-1, 0, 1\}$:

$$z_1 : z_1[-1] = 1, z_1[0] = 0, z_1[1] = 0,$$
$$z_2 : z_2[-1] = 1, z_2[0] = 0, z_2[1] = 1,$$

$$z_3 : z_3[-1] = 0, z_3[0] = 1, z_3[1] = 1,$$
$$z_4 : z_4[-1] = 1, z_4[0] = 1, z_4[1] = 1. \quad (43)$$

Because Ψ operates only over a three-point window, values of z outside $\{-1, 0, 1\}$ play no role. Thus when applying expansion (41) we need consider only maximal closed intervals formed from the four three-point signals in (43). There are three of these: $[z_1, z_2]$, $[z_2, z_4]$, $[z_3, z_4]$. Thus representation (41) yields

$$\Psi(x)[0] = x[-1]x[0]' + x[-1]x[1] + x[0]x[1], \quad (44)$$

which is equivalent to the definition of Ψ in equation (42). Next consider

$$\Psi(x)[0] = x[-1]'x[0]'x[1]' + x[1]' + x[-1]x[0]x[1]. \quad (45)$$

To lie in Ker[Ψ] a signal z must be defined in one of the following five ways over $\{-1, 0, 1, \}$:

$$z_1 : z[-1] = 0, z_1[0] = 0, z_1[1] = 0,$$
$$z_2 : z[-1] = 1, z_2[0] = 0, z_2[1] = 0,$$
$$z_3 : z[-1] = 0, z_3[0] = 1, z_3[1] = 0,$$
$$z_4 : z[-1] = 1, z_4[0] = 1, z_4[1] = 0,$$
$$z_5 : z[-1] = 1, z_5[0] = 1, z_5[1] = 1. \quad (46)$$

There are only two maximal closed intervals formed from the five three-point signals of equation (46): $[z_1, z_4]$, $[z_4, z_5]$. Thus representation (41) yields

$$\Psi(x)[0] = x[1]' + x[-1]x[0], \quad (47)$$

which is equivalent to the original expression for Ψ in (45).

References

1. S. Sternberg, "Image algebra," unpublished notes, 1983.
2. T. Crimmons and W. Brown, "Image algebra and automatic shape recognition," *IEEE Trans. Aerospace Electron. Systems*, vol. 21, 1985.
3. G.X. Ritter and P.D. Gader, "Image algebra techniques for parallel image processing," *Parallel Distrib. Comput.*, vol. 4, 1987.
4. G.X. Ritter, J.N. Wilson, and J.L. Davidson, "Image algebra: an overview," *Comput. Vis., Graph., Image Process.* vol. 49, 1990.
5. E.R. Dougherty and C.R. Giardina, *Mathematical Methods for Artificial Intelligence and Autonomous Systems* Prentice-Hall: Englewood Cliffs, NJ, 1988. pp. 399–414
6. E.R. Dougherty, "A homogeneous unification of image algebra, part I: the homogeneous algebra," *Imaging Sci.* vol. 33, 1989.
7. E.R. Dougherty, "A homogeneous unification of image algebra, part II: unification of image algebra," *Imaging Sci.*, vol. 33, 1989.
8. E.R. Dougherty and C.R. Giardina, "Image algebra—induced operators and induced subalgebras," *Proc. Soc Photo-Opt. Instrum. Eng.*, vol. 845, 1987.
9. G. Matheron, *Random Sets and Integral Geometry*, John Wiley: New York, 1975.
10. J. Serra, "Mathematical morphology for complete lattices," in *Image Analysis and Mathematical Morphology* vol. 2, J. Serra, ed., Academic Press: New York, 1988.
11. J. Serra, "Introduction to morphological filters," in *Image Analysis and Mathematical Morphology*, vol. 2, J. Serra ed., Academic Press: New York, 1988.
12. G. Matheron, "Filters and lattices," in *Image Analysis and Mathematical Morphology*, vol. 2, J. Serra, ed., Academic Press: New York, 1988.
13. H. Heijmans and C. Ronse, "The algebraic basis of mathematical morphology, I: dilations and erosions," *Comput. Vis., Graph., Image Process.*, vol. 50, 1990.
14. H. Heijmans and C. Ronse, "The algebraic basis of mathematical morphology, II: openings and closings, *Comput. Vis., Graph., Image Process.*," vol. 54, 1990.
15. H. Heijmans, "Theoretical aspects of gray-level morphology," *IEEE Trans. Patt. Anal. Mach. Intell.* vol. 13, 1991.
16. C. Ronse, "Why mathematical morphology needs complete lattices," *Signal Process.*, vol. 21, 1990.
17. G.J. Banon and J. Berrera, "Minimal representations for translation invariant set mappings by mathematical morphology," *SIAM J. Appl. Math.*, vol. 51, 1991.
18. P. Maragos and R. Schafer, "Morphological filters—part I: their set-theoretic analysis and relations to linear shift-invariant filters," *IEEE Trans. Acoust., Speech, Signal Process.*, vol. 35, 1987.
19. E.R. Dougherty and C.R. Giardina, "A digital version of the Matheron representation theorem for increasing tau-mappings in terms of a basis for the kernel," in *Proc IEEE Conf. on Computer Vision and Pattern Recognition*, July 1986.
20. C.R. Giardina and E.R. Dougherty, *Morphological Methods in Image and Signal Processing*, Prentice-Hall: Englewood Cliffs, NJ, 1988.
21. J. Serra, *Image Analysis and Mathematical Morphology*, Academic Press: New York, 1982.
22. G. Birkhoff, *Lattice Theory*, American Mathematical Society: Providence, RI, 1967.

Edward R. Dougherty is an associate professor at the Center for Imaging Science of the Rochester Institute of Technology and also serves as an industrial consultant. He holds an M.S. in computer science from Stevens Institute of Technology and a Ph.D. in mathematics from Rutgers University. He has written numerous papers in the area of mathematical morphology and image algebra and has authored or coauthored six books: *An Introduction to Morphological Image Processing, Matrix Structured Image Processing, Image Processing—Continuous to Discrete, Morphological Methods in Image and Signal Processing, Mathematical Methods for Artificial Intelligence and Autonomous Systems*, and *Probability and Statistics for the Engineering, Computing, and Physical Sciences.* He regularly teaches conference short courses in morphological image processing, including those for SPIE, and serves as a chair for two SPIE conferences, *Image Algebra and Morphological Image Processing* and *Nonlinear Image Processing.* His current research involves several principal areas: design of statistically optimal morphological filters, development of model-based approaches to morphological image analysis, granulometric characterization of texture, and development of intrinsically fuzzy approaches to mathematical morphology.

Robert M. Haralick is the Boeing Clairmont Egtvedt Professor in Electrical Engineering at the University of Washington. His recent work is in shape analysis and extraction using mathematical morphology, robust pose estimation, and techniques for making geometric inference from perspective projection information.

Professor Haralick has made a series of contributions in the high-level vision area, specifically dealing with the consistent labeling problem, arrangements, relation homomorphism, matching, and tree search. In the low-level and mid-level areas, Professor Haralick has worked in image texture analysis using spatial gray-tone co-occurrence texture features, and in the feature detection area he has developed the facet model for image processing.

Professor Haralick is a Fellow of IEEE for his contributions in computer vision and image processing. He serves on the Editorial Board of *IEEE Transactions on Pattern Analysis and Machine Intelligence.* He is the computer vision area editor for *Communications of the ACM*. He also serves as associate editor for *Computer Vision, Graphics, and Image Processing and Pattern Recognition.*

He received a B.A. degree in mathematics in 1964, a B.S. degree in electrical engineering in 1966, an M.S. degree in electrical engineering in 1967, and his Ph.D. in 1969, all from the University of Kansas.

Journal of Mathematical Imaging and Vision 2, 185–192 (1992).

Unification of Nonlinear Filtering in the Context of Binary Logical Calculus, Part II: Gray-Scale Filters

EDWARD R. DOUGHERTY
Center for Imaging Science, Rochester Institute of Technology, Rochester, NY 14623-0887

Abstract. This second part of a two-part study concerning the logical structure of nonlinear filters treats gray-scale filters. The algebraic framework of threshold decomposition is dedscribed in terms of the appropriate underlying commuting diagram, along with the manner in which generalized stack filters fall out of the framework when it is interpreted in the context of logical calculus. Relationships between representations for morphological and generalized stack filters are expressed in cellular logic.

Key words. nonlinear filter, morphological filter, stack filter, image algebra, representation, cellular logic

1 Introduction

This, the second part of a two-part study, treats gray-scale filtering by means of logical calculus by first placing the threshold logic used in nonlinear filtering into its natural encompassing mathematical framework. Not only is the logical context useful as an *image algebra*, if you will, but also its explicit use for morphological filtering makes transparent two fundamental relationships of nonlinear image processing: (i) the equivalence of stack filters (Wendt et al. [1]) and cellular-logic flat morphological filters (see Serra [2] for an historical overview of flat operators); and (ii) the representation of generalized stack filters (Lin and Coyle [3]) as a subclass of morphological filters, where by a *morphological filter* we mean a supremum of erosions, or, equivalently, a translation-invariant, increasing mapping (in contrast to Serra [4], who requires idempotence for a filter to be termed *morphological*). Although both relationships are well known, they have historically been framed in the context of mathematical morphology, i.e., representation of generalized stack filters as suprema of erosions. This can be accomplished either, as by Maragos and Schafer [5], in the framework of upper semicontinuous increasing mappings or in the more general lattice-theoretic framework. It is interesting that Maragos and Schafer [5] note that logical representations can be used in place of morphological representations as suprema of erosions. One aspect of the present paper is to carry out precisely such a program with the intent of providing a digitally faithful image algebra as well as a practical means for filter analysis and design.

2 Inducement of Gray-Scale Signals and Threshold Sets

The prominence of flat morphological filters for gray-scale processing, together with the desire to view these filters by means of cellular-logic binary filters, has resulted in widespread use of threshold decomposition (Serra [6], Wendt et al. [1], Shih and Mitchell [7]). Sternberg's conceptualization [8] of gray-scale morphology by the umbra transform can, in fact, be viewed as a generalization of threshold decomposition since the umbra transform for flat filters is algebraically equivalent to threshold decomposition. Moreover, both the umbra and threshold decomposition derive from the order structure of Z and, as shown by Heijmans [9], have direct lattice interpretations. Our goals here are limited: (i) to clearly express the manner in which threshold decomposition is grounded algebraically in a commuting diagram (not necessarily having anything to do with nonlinear image processing); (ii) to deduce the cellular-logic algebraic structure of nonlinear gray-scale digital image processing from the diagram; and (iii) to point out the manner in which binary logical calculus serves to unify nonlinear gray-scale digital image processing and, in particular, to express directly in the context of logic the relationship between morphological and generalized stack filtering.

Let **S** denote the class of signals defined on some set V and possessing gray-range Z, the set of integers, and let **A** denote the class of all denumerable vectors

$$A = [\ldots, A_{-2}, A_{-1}, A_0, A_1, A_2, \ldots] \quad (1)$$

of integer sets that are nested (decreasing), the latter meaning that, for $j > k, A_j < A_k$. We write $A = [A_k]$ to denote a set-vector in **A**. Define the mapping $\Pi : \mathbf{S} \to \mathbf{A}$ by $\Pi(f) = [\Pi_k(f)]$, where $\Pi_k(f)$ is the threshold set

$$\Pi_k(f) = \{x : f(x) \geq k\}. \quad (2)$$

It is immediate that $k \leq j$ implies $\Pi_k(f) > \Pi_j(f)$, so that Π is well defined, that is, $\Pi(f)$ lies in **A**. Π is a bijection (one-to-one and onto mapping) with $\Pi^{-1} : \mathbf{A} \to \mathbf{S}$ being defined by

$$\Pi^{-1}(A)(x) = \max\{k : x \in A_k\}. \quad (3)$$

for $A = [A_k] \in \mathbf{A}$. The upshot of the matter is that, as classes, the class **S** of signals is isomorphic to the class **A** of families of descending sets. Fundamentally, the isomorphism between **S** and **A** is a direct consequence of the definition of a signal in **S** as a subset of the Cartesian product $V \times Z$ for which $(v, z) = (v', z)$ implies $v = v'$, together with the lattice (order) structure on Z.

Consider any signal operator (filter) $\Psi : \mathbf{S} \to \mathbf{S}$. Owing to the bijection between **S** and **A**, Ψ induces an operator $\Phi : \mathbf{A} \to \mathbf{A}$ defined by the following commuting diagram:

$$\begin{array}{ccc} & \Psi & \\ \mathbf{S} & \rightarrow & \mathbf{S} \\ \Pi \downarrow & & \downarrow \Pi. \\ \mathbf{A} & \rightarrow & \mathbf{A} \\ & \Phi & \end{array} \quad (4)$$

Specifically, $\Phi = \Pi\Psi\Pi^{-1}$. Because the output of Φ is an element of **A**, Φ actually consists of a vector of mappings

$$\Phi(A) = [\ldots, \Phi_{-2}(A), \Phi_{-1}(A), \Phi_0(A), \Phi_1(A), \Phi_2(A), \ldots], \quad (5)$$

which we write as $\Phi(A) = [\Phi_k(A)]$. Each Φ_k is called a *section* of Φ and requires all components of a vector in **A** to produce a single output set. Because $\Phi(A) \in \mathbf{A}$, there is an ordering between the outputs of the Φ_k, namely,

$$\cdots > \Phi_{-2}(A) > \Phi_{-1}(A) > \Phi_0(A) > \Phi_1(A) > \Phi_2(A) > \cdots. \quad (6)$$

The inducement of Φ from the commuting diagram (4) means that each signal-to-signal operator can be realized as an operator on **A**. Also from the diagram we obtain inducement of a signal operator given an operator on **A**, namely, $\Psi = \Pi^{-1}\Phi\Pi$. If we denote the collection of all operators $\mathbf{S} \to \mathbf{S}$ by $\mathbf{S}^{\mathbf{S}}$ and the collection of all operators $\mathbf{A} \to \mathbf{A}$ by $\mathbf{A}^{\mathbf{A}}$, then the commuting diagram (4) provides an isomorphism between $\mathbf{S}^{\mathbf{S}}$ and $\mathbf{A}^{\mathbf{A}}$.

A basic question arises regarding inducement of signal operators from vectors of set mappings Φ by means of $\Psi = \Pi^{-1}\Phi\Pi$: if $\Phi = [\Phi_k]$ is a vector of set mappings defined on **A**, when does Φ induce an operator on **S**? To grasp the import of this question, consider an arbitrary mapping $\Phi = [\Phi_k]$ defined on $\mathbf{A}, \Phi : \mathbf{A} \to \mathbf{P}$, where **P** is the collection of denumerable vectors of subsets from the domain space. Φ is an operator on **A** if and only if the output vectors of Φ lie in **A**, so that $\Phi : \mathbf{A} \to \mathbf{A}$. The condition under which this occurs is defined by the decreasing (stacking) constraint (6), which says that the output vector must be descending in the same order as the input vector. For the moment, consider a general mapping Φ on **A** that may or may not be an operator on **A** (lie in $\mathbf{A}^{\mathbf{A}}$). From equation (5) we see that each Φ_k is itself defined on **A**. But each element $A = [A_k]$ in **A** is a set-vector indexed over Z, and therefore Φ_k is a function of a denumerable infinity of variables, each of which is a set. As with any function of several variables, any specific Φ_k may or may not be dependent on all variables. We may have $\Phi_k = \Phi_k(A_{k,1}, A_{k,2}, \ldots)$, where $\{A_{k,1}, A_{k,2}, \ldots\}$ is a subcollection of $\{A_k\}$, either finite or infinite. A special case arises when each section Φ_k depends only on a band of threshold sets about A_k. Then there exists $s \geq 0$ such that

$$\Phi_k = \Phi_k(A_{k-s}, A_{k-s+1}, \ldots, A_k, \ldots, A_{k+s}), \quad (7)$$

and we call Φ an *s-band threshold mapping*. If $s = 0$, then Φ_k depends only A_k and we call Φ a *threshold mapping*. For $s = 0$, the decreasing constraint (6) becomes $\Phi_j(A_j) < \Phi_k(A_k)$ for $j > k$. If Φ satisfies this constraint, we say it is *order preserving*. The key point is that a threshold mapping on **A** is an operator on **A** if and only if it is order preserving. An order-preserving threshold mapping on **A** will be called a *threshold filter*, as will its equivalent signal operator defined by $\Psi = \Pi^{-1}\Phi\Pi$.

3 Moving-Window Gray-Scale Operators

The preceding remarks on gray-scale mappings are independent of the domain space and the nature of the mappings: they are simply a commentary on the consequences of the isomorphism between **S** and **A**. In nonlinear digital filtering our main concern is with translation-invariant, moving-window operators on discrete space, which for algebraic purposes can be assumed to be Z. For the window $W\langle 0\rangle$ and signal f, such an operator Ψ is defined at m by

$$\Psi(f)[m] = \Psi(f[m-M], f[m-M+1], \ldots, f[m+M]), \tag{8}$$

where we do not distinguish between the filter Ψ and the function rule defining Ψ. Translation invariance is manifested by the fact that the function rule is independent of the point m, and hence $\Psi(f)[m]$ is defined by means of $\Psi(f)[0]$. Henceforth, unless otherwise indicated, it will be assumed that we are dealing with translation-invariant moving-window operators over $W\langle m\rangle$.

If $\Phi = [\Phi_k]$ is induced by an arbitrary operator on **S**, then each $\Phi_k(A)$ is a set depending on A, and it can be defined by its characteristic function (for which we use the same notation): $\Phi_k(A)[m] = 1$ if $m \in \Phi_k(A)$, and $\Phi_k(A)[m] = 0$ if $m \notin \Phi_k(A)$. Because of the descending constraint, $\Phi(A)[m]$ is a vector of the form $(\ldots, 1, 1, 0, 0, \ldots)$. But $\Phi = \Pi\Psi\Pi^{-1}$. If we let $g = \Pi^{-1}(A)$ be the signal defined by A, then $\Phi(A)[m] = \Pi\Psi(g)[m]$. $\Psi(g)$ depends only on the values of g in the window $W\langle m\rangle$, and therefore the only part of A required to find $\Psi(g)$ is that portion of A lying over the window $W\langle m\rangle$ in the domain. Consequently, the operators on **A** corresponding to translation-invariant moving-window operators are of the form $\Phi(A)[m] = \Phi(A \cap W\langle m\rangle)[m]$, and each section Φ_k is of the form $\Phi_k(A)[m] = \Phi_k(A \cap W\langle m\rangle)[m]$.

$A \cap W\langle m\rangle$ is itself a set vector: $A \cap W\langle m\rangle = [A_k \cap W\langle m\rangle]$. In characteristic-function format each $A_k \cap W\langle m\rangle$ is a $(2M+1)$-vector of zeros and ones, so that $A \cap W\langle m\rangle$ is a matrix with an infinite number of rows and $2M + 1$ columns. Because the A_k are nested, each column consists of an infinite column of zeros over an infinite column of ones. In mathematical morphology, such an infinite matrix is called an *umbra* (Sternberg [8]). If Φ is induced by Ψ, then $\Phi_k(A)[m]$ is a function of the umbra.

If $\Phi = [\Phi_k]$ happens to be an s-band threshold operator, then $\Phi_k(A)[m]$ depends only on that portion of $A \cap W\langle m\rangle$ that lies in the s-band about A_k. That portion, to be denoted by $[A \cap W\langle m\rangle]_{k,s}$ is a $2s + 1$ by $2M + 1$ matrix. Once a one appears in a column, the entries below it are one. Such a matrix is called an *umbra matrix* (Dougherty and Giardina [10], [11]). Regarding Φ, each section Φ_k is of the form $\Phi_k([A \cap W\langle m\rangle]_{k,s})[m]$, where $\Phi_k([A \cap W\langle m\rangle]_{k,s})[m]$ is either zero or one. Owing to translation invariance, $\Phi_k(A)$ is determined by its definition at $0, \Phi_k(A)[0]$. As a function of 0–1 logical variables, Φ_k is representable by a Boolean expression in $(2M+1)(2s+1)$ variables and hence Φ is representable by a set of Boolean expressions. Owing to the bijection defined by the commuting diagram (4), Ψ itself can be considered to be defined by Boolean expressions, each having $(2M + 1)(2s + 1)$ variables. Logically, Φ (and hence Ψ) is a collection of truth tables. Moving-window s-band threshold operators are known as *generalized stack filters* (Lin and Coyle [3]).

If we assume (as we must for digital processing) that the range space is a finite subset of Z, say, $\{0, 1, \ldots, N-1\}$, then Φ is composed of N truth tables, each possessing $(2M + 1)(2s + 1)$ variables and admitting a minimal-gate logical representation. If we write $\Phi = [\Phi_0, \Phi_1, \ldots, \Phi_{N-1}]$, we see that Φ is nothing more than N cellular-logic arrays. Moreover, the finiteness of the range space ensures that all filters are s-band (for some s). This last point appears in [3] in the context of generalized stack filters; however, whereas there it entails a rather lengthy proof, here we see that it derives at once from the inducement corresponding to the commuting diagram (4).

For the special case in which Φ is a threshold filter, Φ_k is a logical expression depending on only $2M + 1$ variables,

$$\Phi_k(A)[m] = \Phi_k(x[m - M], x[m - M + 1], \ldots, x[m + M]), \quad (9)$$

the right-hand side being a (minimal-gate) sum of products over the window $W\langle m\rangle$. Owing to the isomorphism defined by the bijection we can write $\Psi_k(f)[m]$ in place of $\Phi_k(A)[m]$.

In the preceding analysis for the threshold-filter case we did not concern ourselves with whether we had a threshold mapping on **A** that was or was not order preserving, since it was assumed to correspond to a signal operator and was therefore ipso facto an operator on **A**. In the other direction, suppose we have a threshold mapping $\Phi = [\Phi_k]$ defined on **A** that is given in the sum-of-products form of equation (9). Because $\Phi(A)$ depends only on $A \cap W\langle m\rangle$, any signal operator it induces must be a moving-window operator. Moreover, it induces a signal operator if and only if it is order preserving.

We cannot conclude from order preservation that Φ_k is monotonic as a Boolean function and therefore possesses a positive sum-of-products representation. To see this, consider a three-point window and let $\Phi_k(A)[0] = x[-1]$ for $k > 0, \Phi_0(A)[0] = x[-1] + x[0]x[1]'$, and $\Phi_k(A)[0] = x[-1] + x[0]$ for $k < 0$ (see [3] for a different counterexample).

There is one case in which the sections of threshold filters must be monotonic and therefore possess sum-of-products representations having no complements. This is the case for *flat-morphological*, or *stack*, filters. As is well known, for these each section is identical, each identical section must be a positive Boolean function, and as a signal operator the filter is monotonic: if $f \leq g$, then $\Psi(f) \leq \Psi(g)$. The reason for the *flat* terminology is that each section, as a positive Boolean function, is, by the Matheron representation, a union of binary erosions. Thus $\Phi_k(A)[0] = 1$ if and only if Φ_k possesses a product term $x[j_1]x[j_2]\cdots x[j_r]$ for which each j_i lies in A_k, which in turn is equivalent to $f[j_i] \geq k$ for $i = 1, 2, \ldots, r$. Hence $\Phi_k(A)[0] = 1$ if and only if there exists a product term such that the minimum of f over the indices of the product term is greater than or equal to k. But the latter is equivalent to the maximum of the signal minima over the various products being greater than or equal to k. Since

$$\Psi(f)[0] = \max\{k : \Phi_k(A)[0] = 1\} \quad (10)$$

we conclude that $\Psi(f)[0]$ is precisely the maximum of the signal minima over the various products. Since each minimum of f over a set of indices is a gray-scale erosion by the set determined by those indices, which itself is a gray-scale erosion by a flat structuring element, Ψ is a maximum of erosions by flat structuring elements. As originally proposed by Serra [6], gray-scale morphological filters were assumed to be flat; it was Sternberg [8] who first considered more general structuring elements.

More general than the flat morphological filters are the threshold filters $\Phi = [\Phi_k]$ where, for all k, Φ_k is increasing (and therefore represented by a positive Boolean function) and, for $k < j$, $\Phi_k \geq \Phi_j$. In fact, together these two properties imply order preservation: for $k < j$, if $A_k > A_j$ then $\Phi_k(A_k) > \Phi_k(A_j) > \Phi_j(A_j)$. In particular, such filters are increasing.

4 Gray-Scale Morphological Filters

So far we have considered only spatial translation invariance; once again considering the range space Z, we say that a signal operator Ψ is *offset invariant (range translation invariant)* if for any signal f and integer k, $\Psi(f + k) = \Psi(f) + k$. An operator that is both spatially translation invariant and offset invariant is said to be *morphologically translation invariant*. If Ψ is both spatially translation and offset invariant and is also increasing, we call it a *morphological filter*. The Matheron representation for gray-scale operators states that any such filter can be expressed as a maximum of erosions over its *kernel*, wher

$\mathrm{Ker}[\Psi] = \{g : \Psi(g) \geq 0\}$ (see [11] or Maragos and Schafer [12]). Specifically,

$$\Psi(f) = \vee\{f \ominus g : g \in \mathrm{Ker}[\Psi]\}. \quad (11)$$

A subclass of the kernel, $\mathrm{Bas}[\Psi]$, is the basis for Ψ if (i) for any $g \in \mathrm{Ker}[\Psi]$ there exists $h \in \mathrm{Bas}[\Psi]$ such that $h \leq g$ and (ii) there does not exist a pair of signals in $\mathrm{Bas}[\Psi]$ properly related by $\leq$. When a basis exists (as is usually the case in practice), the representation (11) can be taken over $\mathrm{Bas}[\Psi]$.

If Ψ is a moving-window morphological filter over the window $W\langle 0\rangle$, then the basis representation takes the form

$$\begin{aligned}\Psi(f)[m] = \max\{\min\{f(j) - T[g,m](j) \\ : j \in W\langle m\rangle\} \\ : g \in \mathrm{Bas}[\Psi]\},\end{aligned} \quad (12)$$

where $T[g,m]$, the translation of g by m, is defined by $T[g,m](j) = g(j-m)$. If all basis elements are flat, then the representation (12) reduces to a maximum of signal minima over a collection of translated index sets, which is precisely what is meant by a flat-morphological (stack) filter or, equivalently, a monotonically increasing threshold filter having identical sections. If an increasing threshold filter does not possess identical sections, then it is not offset invariant and hence is not a morphological filter, even though it is spatially translation invariant.

More generally, an increasing, moving-window, s-band threshold filter Ψ must be spatially translation invariant, but it need not be offset invariant. It is offset invariant if and only if the section operators are identical, in which case it is a generalized stack filter with identical sections. If such is the case, then Ψ is a morphological filter and it has a Matheron representation. Moreover, all structuring elements in its basis have domains in $W\langle 0\rangle$ and gray range $[-s, s]$. This conclusion follows from the umbra formulation of erosion. To wit, if g is a structuring element in the Matheron representation and if $\Pi(f) \cap W\langle 0\rangle$ is viewed as an umbra, $\Phi_k(\Pi(f) \cap W\langle 0\rangle)[0] = 1$ if and only if $\Pi(g+k) \cap W\langle 0\rangle$ is a subset of $\Pi(f) \cap W\langle 0\rangle$. Since Ψ depends only on the threshold sets $\Pi_{k+i}(f)$, $-s \leq i \leq s$, the subset relation between $\Pi(f) \cap W\langle 0\rangle$ and $\Pi(g+k) \cap W\langle 0\rangle$ must depend only on these threshold sets. This restricted dependence can occur only if the graph of g is constrained to the rectangle $W\langle 0\rangle \times [-s, s]$, which is precisely the claim. Although we have considered only $W\langle 0\rangle$, the argument applies to any $W\langle m\rangle$ because of translation invariance.

Given an increasing s-band threshold filter Ψ with identical sections (increasing generalized stack filter with identical sections), we can find the Matheron basis for Ψ, say, $\mathrm{Bas}[\Psi] = \{g_1, g_2, \ldots, g_p\}$. If the domain of $g_r, r = 1, 2, \ldots, p$ is denoted by $D[g_r]$, the representation becomes

$$\begin{aligned}\Psi(f)[0] = \max\{\min\{f(j) - g_r(j) \\ : j \in D[g_r]\} \\ : r = 1, 2, \ldots, p\}.\end{aligned} \quad (13)$$

The graph of g_r is a subset of $W\langle 0\rangle \times [-s, s]$ and can be viewed as a $(2s+1)$ by $(2M+1)$ matrix of zeros and ones having at most a single one in each column: specifically, if row–column indexing is used, the graph of g_r can be viewed as being defined over a $(2s+1)$ by $(2M+1)$ array z of points (i_r, j_r), where $z(i_r, j_r) = 1$ if and only if $i_r = g(j_r)$. Representation (13) then takes the form

$$\begin{aligned}\Psi(f)[0] = \max\{\min\{f(j_r) - i_r \\ : z(i_r, j_r) = 1\} \\ : r = 1, \ldots, p\},\end{aligned} \quad (14)$$

which is precisely the form given by Lin and Coyle [3], who do not appeal to the Matheron representation.

Going further, we can stand the problem on its head: given $\mathrm{Bas}[\Psi] = \{g_1, g_2, \ldots, g_p\}$, find the section Φ_0 (and hence all others since they are assumed to be identical). First, $\Psi(f)[0]$ is the maximum k for which $\Phi_k[\Pi(f) \cap W\langle 0\rangle]_{k,s} = 1$. Second, the Matheron representation (and the definition of erosion) implies that, for any $k \leq \Psi(f)[0]$, there exists r such that $z(i_r + k, j_r) = 1$ implies $f(j_r) \geq i_r + k$. Hence $\Phi_0(\Pi(f) \cap W\langle 0\rangle) = 1$ if and only if at least one of the products $\Pi\{x[i_r, j_r] : z(i_r, j_r) = 1\}$ yields 1 when applied to $[\Pi(f) \cap W\langle 0\rangle]_{0,s}$, so that

$$\Phi_0([\Pi(f) \cap W\langle 0\rangle]_{0,s}) = \sum\{\Pi\{x[i_r, j_r]$$

$$: z(i_r, j_r) = 1\} \\ : r = 1, \ldots, p\}. \quad (15)$$

Since all sections are identical, the sum of products on the right side of equation (15) provides the common section expression in terms of logical variables and hence each Φ_k can be evaluated by applying the expression to the appropriately indexed s-band matrix.

To illustrate the preceding notions, consider the morphological filter defined by Bas$[\Psi] = \{g_1, g_2\}$, where g_1 and g_2 are defined by

$$g_1 : g_1[-1] = 0, g_1[0] = 1, g_1[1] = 2, \\ g_2 : g_2[-1] = 2, g_2[0] = 0, g_2[1] = -2. \quad (16)$$

Then $s = 2$ and the arrays $A_1 = z(i_1, j_1)$ and $A_2 = z(i_2, j_2)$ are given by

$$A_1 = \begin{pmatrix} 0 & 0 & 1 \\ 0 & 1 & 0 \\ 1 & 0 & 0 \\ 0 & 0 & 0 \\ 0 & 0 & 0 \end{pmatrix}, \quad A_2 = \begin{pmatrix} 1 & 0 & 0 \\ 0 & 0 & 0 \\ 0 & 1 & 0 \\ 0 & 0 & 0 \\ 0 & 0 & 1 \end{pmatrix}. \quad (17)$$

The Boolean expression for the section is

$$\Phi_0 = x[0, -1]x[1, 0]x[2, 1] \\ +x[2, -1]x[0, 0]x[-2, 1]. \quad (18)$$

Suppose f is given by $f[-1] = 2, f[0] = 5, f[1] = 7$. Then

$$[\Pi(f) \cap W\langle 0\rangle]_{1,2} = \begin{pmatrix} 0 & 1 & 1 \\ 1 & 1 & 1 \\ 1 & 1 & 1 \\ 1 & 1 & 1 \\ 1 & 1 & 1 \end{pmatrix}. \quad (19)$$

Treating Φ_0 as an operator on $[\Pi(f) \cap W\langle 0\rangle]_{1,2}$ yields

$$\Phi_1(\Pi(f)) = (1)(1)(1) + (0)(1)(1) = 1. \quad (20)$$

Next,

$$[\Pi(f) \cap W\langle 0\rangle]_{2,2} = \begin{pmatrix} 0 & 1 & 1 \\ 0 & 1 & 1 \\ 1 & 1 & 1 \\ 1 & 1 & 1 \\ 1 & 1 & 1 \end{pmatrix}. \quad (21)$$

so that the expression for $\Phi_2(\Pi(f))$ is identical to that for $\Phi_1(\Pi(f))$. Finally,

$$[\Pi(f) \cap W\langle 0\rangle]_{3,2} = \begin{pmatrix} 0 & 1 & 1 \\ 0 & 1 & 1 \\ 0 & 1 & 1 \\ 1 & 1 & 1 \\ 1 & 1 & 1 \end{pmatrix}. \quad (22)$$

Hence, $\Phi_3(\Pi(f)) = (0)(1)(1) + (0)(1)(1) = 0$, and taking the maximum k such that $\Phi_k(\Pi(f)) = 1$ yields $\Psi(f)[0] = 2$.

5 Umbra Relationship

The strong relationship between generalized stack formulation and both umbrae and umbra matrices [10] is evident in the preceding example, and we explore this matter further. Regarding the umbra transform, we follow [4] and we refer to it and to Dougherty [13] for the details of operator inducement by means of the umbra transform.

The *umbra* of a subset A of Z^2 is defined by

$$U[A] = \{(j, k) \in Z^2 : \text{there exists} \\ (j, i) \in A \text{ with } i \geq k\}. \quad (23)$$

In this, the digital setting, if we assume that A is bounded from above, then the *surface* of A is the signal defined by

$$S[A][j] = \max\{k : (j, k) \in A\}, \quad (24)$$

where it is understood that $S[A]$ is defined only at those j for which there exists at least one pair $(j, k) \in A$. If g is a signal with domain $D[g]$, the graph of g, denoted $G[g]$, is a subset of $D[g] \times Z$ and the umbra of g is defined by $U[g] = U[G[g]]$, the umbra of the graph of g. The graph of g is recoverable from the umbra of g by applying the surface operator: $G[g] = S[U[g]]$. Identifying g with its graph allows us to write $g = S[U[g]]$. For erosions the basic umbra relation is

$$f \ominus g = S[U[f] \ominus U[g]], \quad (25)$$

so that the gray-scale erosion is found as the surface of a binary erosion of umbrae.

Referring back to representation (14), we find that the arrays $A_r = z(i_r, j_r)$ are actually representations of the graphs of the basis signals g_r. In particular, $U[A_r] = U[g_r]$, where A_r is treated as a subset of Z^2. For the preceding example this is evidenced by equation (17).

The umbra of a signal f is given in terms of $\Pi(f)$ by stacking the $\Pi(f)_k$ and viewing the stack as an infinite matrix in Z^2, thereby yielding

$$U[f] = \begin{pmatrix} \vdots \\ \Pi(f)_1 \\ \Pi(f)_0 \\ \Pi(f)_{-1} \\ \vdots \end{pmatrix}. \qquad (26)$$

As an element of $\mathbf{A}$, $\Pi(f) = [\Pi(f)_k]$ is a vector of the form (1), so that $U[f]$ is the transpose of $\Pi(f)$ written in reverse order. Denoting such a transpose of a vector $A \in \mathbf{A}$ by $A^\#$, we can write

$$U[f] = \Pi(f)^\#, \qquad (27)$$

a key point being that $U[f]$ and $\Pi[f]$ are the same up to format.

According to equation (25), when finding $(f \ominus g)[0]$ by means of umbrae, we are concerned with the umbra of f restricted to the domain of g. This restriction is given by

$$\begin{aligned} U[f]|_{D[g]} &= U[f] \cap (D[g] \times Z) \\ &= \Pi(f)^\# \cap (D[g] \times Z). \end{aligned} \qquad (28)$$

As a subset of Z^2, the matrix $[\Pi(f) \cap W\langle 0\rangle]_{k,s}$ is given by

$$[\Pi(f) \cap W\langle 0\rangle]_{k,s} = U[f]|_{D[g]} \cap (Z \times [k-s, k+s]). \qquad (29)$$

In the above example this is evidenced by equations (20), (21), and (22). By construction, $[\Pi(f) \cap W\langle 0\rangle]_{k,s}$ possesses the same dimensions as the A_r.

According to representation (15), $\Phi_k(\Pi(f))[0] = 1$ if and only if there exists some basis signal g_r for which the entries of the matrix $[\Pi(f) \cap W\langle 0\rangle]_{k,s}$ are ones whenever the corresponding entries of A_r are ones. Letting $A_{r,k}$ denote the matrix in Z^2 that is the upward translation of A_r by k, we can restate this by saying that $\Phi_k(\Pi(f))[0] = 1$ if and only if there exists r such that $A_{r,k} < [\Pi(f) \cap W\langle 0\rangle]_{k,s}$, the latter being treated as a subset within Z^2 in the manner of equation (29). Owing to the dimensions of $A_{r,k}$ and to equation (29), this is equivalent to saying there exists r such that $A_{r,k} < U[f]|_{D[g]}$, which means that $U[g_r]_k < U[f]|_{D[g]}$, where in the case of $U[g_r]_k$ we have again used the subscript k to denote vertical translation by k. But $U[g_r]_k < U[f]|_{D[g]}$ is equivalent to the point $(0, k)$ lying in $U[f] \ominus U[g_r]$. Hence, according to the definition of $\Psi(f)[0]$ as the maximum k such that $\Phi_k(\Pi(f)) = 1$, representation (15) is equivalent to

$$\begin{aligned} \Psi(f)[0] = \max\{k &: \text{there exists } r \text{ with} \\ &(0, k) \in U[f] \ominus U[g_r]\}. \end{aligned} \qquad (30)$$

Equation (30) is simply an umbral restatement of the Matheron representation. Indeed, it is directly equivalent to

$$\begin{aligned} \Psi(f)[0] = \max\{(S[U[f] \\ \ominus U[g_r]])[0] \\ : r = 1, 2, \ldots, p\}, \end{aligned} \qquad (31)$$

which is the Matheron representation. In sum, the generalized-stack representation (15) can be viewed as an umbral form of the Matheron representation.

To illustrate the umbra representation, we again consider the example of equations (16) through (22). There,

$$U[g_1] = U[A_1] = \begin{pmatrix} 0 & 0 & 0 \\ 0 & 0 & 1 \\ 0 & 1 & 1 \\ 1 & 1 & 1 \\ 1 & \mathbf{1} & 1 \\ 1 & 1 & 1 \\ & \vdots & \end{pmatrix}, \qquad (32)$$

$$U[g_2] = U[A_2] = \begin{pmatrix} 0 & 0 & 0 \\ 1 & 0 & 0 \\ 1 & 0 & 0 \\ 1 & \mathbf{1} & 0 \\ 1 & 1 & 0 \\ 1 & 1 & 1 \\ & \vdots & \end{pmatrix}, \qquad (33)$$

$$U[f] = \begin{pmatrix} 0 & 0 & 0 \\ 0 & 0 & 1 \\ 0 & 0 & 1 \\ 0 & 1 & 1 \\ 0 & 1 & 1 \\ 0 & 1 & 1 \\ 1 & 1 & 1 \\ 1 & 1 & 1 \\ 1 & \mathbf{1} & 1 \\ & \vdots & \end{pmatrix}, \qquad (34)$$

where we have used a bold character to denote the origin. Clearly, $U[A_1]_2$ is a subset of $U[f]$ but $U[A_1]_1$ is not, and $U[A_2]_0$ is a subset of $U[f]$ but $U[A_2]_1$ is not. Thus the umbral representation (30) yields $\Psi(f)[0] = \max\{2,0\} = 2$, as did representation (15).

A straightforward adaptation of the umbral representation (30) yields an umbra-matrix representation. This follows at once from its equivalence to representation (31) in terms of erosions and the fact that the surface operator S can be used in conjunction with morphology on umbra matrices to produce gray-scale morphology [10]. Here we note only the similarity between the stack representation matrices A_1 and A_2 of equation (17) for g_1 and g_2 and the corresponding umbra matrices

$$\overline{U}[g_1] = \begin{pmatrix} 0 & 0 & 1 \\ 0 & 1 & 1 \\ 1 & \mathbf{1} & 1 \\ 1 & 1 & 1 \\ 1 & 1 & 1 \end{pmatrix}, \quad \overline{U}[g_2] = \begin{pmatrix} 1 & 0 & 0 \\ 1 & 0 & 0 \\ 1 & \mathbf{1} & 0 \\ 1 & 1 & 0 \\ 1 & 1 & 1 \end{pmatrix}. \qquad (35)$$

From these it should be evident that the umbral argument leading up to equation (30) can be equivalently carried out by using umbra matrices instead of umbrae.

References

1. P.D. Wendt, E.J. Coyle, and N.C. Gallagher, "Stack filters," *IEEE Trans. Acoust., Speech, Signal Process.*, vol. 34, 1986.
2. J. Serra, "Anamorphoses and function lattices," in *Mathematical Morphology in Image Processing*, E. Dougherty, ed., Marcel Dekker: New York, 1992.
3. J.H. Lin and E.J. Coyle, "Minimum mean absolute error estimation over the class of generalized stack filters," *IEEE Trans. Acoust., Speech, Signal Process.*, vol. 38, 1990.
4. J. Serra, *Image Analysis and Mathematical Morphology*, vol. 2, Academic Press: New York, 1988.
5. P. Maragos and R. Schafer, "Morphological filters—part II: their relations to median, order-statistic, and stack filters," *IEEE Trans. Acoust., Speech, Signal Process.*, vol. 35, 1987.
6. J. Serra, *Image Analysis and Mathematical Morphology*, Academic Press: New York, 1982.
7. F. Shih and O. Mitchell, "Threshold decomposition of gray-scale morphology into binary morphology," *IEEE Trans. Patt. Anal. Mach. Intell.*, vol. 11, 1989.
8. S. Sternberg, "Grayscale morphology," *Comput. Vis., Graph., Image Process.*, vol. 35, 1986.
9. H. Heijmans, "Theoretical aspects of gray-level morphology," *IEEE Trans. Patt. Anal. Mach. Intell.*, vol. 13, 1991.
10. E.R. Dougherty and C.R. Giardina, "Morphology on umbra matrices," *Internat. J. Patt. Recog. Art. Intell.*, vol. 2, 1988.
11. C.R. Giardina and E.R. Dougherty, *Morphological Methods in Image and Signal Processing*, Prentice-Hall: Englewood Cliffs, NJ, 1988.
12. P. Maragos and R. Schafer, "Morphological filters—part 1: their set-theoretic analysis and relations to linear shift-invariant filters," *IEEE Trans. Acoust., Speech, Signal Process.*, vol. 35, 1987.
13. E.R. Dougherty, "Euclidean gray-scale granulometries: representation and umbra inducement," *J. Math. Imag. Vis.*, vol. 1, 1992.

Journal of Mathematical Imaging and Vision 2, 193–215 (1992).

Morphological Analysis of Discrete Random Shapes

JOHN GOUTSIAS
Image Analysis and Communications Laboratory, Department of Electrical and Computer Engineering, Johns Hopkins University, Baltimore, MD 21218

Abstract. In this paper I discuss a number of theoretical issues regarding the morphological analysis of discrete random shapes by means of Matheron's random set theory. I revisit this theory by limiting myself to the discrete case, since most image data are available in a discrete form. Although it may seem that the transition from the continuous to the discrete case is straightforward (since most of Matheron's theory is general enough to incorporate the discrete case as a special case), this transition is often challenging and full of exciting and, surprisingly, pleasant results. I introduce the concept of the cumulative-distribution functional of a discrete random set and review some fundamental properties of the capacity functional (a fundamental statistical quantity that uniquely defines a random set and relates random set theory to mathematical morphology). In analogy to a recent result and under a natural boundness condition, I show that there exists a one-to-one correspondence between the probability-mass function of a discrete binary random field and the corresponding cumulative-distribution functional. The relationship between the cumulative-distribution functional and the capacity functional of a discrete random set is also established. The cumulative-distribution and capacity functionals are related to the higher-order moments of a discrete binary random field, and, therefore, their computation is equivalent to computing these moments. A brief discussion of how to perform such computations for a certain class of discrete random sets is provided. The capacity functional of a morphologically transformed, continuous random set cannot be associated to the capacity functional of the random set itself, except in the case of dilation. I show that the derivation of such an association is possible in the discrete case and for the cases of dilation and erosion and more complicated morphological transformations, such as opening and closing. These relationships are then used to derive a fundamental result regarding the statistical behavior of opening and closing morphological filters. I also show that the probability-mass function of a discrete binary random field may be expressed in terms of the cumulative-distribution functional or the capacity functional of a morphologically transformed discrete random set by means of a hit-or-miss transformation. I also introduce moments for discrete random sets, which permit generalization of the concepts of autocorrelation and contact distribution. Furthermore, I demonstrate the fact that the class of opening-based size distributions, introduced axiomatically by Matheron, are higher-order moments of a discrete random set, therefore statistically demonstrating that size distributions are good statistical summaries for shape. Finally, convex random sets are viewed in the discrete domain. My final result regarding convexity, is similar to Matheron's. However, the tools used here for the derivation of such a result are different from the ones used by Matheron, whose approach to this subject is limited to the continuous case.

Key words. binary random fields, discrete random shapes, mathematical morphology, random set theory, shape analysis

1 Introduction

One of the most important problems in modern image-analysis and computer-vision applications is the development of efficient and robust techniques for the synthesis and analysis of shape [1]–[3]. A careful consideration of the available literature on this subject, which is too vast to enumerate here, reveals the classification of shape synthesis and analysis techniques into two categories: *structural* and *stochastic*. Structural techniques explore the geometric nature of shape, whereas stochastic techniques explore its random character.

Stochastic approaches to shape have been receiving special attention because of their robustness to noise contamination and shape degradation [1], [3]. A wide variety of stochastic techniques have been proposed in the literature. The most prominent techniques are based on random point fields (processes) [4]–[11], tessellations and mosaics [12]–[14], coverage models [13], Markov random fields [1], [3], [10], [11], [15]–[22], deformable templates [23], [24], Kendall's polygonal theory of shape [25], [26], and random set theory [2], [5], [7]–[9], [11], [19], [20], [27]–[31]. Of all these techniques, the most widely used are the ones based on random fields. Such techniques may not be suitable for random shape modeling and analysis because they depend solely on probability distribution functions (in the continuous case) or probability-mass functions (in the discrete case), which usually ignore geometric structure. Shape-analysis techniques based on random fields are appropriate only if structural constraints are incorporated, explicitly or implicitly, in the underlying probability model. This is the main theme in [5], [6], [9], [12]–[14], [21], and [23]–[26], for example.

A promising way to consider geometric structure in images is in terms of mathematical morphology [2], [28], [32], [33]. In the theory of mathematical morphology a shape is represented by a closed set X in the two-dimensional Euclidean space R^2. Image-analysis algorithms are then obtained by designing sequences of nonlinear transformations, which are defined in terms of a compact set B in R^2 known as the *structuring element*. The structuring element is a simple shape, which interacts with the image under consideration in order to extract some useful geometric information from it. The type of interaction between X and B is determined by the hit-or-miss topology generated on the space of all closed sets by the two classes $\{X : X \cap B \neq \emptyset\}$ and $\{X : X \cap B = \emptyset\}$. The success of mathematical morphology in a wide variety of interesting problems is discussed in [2], [32], [34], and [35].

Most recently developed morphological techniques focus on the analysis of the structural characteristics of shape. Since most shapes of interest are formed by a combination of structural and stochastic components, it has been suggested in [2], [20], [30], [31], and [36]–[40] (see also [8], [9], [11], [19], and [41]) that the incorporation of morphological transformations within a stochastic formulation may result in a powerful tool for the non-linear analysis of shape.[1] The main approach suggested in [2], [20], [30] and [31] is based on the exploitation of the relationship between mathematical morphology and random set theory, whereas the approach suggested in [37]–[40] explores the connection between Markov random fields and the morphological operations of erosion, dilation, opening, and closing. Finally, the approach suggested in [36] is based on the exploitation of the relationship between point processes, random sets, and morphological skeletons. This paper focuses exclusively on the random-set-theory approach.

Random set theory developed independently by Kendall [27] and Matheron [28] in the mid 1960's, is based on earlier results by Choquet [44]. Although Kendall's theory is very similar to Matheron's, the latter is more appropriate in practical applications since it is directly related to mathematical morphology. According to Matheron's theory, a shape is considered to be a realization from a class of random closed sets X, whose properties are uniquely determined by the hit probability $\Pr[X \cap B \neq \emptyset]$ or, equivalently, by the miss probability $\Pr[X \cap B = \emptyset]$, where B is a compact structuring element known as the *test set*. Matheron's random set theory is also related to *stereology*, a mathematical tool used for the extraction of three-dimensional geometric information from lower-dimensional measurements, such as plane or line measurements, or projections [2], [9], [28], [45]–[47].

[1] An alternative approach to shape analysis suggested in [42], and recently in [43], leads to the so-called *fuzzy morphology*. The fuzzy-morphology approach is quite different, in principle, from the stochastic approaches suggested in [2], [20], [30], [31], and [36]–[40]. There exists a general lack of results regarding the relative merits of these two approaches.

Random set theory provides a promising theoretical foundation for the modeling and analysis of random shapes because it allows the exploitation of both their structural and statistical characteristics. It has proven to be a useful tool in a variety of theoretical developments [2], [9], [28], [30], [31], [48]–[50], in general shape-analysis applications [2], [8], [19], [30]–[32], [48]–[51], and in some interesting applications in biology [2], [11], [30]–[32], [52], ecology [2], [53], materials science [2], [9], [30], [54]–[56], quantitative microscopy [2], [30], [32], [51], [53], geology [2], [9], [30], [51], [57], and physics [58]. Recently, the theory of random sets has been extended to the problem of synthesis and analysis of random gray-level images with encouraging results [59]–[63]. All these applications have identified random set theory as a promising tool for summarizing and analyzing a wide variety of image data and have demonstrated the potential of random set models in simulating complex geometric patterns.

Most often, automated shape synthesis, analysis, and recognition tasks operate on discrete data. Therefore it is quite important to consider random sets defined only on discrete spaces. Discrete shapes may be obtained as a result of a discretization procedure applied to a continuous shape, or they may be directly defined on a discrete space. Shape discretization is a complicated issue [2], [6], [30], [64]–[66], and I shall not discuss it here; rather, I shall consider shapes defined directly on Z^2, where Z is the set of all integers, and I shall concentrate on discussing various aspects of discrete random set theory. For more information on shape discretization refer to [2], [64], and [66].

As was previously discussed, the two popular approaches to modeling random shapes use either a random point field, specified in terms of a probability-mass function over a random counting measure, or a Markov random field, specified in terms of a Gibbsian probability-mass function. In [67]–[70] (see also [9] and [11]) a theoretical foundation is provided for the relationship between Matheron's theory of random sets, random point fields on R^n, and stochastic geometry [5]–[9]. In [71] the relationship between Matheron's theory of random sets and general random fields on R^n is also discussed. These relationships may be effectively used as connecting nodes between these seemingly different approaches for modeling and analyzing discrete random shapes. In fact, by using the results in [9], [46], [67], and [69]–[71] it can be shown that random point fields, discrete binary random fields, and discrete random sets on Z^2 are all equivalent. However, as I shall discuss shortly, the tools resulting from a random-set-theory approach to shape analysis are quite different from the ones used in the other two approaches.

This paper is organized as follows. In section 2 I review some fundamental concepts of Matheron's random set theory, define discrete random sets, exploit their relationship with discrete binary random fields, and establish the necessary notation. I introduce the cumulative-distribution functional of a discrete random set as a natural analogue of the cumulative-distribution function of a random variable. The connections among the probability-mass function of a discrete binary random field, the cumulative-distribution functional, and the capacity functional of a discrete random set are established. The relationship between a random point field and a discrete random set is briefly discussed, and stationarity of discrete random sets is reviewed. Finally, I provide a few preliminary suggestions on how to compute the cumulative-distribution functional and the capacity functional of a certain class of discrete random sets. In section 3 I consider four basic morphological transformations of discrete random sets and derive the mathematical relationship between their cumulative-distribution and capacity functionals. I illustrate the importance of such relationships by means of a few examples, including the derivation of a fundamental property regarding the statistical behavior of openings and closings. Finally, I show that the probability-mass function of a discrete binary random field may be expressed in terms of the cumulative-distribution functional or the capacity functional of a morphologically transformed discrete random set by means of a hit-or-miss transformation. Moments for discrete random sets are introduced in section 4. The discussion demonstrates the fact that the proposed moments generalize the concepts of autocorrelation and contact distribution. I define the autocorrelation functional of a discrete random set and study its properties. Finally, I show that a certain class of size distributions, introduced in terms of an axiomatic approach by Matheron [28], is a special case of the proposed moments, thus statistically supporting the importance of such distributions as shape summaries. In the last section I study convex-connected discrete random sets. The goal here is to verify the validity of Matheron's results regarding convexity in the discrete case. My analysis is similar to Matheron's; however, Matheron has worked out only the continuous case. No indication on the validity of his results in the discrete case appears in [28]. My analysis is quite involved, requiring tools from digital topology and geometry.

2 Discrete Random Sets

Consider the two-dimensional discrete space Z^2. At every point $(i,j) \in Z^2$ we assign a binary random variable $\mathbf{x}_{ij}$ that takes values $x_{ij} \in \{0,1\}$.[2] The collection $\mathbf{x} = \{\mathbf{x}_{ij}, (i,j) \in Z^2\}$ of all random variables on Z^2, together with a probability-mass function $\Pr[\mathbf{x} = x]$, for every $x \in \{0,1\}^{Z^2}$, whose corresponding finite-dimensional cumulative-distribution functions satisfy Kolmogorov's conditions of symmetry and consistency [71], [72], defines a discrete *binary random field* (BRF) on Z^2 [11], [71]. A *discrete random set* (DRS) $\mathbf{X}$ on Z^2 is now defined by (see also [11], [31], [41], [46], [51] and [71])

$$\mathbf{X} \triangleq \{(i,j) \in Z^2 : \mathbf{x}_{ij} = 1\}, \qquad (1)$$

where $\triangleq$ denotes *equal by definition*. Observe that

$$\mathbf{x}_{ij} = \mathbf{I}_{\mathbf{X}}(i,j) \quad \text{for every } (i,j) \in Z^2, \qquad (2)$$

where $I_A(i,j)$ is the *indicator function* of set A, i.e.,

$$I_A(i,j) \triangleq \begin{cases} 1 & \text{if } (i,j) \in A \\ 0 & \text{otherwise.} \end{cases}$$

[2] Throughout the paper, random quantities are denoted by boldface

A DRS $\mathbf{X}$ on Z^2 is statistically equivalent to a BRF $\mathbf{x}$ on Z^2 (i.e., $\Pr[\mathbf{x} = x] = \Pr[\mathbf{X} = X]$ for every $x \in \{0,1\}^{Z^2}$), their equivalence being determined by (1) and (2); therefore, why redefine a BRF in terms of a DRS? As discussed in section 1, traditional random field theory has proven itself useful in many image-processing and image-analysis problems (e.g., image classification, restoration, smoothing). However, this theory is limited by the lack of efficient tools for summarizing, processing, and analyzing geometric shape information. As we shall see in the following, (1), (2), and the development in [2] and [28] provide us with promising tools for the morphological processing and analysis of discrete random shapes.

It is a basic fact of elementary probability theory [72] that a discrete-valued random variable $\mathbf{z}$ that takes values in a bounded complete sublattice $(Z_b, \leq)$ of the complete lattice $(Z, \leq)$, with least element $z_{\min}$ and greatest element $z_{\max}$ [73], is uniquely characterized in terms of its cumulative-distribution function[3]

$$\begin{aligned} F_{\mathbf{z}}(z) &\triangleq \Pr[z_{\min} \leq \mathbf{z} \leq z - 1] \\ &= 1 - \Pr[z \leq \mathbf{z} \leq z_{\max}] \\ &\qquad \text{for every } z \in Z_b. \end{aligned} \qquad (3)$$

Recall that $0 \leq F_{\mathbf{z}}(z) \leq 1$ for every $z \in Z_b$, $F_{\mathbf{z}}(z_{\min}) = 0$, and $F_{\mathbf{z}}(z_1) \leq F_{\mathbf{z}}(z_2)$ for every $z_1 \leq z_2$, $z_1, z_2 \in Z_b$. Furthermore,

$$F_{\mathbf{z}}(z) = 1 - \sum_{z \leq z' \leq z_{\max}} \Pr[\mathbf{z} = z'] \qquad (4a)$$

and

$$\Pr[\mathbf{z} = z] = \begin{cases} F_{\mathbf{z}}(z+1) - F_{\mathbf{z}}(z) \\ \quad \text{for } z_{\min} \leq z \leq z_{\max} - 1 \\ 1 - F_{\mathbf{z}}(z_{\max}) \\ \quad \text{for } z = z_{\max} \end{cases}. \qquad (4b)$$

Let P and S denote the collection of all subsets, and all bounded subsets of Z^2, respectively. The fundamental algebraic structure of sets defined on Z^2 is determined by the Boolean lattice $(P, \subseteq)$, with least element $\emptyset$ and greatest element Z^2 itself, furnished with the partial order relation $\subseteq$ [73]. It is now natural to

[3] The definition of the cumulative-distribution function used here is a shifted version of the usual one [72], the shift being imposed for convenience.

seek extension of the concept of the cumulative-distribution function to the case of DRS's. Let us define the functional $F_{\mathbf{X}}(B)$ by [compare with (3)]

$$F_{\mathbf{X}}(B) \triangleq 1 - \Pr[\mathbf{X} \supseteq B] \quad \text{for every } B \in S. \quad (5)$$

It is an immediate consequence of (5) and the fact that $\emptyset \subseteq A$ for any set A that

$$0 \leq F_{\mathbf{X}}(B) \leq 1 \quad \text{for every } B \in S, \quad (6a)$$

$$F_{\mathbf{X}}(\emptyset) = 0 \quad (6b)$$

and

$$F_{\mathbf{X}}(B_1) \leq F_{\mathbf{X}}(B_2) \quad \text{for every } B_1 \subseteq B_2, B_1, B_2 \in S; \quad (6c)$$

therefore, $F_{\mathbf{X}}(B)$, $B \in S$, is the *cumulative-distribution functional* of the DRS $\mathbf{X}$. From (1) and (5) observe that

$$\begin{aligned} F_{\mathbf{X}}(B) &= 1 - \sum_{x: X \supseteq B} \Pr[\mathbf{x} = x] \\ &= 1 - \Pr[\Lambda_{(i,j) \in B}\{\mathbf{x}_{ij} = 1\}] \\ &\qquad \text{for every } B \in S, \end{aligned} \quad (7)$$

where Λ denotes logical AND, which is the analogue of (4a). The derivation of an expression similar to (4b) is more complicated, primarily because of the lack of a total ordering in $(P, \subseteq)$. Let S_W denote the collection of all bounded subsets of Z^2 that are included in a finite-size window W, $|W| < +\infty$, where $|A|$ denotes the cardinality of set A (i.e., the total number of its elements). A DRS $\mathbf{X}$ will be called an *almost surely* (a.s.) *W-bounded DRS* if $\Pr[\mathbf{X} \in S_W] = 1$. We now have the following proposition (a similar result first appeared in [50]).

PROPOSITION 1. For every BRF $\mathbf{x}$ such that $\mathbf{X}$, given by (1), is an a.s. W-bounded DRS, we have that

$$\begin{aligned} \Pr[\mathbf{x} = x] &= \Pr[\mathbf{X} = X] \\ &= \sum_{X' \subseteq X^c \cap W} (-1)^{|X'|}[1 - F_{\mathbf{X}}(X \cup X')], \end{aligned} \quad (8)$$

for every $X \in S_W$, where X^c is the set complement of X in P.

Proof. From (5) and (7) it is clear that

$$\begin{aligned} \Pr[\mathbf{x} = x] &= \Pr[\mathbf{X} = X] \\ &= \Pr[\mathbf{X} \supseteq X] - \Pr[\mathbf{X} \supset X] \\ &= 1 - F_{\mathbf{X}}(X) - \Pr[\mathbf{X} \supset X]. \end{aligned} \quad (9)$$

If $\mathbf{X}$ is an a.s. W-bounded DRS, then, as a result of the inclusion–exclusion formula [72] and the fact that $\Pr[\mathbf{X} = X] = 0$ for every X not included in W (in which case, and together with (7), we have that $F_{\mathbf{X}}(B) = 1$ for every B not included in W), we have that (see also (5))

$$\begin{aligned} \Pr[\mathbf{X} \supset X] &= \Pr[V_{x_i \in X^c}\{\mathbf{X} \supseteq X \cup \{x_i\}\}] \\ &= -\sum_{X' \neq \emptyset} (-1)^{|X'|}\Pr[\mathbf{X} \supseteq X \cup X'] \\ &= -\sum_{X' \neq \emptyset} (-1)^{|X'|}[1 - F_{\mathbf{X}}(X \cup X')] \\ &= -\sum_{\substack{X' \subseteq X^c \cap W, \\ X' \neq \emptyset}} (-1)^{|X'|}[1 - F_{\mathbf{X}}(X \cup X')] \end{aligned}$$

for every $X \in S_W$, where V denotes logical OR, which, together with (9) and the fact that $|\emptyset| = 0$, shows (8).

Equations (7) and (8) establish a one-to-one correspondence between the probability-mass function of a BRF $\mathbf{x}$ on Z^2 and the cumulative-distribution functional of an a.s. W-bounded DRS $\mathbf{X}$. The sum in (8) will always converge in this case. The a.s. W-boundness assumption is not limiting since in most practical applications $\mathbf{X}$ is observed through a finite-size window W, i.e., we are mostly interested in $\mathbf{X} \cap W$, with $|W| < +\infty$.

It is now worthwhile to notice that (see (7))

$$F_{\mathbf{X}}(B) = 1 - E\left[\prod_{(i,j) \in B} \mathbf{I}_{\mathbf{X}}(i,j)\right] \quad \text{for every } B \in S,$$

for any DRS $\mathbf{X}$, where $E[\bullet]$ denotes expectation; therefore $1 - F_{\mathbf{X}}(B)$ is a typical $|B|$-order moment of the BRF $\mathbf{x}$ and (8) is simply a moment-inversion formula.

The definition of the cumulative-distribution functional $F_{\mathbf{X}}(\bullet)$ of a DRS $\mathbf{X}$ is a direct result of the Boolean lattice structure of P, which forms the foundation of set morphology [74]. Further connections to mathematical morphology can be established by studying the relationship between $\Pr[\mathbf{X} \cap B \neq \emptyset]$ and $F_{\mathbf{X}}(B)$. The *discrete-capacity functional* $T_{\mathbf{X}}(B)$ of a DRS $\mathbf{X}$ is defined by [28]

$$T_{\mathbf{X}}(B) \stackrel{\Delta}{=} \Pr[\mathbf{X} \cap B \neq \emptyset] \quad \text{for every} B \in S. \tag{10}$$

From (10) it is easy to see [28] that (compare with (6))

$$0 \leq T_{\mathbf{X}}(B) \leq 1 \quad \text{for every } B \in S, \tag{11a}$$

$$T_{\mathbf{X}}(\emptyset) = 0 \tag{11b}$$

and

$$T_{\mathbf{X}}(B_1) \leq T_{\mathbf{X}}(B_2) \text{ for every } B_1, B_2 \in S \quad \text{such that } B_1 \subseteq B_2. \tag{11c}$$

We may decide to assign the role of the cumulative-distribution functional to $T_{\mathbf{X}}(B)$ instead of $F_{\mathbf{X}}(B)$ since equations (11) are satisfied (it can also be shown that equations similar to (7) and (8) are satisfied by $T_{\mathbf{X}}(B)$ [50]); however, our previous discussion shows that this role should be assigned to $F_{\mathbf{X}}(B)$. In fact, it is easy to show (see section 3) that $T_{\mathbf{X}}(B) = F_{\mathbf{X}^c}(B)$ for every $B \in S$; therefore $T_{\mathbf{X}}(B)$ may be viewed as the cumulative-distribution functional of the DRS $\mathbf{X}^c$. Finally, observe that (compare with (7))

$$T_{\mathbf{X}}(B) = 1 - \Pr[\Lambda_{(i,j)\in B}\{\mathbf{x}_{ij} = 0\}] \quad \text{for every } B \in S \tag{12}$$

or

$$T_{\mathbf{X}}(B) = 1 - E\left[\prod_{(i,j)\in B} [1 - \mathbf{I}_{\mathbf{X}}(i,j)]\right]. \quad \text{for every } B \in S.$$

Therefore $1 - T_{\mathbf{X}}(B)$ is a generalized $|B|$-order moment of the BRF $\mathbf{x}$.

Let us now define the probabilities $Q_{\mathbf{X},n}(B; B_1, B_2, \ldots, B_n)$ by (see also [28] and [44])

$$Q_{\mathbf{X},0}(B) \stackrel{\Delta}{=} \Pr[\mathbf{X} \cap B = \emptyset] \tag{13a}$$

and

$$Q_{\mathbf{X},n}(B; B_1, B_2, \ldots, B_n) \stackrel{\Delta}{=} \Pr[\mathbf{X} \cap B = \emptyset, \Lambda_{i=1}^{n}\{\mathbf{X} \cap B_i \neq \emptyset\}], \tag{13b}$$

for $B, B_1, B_2, \ldots, B_n \in S$, $n = 1, 2, \ldots$. These probabilities will be useful in section 4, where moments are introduced for a DRS. We now have the following proposition [28], [44]:

PROPOSITION 2. In general,

$$Q_{\mathbf{X}}(B) \stackrel{\Delta}{=} Q_{\mathbf{X},0}(B) = 1 - T_{\mathbf{X}}(B) \geq 0 \quad \text{for every } B \in S \tag{14a}$$

and

$$Q_{\mathbf{X},n}(B; B_1, B_2, \ldots, B_n) = Q_{\mathbf{X},n-1}(B; B_1, B_2, \ldots, B_{n-1}) - Q_{\mathbf{X},n-1}(B \cup B_n; B_1, B_2, \ldots, B_{n-1}) \geq 0, \tag{14b}$$

for every $B, B_1, B_2, \ldots B_n \in S$, $n = 1, 2 \ldots$.

Proof. Observe that (see (10) and (13a))

$$Q_{\mathbf{X},0}(B) = \Pr[\mathbf{X} \cap B = \emptyset] = 1 - \Pr[\mathbf{X} \cap B \neq \emptyset] = 1 - T_{\mathbf{X}}(B),$$

which shows (14a). We also have that

$$\Pr[\mathbf{X} \cap B = \emptyset, \Lambda_{i=1}^{n-1}\{\mathbf{X} \cap B_i \neq \emptyset\}, \mathbf{X} \cap B_n = \emptyset] + \Pr[\mathbf{X} \cap B = \emptyset, \Lambda_{i=1}^{n}\{\mathbf{X} \cap B_i \neq \emptyset\}] = \Pr[\mathbf{X} \cap B = \emptyset, \Lambda_{i=1}^{n-1}\{\mathbf{X} \cap B_i \neq \emptyset\}]. \tag{15}$$

Equation (14b) is a direct consequence of (13b), (15), and the fact that

$$Q_{\mathbf{X},n-1}(B \cup B_n; B_1, B_2, \ldots, B_{n-1}) = \Pr[\mathbf{X} \cap (B \cup B_n) = \emptyset, \Lambda_{i=1}^{n-1}\{\mathbf{X} \cap B_i \neq \emptyset\}] = \Pr[\mathbf{X} \cap B = \emptyset, \Lambda_{i=1}^{n-1}\{\mathbf{X} \cap B_i \neq \emptyset\}, \mathbf{X} \cap B_n = \emptyset]$$

Non-negativity is a direct consequence of the fact that $Q_{\mathbf{X},n}$ is a probability.

As a direct consequence of the previous proposition, we have the following two corollaries [44].

COROLLARY 1. In general

$$Q_{\mathbf{X},n}(B; B_1, B_2, \ldots, B_n)$$

$$= 1 - T_{\mathbf{X}}(B) - \sum_{k=1}^{n}[1 - T_{\mathbf{X}}(B \cup B_k)] + \sum_{k=1}^{n-1}\sum_{l=k+1}^{n}[1 - T_{\mathbf{X}}(B \cup B_k \cup B_l)] - \cdots + \cdots + (-1)^n[1 - T_{\mathbf{X}}(B \cup B_1 \cup B_2 \cup \ldots \cup B_n)], \quad (16)$$

for every $B, B_1, B_2, \ldots, B_n \in S$ and every $n \geq 1$.

Proof. The proof is a direct consequence of repeated iterations on (14).

COROLLARY 2. In general

$$T_{\mathbf{X}}(B_1 \cup B_2) \leq T_{\mathbf{X}}(B_1) + T_{\mathbf{X}}(B_2) - T_{\mathbf{X}}(B_1 \cap B_2) \leq T_{\mathbf{X}}(B_1) + T_{\mathbf{X}}(B_2), \quad (17)$$

for every $B_1, B_2 \in S$.

Proof. From (16), with $n = 2$, we have that

$$Q_{\mathbf{X},2}(B; B_1, B_2) = T_{\mathbf{X}}(B \cup B_1) - T_{\mathbf{X}}(B) + T_{\mathbf{X}}(B \cup B_2) - T_{\mathbf{X}}(B \cup B_1 \cup B_2) \geq 0, \quad (18)$$

for every $B, B_1, B_2 \in S$. Equation (17) is now a direct consequence of (18), with $B = B_1 \cap B_2$.

A fundamental quantity in the theory of DRS's is $Q_{\mathbf{X}}(B)$, defined by (14a). From (13) and (14) we see that this quantity generates the probabilities $\Pr[\mathbf{X} \cap B = \emptyset, \mathbf{X} \cap B_1 \neq \emptyset, \mathbf{X} \cap B_2 \neq \emptyset, \ldots, \mathbf{X} \cap B_n \neq \emptyset]$, which, as we shall see in section 4, generate moments associated with the DRS $\mathbf{X}$; therefore we shall call this functional the *discrete generating functional* of $\mathbf{X}$. The next proposition establishes the desirable connection between $F_{\mathbf{X}}(B)$ and $T_{\mathbf{X}}(B)$.

PROPOSITION 3. In general, and for every $B \in S$, we have that

$$T_{\mathbf{X}}(B) = \sum_{\substack{B' \subseteq B,\\ B' \neq \emptyset}} (-1)^{|B'|+1}[1 - F_{\mathbf{X}}(B')] \quad (19a)$$

and

$$F_{\mathbf{X}}(B) = \sum_{\substack{B' \subseteq B,\\ B' \neq \emptyset}} (-1)^{|B'|+1}[1 - T_{\mathbf{X}}(B')]. \quad (19b)$$

Proof. Observe that (see also (10))

$$\begin{aligned} T_{\mathbf{X}}(B) &= T_{\mathbf{X}}(\{b_1, b_2, \ldots, b_n\}) \\ &= \Pr[\mathbf{X} \cap \{b_1, b_2, \ldots, b_n\} \neq \emptyset] \\ &= \Pr[V_{i=1}^{n}\{\mathbf{X} \cap \{b_i\} \neq \emptyset\}] \\ &= \Pr[V_{i=1}^{n}\{\mathbf{X} \supseteq \{b_i\}\}] \\ &= -\sum_{\substack{B' \subseteq B,\\ B' \neq \emptyset}} (-1)^{|B'|}\Pr[\mathbf{X} \supseteq B'], \end{aligned}$$

which, together with (5), shows (19a). Equation (19b) is a direct consequence of (19a), with $\mathbf{X} \to \mathbf{X}^c$, and the fact that $T_{\mathbf{X}}(B) = F_{\mathbf{X}^c}(B)$.

From the previous proposition, it is clear that given the discrete-capacity functional $T_{\mathbf{X}}(\bullet)$, the cumulative-distribution functional $F_{\mathbf{X}}(\bullet)$ can be, at least in principle, computed recursively in terms of (19b), whereas given $F_{\mathbf{X}}(\bullet)$, we can compute $T_{\mathbf{X}}(\bullet)$ by means of (19a). It is now a direct consequence of (7), (8), and (19) that given the probability-mass function $\Pr[\mathbf{x} = x]$ of a BRF $\mathbf{x}$ on Z^2 with an a.s. W-bounded $\mathbf{X}$, the discrete-capacity functional $T_{\mathbf{X}}(B)$ of $\mathbf{X}$ can be uniquely computed, and vice versa.[4] Therefore any BRF $\mathbf{x}$ on Z^2 with an a.s. W-bounded $\mathbf{X}$ is uniquely characterized by the discrete-capacity functional of $\mathbf{X}$.[5] The suggested one-to-one relationship between $T_{\mathbf{X}}(B)$ and $\Pr[\mathbf{x} = x]$ is applicable only in the discrete case and under the a.s. W-boundness assumption. We believe that the absence of such a relationship in the continuous case has imposed certain limitations on Matheron's continuous random set theory (see also [11] and [46]). When $\mathbf{X}$ is not a.s. W-bounded, there might exist no one-to-one relationship between $\Pr[\mathbf{x} = x]$ and $T_{\mathbf{X}}(B)$. However, as a special case of the celebrated Choquet–Kendall–Matheron theorem [27], [28], [44] (see also [75]) a DRS $\mathbf{X}$ is uniquely characterized by its discrete-capacity functional $T_{\mathbf{X}}(B)$, defined over all bounded subsets $B \in S$ of Z^2.

[4] This was first observed in [50] by using an analysis similar to the one presented here and is a direct consequence of the moment-inversion formula (8).

[5] This is a strong result that was first proved in [67] under the framework of a more general approach. Our proof is limited to the discrete case and is based on the relationship between the discrete-capacity functional of $\mathbf{X}$ and the moments of $\mathbf{x}$.

Any DRS **X** is a collection of random points on Z^2; therefore **X** can also be considered as a simple random point field on Z^2 [9]–[11]. Indeed, it can be shown (see [9], [11], [67], and [70]) that the statistical properties of a simple random point field **X** on R^n are uniquely determined by the void probability $\Pr[\mathbf{X}(B) = 0]$ for every $B \in S$, where $\mathbf{X}(B)$ is a random counting measure counting the number of points of **X** in $B \in S$. Since (see also (13a) and (14a))

$$\begin{aligned}\Pr[\mathbf{X}(B) = 0] &= \Pr[\mathbf{X} \cap B = \emptyset] \\ &= Q_{\mathbf{X}}(B) \\ &= 1 - T_{\mathbf{X}}(B) \quad \text{for every } B \in S,\end{aligned}$$

a DRS is statistically equivalent to a simple random point field on Z^2. Therefore, statistical techniques based on random point field theory can be used to analyze discrete random shapes. However, these techniques are primarily based on the statistical properties of random measure $\mathbf{X}(B)$ [9]–[11] and are not, in general, compatible with the morphological approach suggested in [2] and [28]. To ensure compatibility we should focus on the statistical behavior of the event $\mathbf{X}(B) = 0$, which naturally leads to the discrete-capacity functional $T_{\mathbf{X}}(B)$ and, therefore, to Matheron's theory of random sets. This is why we prefer here to call **X** a DRS rather that a random point field.

By analogy to the theory of random point fields we say that (see also [9], [11], and [69]) a DRS **X** is *n^{th}-order stationary*, $n \geq 1$, if its cumulative-distribution functional $F_{\mathbf{X}}(B)$ is invariant under any arbitrary translation of **X** for every $B \in S$, such that $|B| = n$, i.e., if

$$F_{\mathbf{X}\oplus\{z\}}(B) = F_{\mathbf{X}}(B) \quad \text{for every } B \in S \text{ such that } |B| = n \text{ and every } z \in Z^2,$$

where $\oplus$ denotes morphological dilation [33]. Since $X \oplus \{z\} \supseteq B \Leftrightarrow X \supseteq B \oplus \{-z\}$, **X** is n^{th}-order stationary if and only if (see also (5))

$$F_{\mathbf{X}}(B \oplus \{z\}) = F_{\mathbf{X}}(B) \quad \text{for every } B \in S \text{ such that } |B| = n \text{ and every } z \in Z^2. \tag{20}$$

It is clear from (19) and (20) that a necessary and sufficient condition for a DRS **X** to be n^{th}-order stationary is [28]

$$T_{\mathbf{X}}(B \oplus \{z\}) = T_{\mathbf{X}}(B) \quad \text{for every } B \in S \text{ such that } |B| = n \text{ and every } z \in Z^2.$$

The DRS **X** will be called *strictly stationary*, if it is n^{th}-order stationary for every $n \geq 1$. The previous exposition is intuitive; therefore we shall not discuss stationarity any further. For more details refer to [9], [11], [28], and [69].

It could be extremely advantageous if given the probability-mass function $\Pr[\mathbf{x} = x]$ of a BRF **x**, we could compute $F_{\mathbf{X}}(B)$ and $T_{\mathbf{X}}(B)$ [67], [70] for some $B \in S$. In this case, instead of using conventional statistical tools for the analysis of **x** (e.g., probability-mass functions or power spectra), we could use the morphological-random-set theoretical tools proposed in [2] and [28], which are better suited for shape analysis, and effectively study their statistical properties (for example, see [50]). The solution to this problem seems quite difficult in general, since computation of $F_{\mathbf{X}}(B)$ and $T_{\mathbf{X}}(B)$ requires computation of

$$\Pr[\Lambda_{(i,j)\in B}\{\mathbf{x}_{ij} = 1\}] \quad \text{and} \quad \Pr[\Lambda_{(i,j)\in B}\{\mathbf{x}_{ij} = 0\}],$$

respectively, of the marginal probability-mass functions of **x** (see (7) and (12)). However, the utility of $F_{\mathbf{X}}(B)$ and $T_{\mathbf{X}}(B)$ should not be underestimated. In some nontrivial (i.e., not independent and identically distributed) BRF models this computation may be possible. One such model is the discrete binary, mutually compatible, Gibbs random field [76], which under certain conditions is a good approximation of a general Markov random field model [77]. It can be shown in a rather straightforward manner that in this case, and under the assumption of stationarity [76], [78]–[80], the computation of $F_{\mathbf{X}}(B)$ and $T_{\mathbf{X}}(B)$ reduces the binary summations over some points in $B^c \cap W_{\text{min}}$, where W_{min} is the smallest possible discrete parallelogram that contains B. The computation of these summations seems feasible, especially for relatively small-size B's. For more general stationary Markov random fields, the Monte Carlo simulation techniques recently proposed in [81] and [82] for calculating partition functions may be modified in order to develop unbiased and consistent estimators for the marginal probability-mass functions and, therefore, for $F_{\mathbf{X}}(B)$ and $T_{\mathbf{X}}(B)$. I shall not discuss this subject any further since it is currently being investigated. However, I should point out that if no particular probability model is assumed for **X**, then under conditions of strict stationarity and ergodicity for the DRS **X** it can be shown [9] (see also [2], [19], [20], [41], [46], and [51]), that the spatial average

$$\hat{F}_{\mathbf{X},n}(B) \triangleq 1 - \frac{|(X \cap W_n) \ominus B|}{|W_n \ominus B|} \quad \text{for } n = 1, 2, \ldots$$

is an unbiased and consistent estimator for $F_{\mathbf{X}}(B)$ for every $B \in S$, where $\ominus$ denotes morphological erosion [33]. In the previous definition $\{W_n, n = 1, 2, \ldots\}$ is a sequence of nonempty, convex (a set $X \in P$ is convex if it is identical to its convex hull [2]—see also section 5), bounded, discrete windows with $W_n \subset W_m$ for $n < m$ and $\lim_{n \to +\infty} |W_n| = +\infty$. Since $T_{\mathbf{X}}(B) = F_{\mathbf{X}^c}(B)$, the spatial average

$$\hat{T}_{\mathbf{X},n}(B) \triangleq 1 - \frac{|(X^c \cap W_n) \ominus B|}{|W_n \ominus B|} \quad \text{for } n = 1, 2, \ldots$$

is also an unbiased and consistent estimator for $T_{\mathbf{X}}(B)$ for every $B \in S$.

3 Morphological Transformations of DRS's

Morphological shape analysis is based on transforming a given shape X by means of a sequence $\{\Psi_i[\bullet], i = 1, 2, \ldots\}$ of morphological transformations. These transformations can be easily designed to fit desirable geometric specifications [2], [32], [34], [35]. However, when X is modeled by means of a DRS **X**, it is absolutely necessary to study both the geometric and the statistical effects of $\Psi[\bullet]$ on **X**. The results of such a study can then be used for designing morphological transformations with the desirable geometric and statistical properties. There is extensive literature on the statistical properties of the output of a linear, shift-invariant system excited by a random input (for example, see [83]). Reference [84] contains relationships between the input and output statistics for the case in which the input shape is modelled by a one-dimensional random process (and, to a limited extent, by a two-dimensional random field) and filtered by a one-dimensional morphological close-opening filter. However, there has been little study of the statistical effect of a morphological transformation $\Psi[\bullet]$ on a DRS **X**, primarily because of the limitations of the mostly continuous random-set-theory approach proposed in [28] (see also [75]). The purpose of this section is to propose tools that may lead to such a study.

If **X** is a DRS, then $\mathbf{Y} = \Psi[\mathbf{X}]$ will also be a DRS since, in the discrete case, $\Psi[\bullet]$ is always measurable [2], [28], [72], [75]. To understand the effect of $\Psi[\bullet]$ on **X** it is desirable that we determine the output cumulative-distribution functional $F_{\mathbf{Y}}(B)$ or the output discrete-capacity functional $T_{\mathbf{Y}}(B)$. The discrete-capacity functional of the DRS **Y** is given by

$$T_{\mathbf{Y}}(B) = \Pr[\Psi[\mathbf{X}] \cap B \neq \emptyset] \quad \text{for every } B \in S. \tag{21}$$

Equation (21) provides no direct relationship between the discrete-capacity functionals $T_{\mathbf{Y}}(B)$ and $T_{\mathbf{X}}(B)$. Although such a relationship is desirable, it is impossible to obtain in general. However, the following two propositions reveal some instances in which obtaining such a relationship is possible. In the following, E^t denotes set transposition (i.e., $E^t = \{e : -e \in E\}$).

PROPOSITION 4. For any DRS **X** and for every $B \in S$ we have that

$$T_{\mathbf{X} \cap W}(B) = T_{\mathbf{X}}(B \cap W) \quad \text{for every } W \in P \tag{22a}$$

and

$$T_{\mathbf{X}^c}(B) = F_{\mathbf{X}}(B), \tag{22b}$$

whereas

$$F_{\mathbf{X} \cup W}(B) = F_{\mathbf{X}}(B \cap W^c) \quad \text{for every } W \in P. \tag{23}$$

Proof. Observe first that $B \cap W \in S$ and that

$$\begin{aligned} T_{\mathbf{X} \cap W}(B) &= \Pr[(\mathbf{X} \cap W) \cap B \neq \emptyset] \\ &= \Pr[\mathbf{X} \cap (W \cap B) \neq \emptyset] \\ &= T_{\mathbf{X}}(W \cap B), \end{aligned}$$

which shows (22a). Observe now that (see also (5))

$$\begin{aligned} T_{\mathbf{X}^c}(B) &= \Pr[\mathbf{X}^c \cap B \neq \emptyset] \\ &= 1 - \Pr[\mathbf{X}^c \cap B = \emptyset] \\ &= 1 - \Pr[\mathbf{X} \supseteq B] \\ &= F_{\mathbf{X}}(B), \end{aligned}$$

which shows (22b). Finally, (23) is a direct consequence of (22a), with $\mathbf{X} \to \mathbf{X}^c$ and $W \to W^c$, and (22b).

PROPOSITION 5. For a DRS $\mathbf{X}$ and for every structuring element $E \in S$ we have that

$$T_{\mathbf{X}\oplus E}(B) = T_{\mathbf{X}}(B \oplus E^t) \tag{24a}$$

and

$$T_{\mathbf{X}\ominus E}(B) = \sum_{\substack{B' \subseteq B, \\ B' \neq \emptyset}} (-1)^{|B'|+1}[1 - F_{\mathbf{X}}(B' \oplus E)], \tag{24b}$$

for every $B \in S$, whereas

$$F_{\mathbf{X}\oplus E}(B) = \sum_{\substack{B' \subseteq B, \\ B' \neq \emptyset}} (-1)^{|B'|+1}[1 - T_{\mathbf{X}}(B' \oplus E^t)] \tag{25a}$$

and

$$F_{\mathbf{X}\ominus E}(B) = F_{\mathbf{X}}(B \oplus E), \tag{25b}$$

for every $B \in S$.

Proof. Observe first that $B \oplus E^t \in S$ and that

$$\begin{aligned} (X \oplus E) \cap B &= [\cup_{e\in E}(X \oplus \{e\})] \cap B \\ &= \cup_{e\in E}[(X \oplus \{e\}) \cap B] \\ &= \cup_{e\in E}[X \cap (B \oplus \{-e\})] \oplus \{e\}, \end{aligned}$$

whereas

$$\cup_{e\in E}[X \cap (B \oplus \{-e\})] \oplus \{e\} = \emptyset$$
$$\text{if and only if } \cup_{e\in E}[X \cap (B \oplus \{-e\})] = \emptyset.$$

But

$$\begin{aligned} \cup_{e\in E}[X \cap (B \oplus \{-e\})] &= X \cap [\cup_{e\in E} B \oplus \{-e\}] \\ &= X \cap (B \oplus E^t), \end{aligned}$$

from which we obtain that

$$(X \oplus E) \cap B = \emptyset$$
$$\text{if and only if } X \cap (B \oplus E^t) = \emptyset.$$

The last statement, together with (13a) and (14a), shows (24a).

From (19b), (22b), and the duality between dilation and erosion (i.e., $(X \oplus E)^c = X^c \ominus E^t$ [33]) we have that

$$\begin{aligned} T_{\mathbf{X}\ominus E}(B) &= T_{(\mathbf{X}^c\oplus E^t)^c}(B) \\ &= F_{\mathbf{X}^c\oplus E^t}(B) \\ &= \sum_{\substack{B' \subseteq B, \\ B' \neq \emptyset}} (-1)^{|B'|+1}[1 - T_{\mathbf{X}^c\oplus E^t}(B')], \end{aligned}$$

which, together with (22b) and (24a), shows (24b). Finally, (25a) and (25b) are a direct consequence of (22b), (24a), and (24b), with $\mathbf{X} \to \mathbf{X}^c$ and $E \to E^t$, and the duality between erosion and dilation.

Relationships (22a) and (24a) are also satisfied in the continuous case [75]. However, (22b), (23), (24b), and (25) are satisfied only in the discrete case. Relationship (22a) is very useful in practice since shapes are usually observed through an observation window W. It simply demonstrates that it suffices to know the discrete-capacity functional of $\mathbf{X}$ for every $B \subseteq W$ to uniquely characterize the DRS $\mathbf{X} \cap W$. Relationship (22b), which provides the connection between the statistical characteristics of a DRS $\mathbf{X}$ and its complement $\mathbf{X}^c$, allows the derivation of (23), (24b), and (25). Finally, (22b), together with (24) and (25), allows the derivation of similar relationships for more complicated morphological transformations, such as openings and closings. Indeed, we have the following corollary to Proposition 5, where $\circ$ and $\bullet$ denote morphological opening and closing, respectively.

COROLLARY 3. For a DRS $\mathbf{X}$, and for every structuring element $E \in S$, we have that

$$T_{\mathbf{X}\circ E}(B) = \sum_{\substack{B' \subseteq B\oplus E^t, \\ B' \neq \emptyset}} (-1)^{|B'|+1}[1 - F_{\mathbf{X}}(B' \oplus E)] \tag{26a}$$

and

$$T_{\mathbf{X}\bullet E}(B) = \sum_{\substack{B' \subseteq B, \\ B' \neq \emptyset}} (-1)^{|B'|+1}[1 - F_{\mathbf{X}\oplus E}(B' \oplus E)], \tag{26b}$$

for every $B \in S$, whereas

$$F_{\mathbf{X} \circ E}(B) = \sum_{\substack{B' \subseteq B, \\ B' \neq \emptyset}} (-1)^{|B'|+1}[1 - T_{\mathbf{X} \ominus E}(B' \oplus E^t)] \tag{27a}$$

and

$$F_{\mathbf{X} \bullet E}(B) = \sum_{\substack{B' \subseteq B \oplus E, \\ B' \neq \emptyset}} (-1)^{|B'|+1}[1 - T_{\mathbf{X}}(B' \oplus E^t)], \tag{27b}$$

for every $B \in S$.

Proof. Recall that $X \circ E = (X \ominus E) \oplus E$ and $X \bullet E = (X \oplus E) \ominus E$ [33]. The proof of (26) is now a direct consequence of (24), whereas the proof of (27) is a direct consequence of (25).

Relationships (19) and (24)–(27) are fundamental in the theoretical and experimental study of DRS's and in the statistical analysis of their morphological transformations. I shall illustrate this with a few examples. First, let me consider a nonempty, symmetric structuring element E that contains the origin. We define the structuring element nE by

$$nE = E \oplus E \oplus \cdots \oplus E \quad (n-1 \text{ dilations}) \text{ for } n = 0, 1, \ldots, \tag{28}$$

where $0E = \emptyset$. Let me also define two sequences $f_{\mathbf{X}}(E,n)$ and $o_{\mathbf{X}}(E,n)$ by

$$f_{\mathbf{X}}(E,n) \triangleq F_{\mathbf{X}}(nE) \quad \text{for } n = 0, 1, \ldots \tag{29a}$$

and

$$o_{\mathbf{X}}(E,n) \triangleq T_{\mathbf{X} \circ nE}(\{o\}) \quad \text{for } n = 0, 1, \ldots. \tag{29b}$$

Sequence $\{f_{\mathbf{X}}(E,n), n = 0, 1, \ldots\}$ is a discrete version of the contact-distribution function (with respect to the structuring element E) of the DRS $\mathbf{X}^c$ (see [2], [9], and (22b)), whereas sequence $\{o_{\mathbf{X}}(E,n), n = 0, 1, \ldots\}$ is a discrete version of an opening-based size-distribution function (with respect to the structuring element E) of the DRS $\mathbf{X}$ [2], [28]. From (24b), with $E \to nE$ and $B = \{o\}$, and from (29a) observe that

$$f_{\mathbf{X}}(E,n) = 1 - T_{\mathbf{X} \ominus nE}(\{o\}), \tag{30}$$

whereas from (19a), (25b), (26a), with $E \to nE$ and $B = \{o\}$, and (29b) (or from (24a), with $\mathbf{X} \to \mathbf{X} \ominus nE$, $E \to nE$, and $B = \{o\}$, and (29b)) we have that

$$o_{\mathbf{X}}(E,n) = T_{\mathbf{X} \ominus nE}(nE), \tag{31}$$

for every $n = 0, 1, \ldots$. It is now clear from (30) and (31) that $o_{\mathbf{X}}(E,n)$ is a better statistical summary for $\mathbf{X}$ than is $f_{\mathbf{X}}(E,n)$, since $o_{\mathbf{X}}(E,n)$ is an $|nE|$-order moment of the BRF $\mathbf{I}_{\mathbf{X} \ominus nE}$, whereas $f_{\mathbf{X}}(E,n)$ is only a first-order moment. This has been experimentally observed in [19] and [20]. The use of (19) and (24)–(27) allows us to justify the superiority of the size distribution and, therefore, statistically justify Matheron's axiomatic development of these distributions [28]. For more discussion on this subject, see section 4.

Let us now assume that $\hat{F}_{\mathbf{X}}(B)$ is an unbiased and consistent estimator for $F_{\mathbf{X}}(B)$; then it can be easily shown that $\hat{T}_{\mathbf{X} \circ E}(B)$, obtained form (26a) by replacing $F_{\mathbf{X}}(\bullet)$ with $\hat{F}_{\mathbf{X}}(\bullet)$, is also an unbiased and consistent estimator for $T_{\mathbf{X} \circ E}(B)$. Knowledge of this quantity may permit experimental study of the statistical behavior of openings in a more general setting than the one proposed in [84], which, in turn, may allow proper filter design (for example, see [49] and [50]). This issue is currently under investigation. Finally, I would like to illustrate the theoretical importance of the previous equations by studying a classical linear-filtering problem under the framework of mathematical morphology.

In many morphological-filtering applications [38], [48]–[50], it is desirable to know under what conditions a shape X is E-open (i.e., $X \circ E = E$) or E-closed (i.e., $X \bullet E = E$) or both since most morphological filtering results in shapes that are either open or closed or both. Since we are interested in modeling X by means of a DRS $\mathbf{X}$, we define $\mathbf{X}$ to be an *a.s. E-open DRS* if $\Pr[\mathbf{X} \in X^E_{\text{opn}}] = 1$, where $X^E_{\text{opn}} \triangleq \{X \in P : X \circ E = X\}$, whereas we define $\mathbf{X}$ to be an *a.s. E-closed DRS* if $\Pr[\mathbf{X} \in X^E_{\text{clo}}] = 1$, where $X^E_{\text{clo}} \triangleq \{X \in P : X \bullet E = X\}$. We now have the following proposition (see also [50] for a similar result):

PROPOSITION 6. If $\mathbf{X}$ is an a.s. E-open DRS,

then (a) $F_{\mathbf{X}\circ E}(B) = F_{\mathbf{X}}(B)$ for every $B \in S$ or, equivalently, (b) $T_{\mathbf{X}\circ E}(B) = T_{\mathbf{X}}(B)$ for every $B \in S$.

Proof. If $\mathbf{X}$ is a.s. E-open, then $\Pr[\mathbf{X} = X] = \Pr[\mathbf{X}\circ E = X]$ for every $X \in P$, which, in turn, implies (a) by virtue of (7). (a) $\Leftrightarrow$ (b) by virtue of (19).

A similar result holds for the case of closing. We have the following proposition, whose proof is similar to the proof of Proposition 6 and therefore is omitted.

PROPOSITION 7. If $\mathbf{X}$ is an a.s. E-closed DRS, then (a) $F_{\mathbf{X}\bullet E}(B) = F_{\mathbf{X}}(B)$ for every $B \in S$ or, equivalently, (b) $T_{\mathbf{X}\bullet E}(B) = T_{\mathbf{X}}(B)$ for every $B \in S$.

Let us now consider a discrete shape X, partially available, by means of an observation $Y = (X - M) \cup N$. The objective here is to design a morphological transformation $\Psi[\bullet]$ that, when applied on Y, results in a shape $\hat{X} = \Psi[(X - M) \cup N]$, which is, in some sense, a close replica of X. This is the classical noise-filtering problem, revisited in [48]–[50], in terms of mathematical morphology and DRS theory, which is a plausible approach when the observation noise has strong geometric character (e.g., shape occlusion or clutter). In [48] the two noise components M and N are simulated by a germ-grain DRS model [2], [9], [11]. In this case M primarily models missing data, whereas N models additive-shape noise.

Before we attempt any solution to this problem by mathematical morphology we should make sure that such a problem has a meaningful solution under some reasonable morphological constraints on X, M, and N. I shall demonstrate this in the following. We shall model X by means of a DRS $\mathbf{X}$, and we shall assume that M and N are two deterministic, but unknown, sets. This assumption is similar to the one in [48] and is necessary here for simplifying our presentation (see [49] and [50] for a more general treatment). The next proposition reveals a fundamental statistical property of a certain class of morphological filters.

PROPOSITION 8. (a) If N is such that for every shape $X \in P$ there exists a structuring element $C_N \in S$ such that

$$X^c \cap (B \oplus C) = \emptyset \Leftrightarrow X^c \cap (B \oplus C) \cap N^c = \emptyset \quad \text{for every } B \in S, C \supseteq C_N, \tag{32a}$$

then

$$T_{(\mathbf{X}\cup N)\circ E}(B) = T_{\mathbf{X}}(B) \quad \text{for every } B \in S, \tag{32b}$$

provided that $\mathbf{X}$ is a.s. E-open for some $E \supseteq C_N$. (b) If M is such that for every shape $X \in P$ there exists a structuring element $D_M \in S$ such that

$$X \cap (B \oplus D^t) \neq \emptyset \Leftrightarrow X \cap (B \oplus D^t) \cap M^c \neq \emptyset \quad \text{for every } B \in S, D \supseteq D_M, \tag{33a}$$

then

$$T_{(\mathbf{X}-M)\bullet E}(B) = T_{\mathbf{X}}(B) \quad \text{for every } B \in S, \tag{33b}$$

provided that $\mathbf{X}$ is a.s. E-closed for some $E \supseteq D_M$.

Proof. From (23) and (26a) observe that

$$\begin{aligned} T_{(\mathbf{X}\cup N)\circ C}(B) &= \sum_{\substack{B' \subseteq B \oplus C^t,\\ B' \neq \emptyset}} (-1)^{|B'|+1}[1 - F_{\mathbf{X}\cup N}(B' \oplus C)] \\ &= \sum_{\substack{B' \subseteq B \oplus C^t,\\ B' \neq \emptyset}} (-1)^{|B'|+1}[1 - F_{\mathbf{X}}((B' \oplus C) \cap N^c)]. \end{aligned} \tag{34}$$

From (32a) we have that

$$\begin{aligned} X \supseteq (B' \oplus C) \cap N^c &\Leftrightarrow X^c \cap (B' \oplus C) \cap N^c = \emptyset \\ &\Leftrightarrow X^c \cap (B' \oplus C) = \emptyset \\ &\Leftrightarrow X \supseteq B' \oplus C, \end{aligned}$$

for every $B' \in S$ and $C \supseteq C_N$, from which we obtain (see also (5))

$$F_{\mathbf{X}}((B' \oplus C) \cap N^c) = F_{\mathbf{X}}(B' \oplus C) \quad \text{for every } B' \in S \text{ and } C \supseteq C_N. \tag{35}$$

From (26a), (34), and (35) we now have that

$$T_{(\mathbf{X}\cup N)\circ C}(B) = T_{\mathbf{X}\circ C}(B) \quad \text{for every } B \in S \text{ and } C \supseteq C_N,$$

and, since $\mathbf{X}$ is a.s. E-open for some $E \supseteq C_N$, we obtain (32b).

Equation (33b) can be shown in a fashion similar to that of (32b) and by using (10), (22a), (27b), (33a), and the fact that $X - M = X \cap M^c$.

According to Proposition 8, if N satisfies condition (32a), then the output of a simple morphological opening filter with structuring element $E \supseteq C_N$ to input $\mathbf{X} \cup N$ will be identically distributed to the noise-free E-open input $\mathbf{X}$, whereas if M satisfies condition (33a), then the output of a simple morphological closing filter with structuring element $E \supseteq D_M$ to input $\mathbf{X} - M$ will be identically distributed to the noise-free E-closed input $\mathbf{X}$. If $\mathbf{X}$ is both a.s. E-open and a.s. E-closed, then it can be easily shown that the output of a simple morphological open-closing filter with structuring element E to input $(\mathbf{X} - M) \cup N$ will be identically distributed to the noise-free input $\mathbf{X}$, provided that (32a) and (33a) are both satisfied and that $E \supseteq C_N \cup D_M$. Conditions (32a) and (33a) are reasonable geometric conditions that are directly related to the size of the noise under consideration. Indeed, let us closely examine (32a). It is clear that

$$X^c \cap (B \oplus C) = \emptyset \\ \Rightarrow X^c \cap (B \oplus C) \cap N^c = \emptyset,$$

for every $X \in P$ and B, $C \in S$. Therefore the actual condition imposed on N is

$$X^c \cap (B \oplus C) \cap N^c = \emptyset, \\ \Rightarrow X^c \cap (B \oplus C) = \emptyset,$$

for every $X \in P$, $B \in S$, and $C \supseteq C_N$. If the size of noise N is limited, and since dilation is an increasing transformation, there exists a structuring element C_N such that $N \not\supseteq X^c \cap (B \oplus C)$ for every $B \in S$ and $C \supseteq C_N$, unless $X^c \cap (B \oplus C) = \emptyset$. Assume now that $X^c \cap (B \oplus C) \cap N^c = \emptyset$ for some $B \in S$ and some $C \supseteq C_N$. In this case, $N \supseteq X^c \cap (B \oplus C)$ and, therefore, $X^c \cap (B \oplus C) = \emptyset$; otherwise, we have a contradiction. A similar observation can be made for condition (33a).

To conclude this section let us consider the hit-or-miss transform $X \circledast (B_1, B_2)$ of a set X by the two structuring elements $B_1, B_2 \in S$ defined by [2]:

$$X \circledast (B_1, B_2) \triangleq X \ominus B_1 - X \oplus B_2^t \\ = (X \ominus B_1) \cap (X^c \ominus B_2). \quad (36)$$

We have the following result.

PROPOSITION 9. For any DRS $\mathbf{X}$ we have that

$$\Pr[\mathbf{X} = B] = 1 - F_{\mathbf{X} \circledast (B, B^c)}(\{o\}) \\ = T_{\mathbf{X} \circledast (B, B^c)}(\{o\}), \quad (37)$$

for every $B \in S$.

Proof. Observe that

$$\begin{aligned} \Pr[\mathbf{X} = B] &= \Pr[\mathbf{X} \supseteq B, \mathbf{X} \subseteq B] \\ &= \Pr[\mathbf{X} \supseteq B, \mathbf{X}^c \supseteq B^c] \\ &= \Pr[\mathbf{X} \supseteq B \oplus \{o\}, \mathbf{X}^c \supseteq B^c \oplus \{o\}] \\ &= \Pr[\mathbf{X} \ominus B \supseteq \{o\}, \mathbf{X}^c \ominus B^c \supseteq \{o\}] \\ &= \Pr[((\mathbf{X} \ominus B) \cap (\mathbf{X}^c \ominus B^c)) \supseteq \{o\}] \\ &= \Pr[((\mathbf{X} \ominus B) \cap (\mathbf{X}^c \ominus B^c)) \cap \{o\} \neq \emptyset], \end{aligned}$$

which, together with (5), (10), and (36), shows (37).

According to (1), (2), and the previous proposition, the probability-mass function $\Pr[\mathbf{x} = x]$ of a BRF $\mathbf{x}$ is given by the cumulative-distribution functional or the discrete-capacity functional of the hit-or-miss transformed DRS $\mathbf{X} \circledast (B, B^c)$ computed at the origin.

4 Moments of DRS's

Moments are very useful tools for summarizing the statistical behaviour of random variables, random processes, and random fields. As expected, the same is also true for random sets. One of the most popular moments in BRF theory is the autocorrelation $r_{\mathbf{x}}(i, j; k, l)$ of a BRF $\mathbf{x}$, given by [2], [9], [20], [51] (see also [19], [28], and [85]):

$$\begin{aligned} r_{\mathbf{x}}(i, j; k, l) &\triangleq E[\mathbf{I}_{\mathbf{X}}(i, j)\mathbf{I}_{\mathbf{X}}(k, l)] \\ &= \Pr[(i, j) \in \mathbf{X}, (k, l) \in \mathbf{X}] \\ &= \Pr[\mathbf{X} \cap \{(i, j)\} \neq \emptyset, \mathbf{X} \cap \{(k, l)\} \neq \emptyset], \end{aligned} \quad (38)$$

for every (i, j), $(k, l) \in Z^2$. The autocorrelation (38) has been extensively used in a variety of applications with moderate success (for example, see [2], [9], [19], [20], and [86]). As is clear from (38), $r_{\mathbf{X}}(i, j; k, l)$ cannot effectively summarize shape information since it provides only a measurement of the proportion of pairs of points $((i, j), (k, l))$ that are in $\mathbf{X}$. However, (38) can be generalized to a more desirable statistical summary for random shape. We have the following definition.

DEFINITION 1. The nth*-order moment* $R_{\mathbf{X}}^{(n)}(B_1, B_2, \ldots, B_n)$ of a DRS $\mathbf{X}$ is defined by

$$R_{\mathbf{X}}^{(n)}(B_1, B_2, \ldots, B_n) \triangleq \Pr[\mathbf{X} \cap B_1 \neq \emptyset, \mathbf{X} \cap B_2, \neq \emptyset \ldots, \mathbf{X} \cap B_n \neq \emptyset], \quad (39)$$

for $B_1, B_2, \ldots, B_n \in S$ and $n \geq 1$.

From (10) and (39), with $n = 1$, we have that

$$R_{\mathbf{X}}^{(1)}(B) = T_{\mathbf{X}}(B) \quad \text{for every } B \in S; \quad (40)$$

therefore the discrete-capacity functional $T_{\mathbf{X}}(B)$ may be viewed as the 1st-order moment of a DRS $\mathbf{X}$ (recall from section 2 that $1 - T_{\mathbf{X}}(B)$ is also a generalized $|B|$-order moment of the BRF $\mathbf{x}$). We now define the *autocorrelation functional* $R_{\mathbf{X}}(B_1, B_2)$ of a DRS $\mathbf{X}$ to be its 2nd-order moment, i.e.,

$$R_{\mathbf{X}}(B_1, B_2) \triangleq R_{\mathbf{X}}^{(2)}(B_1, B_2) = \Pr[\mathbf{X} \cap B_1 \neq \emptyset, \mathbf{X} \cap B_2 \neq \emptyset], \quad (41)$$

for every B_1, $B_2 \in S$. From (38) and (40) we have that

$$r_{\mathbf{X}}(i, j; k, l) = R_{\mathbf{X}}(\{(i, j)\}, \{(k, l)\})$$

for every (i, j), $(k, l) \in Z^2$, therefore the autocorrelation (38) is a special case of the autocorrelation functional $R_{\mathbf{X}}(B_1, B_2)$.

The nth-order moment $R_{\mathbf{X}}^{(n)}(B_1, B_2, \ldots, B_n)$ can be expressed in terms of the functional $Q_{\mathbf{X},n}(B; B_1, B_2, \ldots, B_n)$ defined by (13). This relationship is established in the following proposition.

PROPOSITION 10. In general,

$$R_{\mathbf{X}}^{(n+1)}(B_1, B_2, \ldots, B_{n+1}) = R_{\mathbf{X}}^{(n)}(B_1, B_2, \ldots, B_n) - Q_{\mathbf{X},n}(B_{n+1}; B_1, B_2, \ldots, B_n), \quad (42)$$

for every $B_1, B_2, \ldots, B_{n+1} \in S$ and $n \geq 1$.

Proof. From (13) and (39) we have that

$$\begin{aligned} R_{\mathbf{X}}^{(n+1)}&(B_1, B_2, \ldots, B_{n+1}) \\ &= \Pr[\mathbf{X} \cap B_1 = \emptyset, \mathbf{X} \cap B_2 = \emptyset, \ldots, \mathbf{X} \cap B_{n+1} = \emptyset] \\ &= \Pr[\mathbf{X} \cap B_1 = \emptyset, \mathbf{X} \cap B_2 = \emptyset, \ldots, \mathbf{X} \cap B_n = \emptyset] \\ &\quad - \Pr[\mathbf{X} \cap B_1 = \emptyset, \mathbf{X} \cap B_2 = \emptyset, \\ &\quad \ldots, \mathbf{X} \cap B_n = \emptyset, \mathbf{X} \cap B_{n+1} = \emptyset] \\ &= R_{\mathbf{X}}^{(n)}(B_1, B_2, \ldots, B_n) \\ &\quad - Q_{\mathbf{X},n}(B_{n+1}; B_1, B_2, \ldots, B_n), \end{aligned}$$

which shows (42).

From (14) observe that the functional $Q_{\mathbf{X},n}$, $n \geq 1$, provides a finite-difference approximation of the nth-order "derivative" $Q_{\mathbf{X}}^{(n)}(B; x)$ of $Q_{\mathbf{X}}(B)$ at a point $x \in R^2$ defined over all compact subsets B of R^2. If this "derivative" exists, it is defined by (see also [44])

$$Q_{\mathbf{X}}^{(n)}(B; x) \triangleq \lim_{\rho \to 0} \frac{Q_{\mathbf{X}}^{(n-1)}(B; x) - Q_{\mathbf{X}}^{(n-1)}(B \cup B_{\rho,x}; x)}{\text{Area}(B_{\rho,x})} \quad \text{for } n \geq 1$$

with $Q_{\mathbf{X}}^{(0)}(B; x) \triangleq Q_{\mathbf{X}}(B)$ for every $x \in R^2$ and every compact set $B \subset R^2$, where $B_{\rho,x}$ is any compact subset of the compact disk $D(x, \rho)$ of radius ρ centered at a point $x \in R^2$. From (14a), (40), and (42) it is also clear that the nth-order moment $R_{\mathbf{X}}^{(n)}$ of $\mathbf{X}$ is given in terms of the derivatives $Q_{\mathbf{X},k}, 0 \leq k \leq n-1$, of $Q_{\mathbf{X}}(B)$; therefore $Q_{\mathbf{X}}(B)$ is a moment-generating functional [9], [11], [72]. From (8), (16), and (22b) we have that

$$\begin{aligned} Q_{\mathbf{X}^c,n}&(B; \{b_1\}, \{b_2\}, \ldots, \{b_n\}) \\ &= \sum_{B' \subseteq B^c \cap W} (-1)^{|B'|}[1 - T_{\mathbf{X}^c}(B \cup B')] \\ &= \sum_{B' \subseteq B^c \cap W} (-1)^{|B'|}[1 - F_{\mathbf{X}}(B \cup B')] = \Pr[\mathbf{X} = B \end{aligned}$$

for every $B \in S_W$ such that $B^c \cap W = \{b_1, b_2, \ldots, b_n\}$, which, together with (42), results in

$$\Pr[\mathbf{X} = B] = R^{(n)}_{\mathbf{X}^c}(\{b_1\}, \{b_2\}, \ldots, \{b_n\}) - R^{(n+1)}_{\mathbf{X}^c}(\{b_1\}, \{b_2\}, \ldots, \{b_n\}, B),$$

for every $B \in S_W$ such that $B^c \cap W = \{b_1, b_2, \ldots, b_n\}$. Therefore the probability-mass function of an a.s. W-bounded DRS $\mathbf{X}$ is directly related to the moments of $\mathbf{X}^c$.

The 2nd-order moment $R_{\mathbf{X}}(B_1, B_2)$ of a DRS $\mathbf{X}$ is of a special interest since it generalizes the concept of autocorrelation $r_{\mathbf{X}}$. In the following we shall present various properties of $R_{\mathbf{X}}(B_1, B_2)$ that turn out to be quite similar to the properties of the autocorrelation of a BRF $\mathbf{x}$, thus extending the similarity between $r_{\mathbf{X}}$ and $R_{\mathbf{X}}$. We have the following two propositions.

PROPOSITION 11. In general,

$$R_{\mathbf{X}}(B_1, B_2) = T_{\mathbf{X}}(B_1) + T_{\mathbf{X}}(B_2) - T_{\mathbf{X}}(B_1 \cup B_2), \tag{43}$$

for every $B_1, B_2 \in S$.

Proof. The proof is a direct consequence of (16), (40), and (42).

PROPOSITION 12. In general,

$$R_{\mathbf{X}}(B_1, B_2) = R_{\mathbf{X}}(B_2, B_1), \tag{44a}$$

$$0 \leq T_{\mathbf{X}}(B_1 \cap B_2) \leq R_{\mathbf{X}}(B_1, B_2) \leq \min\{T_{\mathbf{X}}(B_1), T_{\mathbf{X}}(B_2)\} \leq 1, \tag{44b}$$

$$R_{\mathbf{X}}(\emptyset, B) = R_{\mathbf{X}}(B, \emptyset) \qquad R_{\mathbf{X}}(\emptyset, \emptyset) = 0, \tag{44c}$$

for every B, B_1, $B_2 \in S$, whereas

$$R_{\mathbf{X}}(B_1, B_2) \leq R_{\mathbf{X}}(B, B_2) \text{ for every } B, B_1, B_2 \in S \text{ such that } B_1 \subseteq B \tag{44d}$$

and

$$R_{\mathbf{X}}(B_1, B_2) = \begin{cases} T_{\mathbf{X}}(B_1) & \text{for every } B_1, B_2 \in S \text{ such that } B_1 \subseteq B_2 \\ T_{\mathbf{X}}(B_2) & \text{for every } B_1, B_2 \in S \text{ such that } B_2 \subseteq B_1 \end{cases}. \tag{44e}$$

Proof. The proof of (44a) and (44e) is a direct consequence of (43). The proof of (44b) is a direct consequence of (11a), (11c), (17), and (43), whereas (44c) is a direct consequence of (11b) and (43). Finally, (44d) is a direct consequence of (41).

The autocorrelation functional $R_{\mathbf{X}}(B_1, B_2)$ can be used in a manner that is similar to the use of the autocorrelation $r_{\mathbf{x}}(i, j; k, l)$ of a BRF $\mathbf{x}$. For example, $R_{\mathbf{X}}$ can be used as a measure of the statistical dependence between disjoint subsets of $\mathbf{X}$. Indeed, if B_1 and B_2 are two disjoint structuring elements in S, then it can be easily shown that the events $X \cap B_1 = \emptyset$ and $X \cap B_2 = \emptyset$ are statistically independent if and only if $Q_{\mathbf{X}}(B_1 \cup B_2) = Q_{\mathbf{X}}(B_1)Q_{\mathbf{X}}(B_2)$, in which case (see also (14a) and (43)) $R_{\mathbf{X}}(B_1, B_2) = T_{\mathbf{X}}(B_1)T_{\mathbf{X}}(B_2)$. As expected, the inverse of this statement is not true in general.

The autocorrelation functional $R_{\mathbf{X}}$ should be more effective in summarizing random shapes than is $r_{\mathbf{x}}$, primarily because of the flexibility introduced by the choice of the two test sets B_1 and B_2 associated with $R_{\mathbf{X}}$. To make $R_{\mathbf{X}}$ practical we may choose $B_1 = nE \oplus \{(i, j)\}$ and $B_2 = nE \oplus \{(k, l)\}$, where E is a structuring element in S, $(i, j), (k, l) \in Z^2$, and nE is defined by (28). In this case the sequence $r_{\mathbf{X},E,n}(i, j; k, l)$, defined by

$$r_{\mathbf{X},E,n}(i, j; k, l) \triangleq R_{\mathbf{X}}(nE \oplus \{(i, j)\}, nE \oplus \{(k, l)\}), \tag{45}$$

for $n = 1, 2, \ldots$ and $(i, j), (k, l) \in Z^2$ is a combination of the contact-distribution function $f_{\mathbf{X}}(E, n)$ given by (29a) and the autocorrelation $r_{\mathbf{x}}(i, j; k, l)$ given by (38); therefore $r_{\mathbf{X},E,n}$ should perform better than either (29a) or (38).

The functional $r_{\mathbf{X},E,n}$ can be further generalized in terms of the nth-order moment $R^{(n)}_{\mathbf{X}}$, with important consequences. Indeed, consider a set C of m points $\{c_1, c_2, \ldots, c_m\}$ in Z^2. Define the functional $R_{\mathbf{X},C}(E, n)$ by (see (45))

$$R_{\mathbf{X},C}(E, n) \triangleq R^{(n)}_{\mathbf{X}}(nE \oplus \{c_1\}, nE \oplus \{c_2\}, \ldots, nE \oplus \{c_m\}) \tag{46}$$

for $n = 1, 2, \ldots$. From (39) and (46) observe that

$$\begin{aligned} &R_{\mathbf{X},C}(E,n) \\ &= \Pr[\mathbf{X} \cap nE \oplus \{c_1\} \neq \emptyset, \mathbf{X} \cap nE \oplus \{c_2\} \neq \emptyset, \\ &\ldots, \mathbf{X} \cap nE \oplus \{c_m\} \neq \emptyset] \\ &= \Pr[\mathbf{X} \oplus nE^t \supseteq \{c_1\}, \mathbf{X} \oplus nE^t \supseteq \{c_2\}, \ldots, \\ &\quad \mathbf{X} \oplus nE^t \supseteq \{c_m\}] \\ &= \Pr[\mathbf{X} \oplus nE^t \supseteq C] \\ &= \Pr[\mathbf{X} \oplus nE^t \supseteq C \oplus \{o\}] \\ &= \Pr[(\mathbf{X} \oplus nE^t) \ominus C \supseteq \{o\}], \end{aligned}$$

which, together with (5), results in

$$R_{\mathbf{X},C}(E,n) = 1 - F_{(\mathbf{X} \oplus nE^t) \ominus C}(\{o\}). \quad (47)$$

Consider now a nonempty, symmetric structuring element E that contains the origin, and take $C = nE$. From (22b), (47), and the duality between opening and closing we can easily show that

$$R_{\mathbf{X}^c,nE}(E,n) = 1 - T_{\mathbf{X} \circ nE}(\{o\}),$$

which, together with (29b), shows that $1 - R_{\mathbf{X}^c,nE}(E,n) = o_{\mathbf{X}}(E,n)$, i.e., $1 - R_{\mathbf{X}^c,nE}(E,n)$ is the opening-based size-distribution function of $\mathbf{X}$. Although Matheron has chosen an axiomatic approach to size distributions [28] (see also [2]), the previous discussion demonstrates that size distributions are special cases of the higher-order moments of a DRS $\mathbf{X}$ as defined by (39). As a direct consequence of this result, it is now clear that, since size distributions are higher-order moments, they will be more effective in summarizing shape information than either the contact-distribution function (see also the discussion in section 3) or the autocorrelation function and are even more effective than (45).

5 Almost Surely Convex-8-Connected DRS's

Convex sets are fundamental in many scientific disciplines, including integral geometry, stereology, and mathematical morphology [2], [5], [7], [28], [30], [45], [47]. It is therefore important to study the class of convex DRS's. As we have discussed in section 1, although a theory for convex random sets defined on the Euclidean space R^2 exists [28], because of some peculiarities of discrete topology no convexity results are known for random sets defined on Z^2. The purpose of this section is to show that Matheron's results on convexity of random sets on R^2 are still valid in the discrete case and under some additional assumptions on connectivity. Before we proceed with our analysis we shall summarize some useful results from discrete topology. More details on this subject may be found in [2] and [87].

The polytope pol (X) of a set of points $X \in P$ is defined to be the intersection of all closed half-planes (i.e., the part of R^2 lying on and on one side of a line) that contain X. The set $H(X) \triangleq \mathrm{pol}(X) \cap Z^2$ of all points in pol(X) is the convex hull of X. A set $X \in P$ is a convex set if it is identical to its convex hull, i.e., if $X = H(X)$. The polytope pol(X) of a set of points $X \in P$ is always a convex set in R^2, i.e., for every pair of points x_1, $x_2 \in \mathrm{pol}(X)$ the line segment

$$L_{x_1x_2} = \{x \in \mathbf{R}^2 : x = \lambda x_1 + (1-\lambda)x_2, 0 \leq \lambda \leq 1\} \subseteq \mathrm{pol}(X).$$

The vertices of the pol(X) are points in X, and therefore every edge of the polytope connects a pair of points $(x_1, x_2) \in X \times X$. A line L passing through two points in pol(X) divides it into two polytopes $\mathrm{pol}_1(X)$ and $\mathrm{pol}_2(X)$ such that $\mathrm{pol}_1(X) \cap \mathrm{pol}_2(X) = L \cap \mathrm{pol}(X)$ and $\mathrm{pol}_1(X) \cup \mathrm{pol}_2(X) = \mathrm{pol}(X)$. We have the following lemma.[6]

LEMMA 1. If X is a convex set in P, then the sets $X_1 = \mathrm{pol}_1(X) \cap Z^2$ and $X_2 = \mathrm{pol}_2(X) \cap Z^2$ are also convex sets in P such that $X_1 \cap X_2 = L \cap X$ and $X_1 \cup X_2 = X$.

The 4-neighbors of a point $(i,j) \in Z^2$ are the four points $(i \pm 1, j)$ and $(i, j \pm 1)$ in Z^2, whereas the 8-neighbors of a point $(i,j) \in Z^2$ are the eight points $(i \pm 1, j)$, $(i, j \pm 1)$, and $(i \pm 1, j \pm 1)$ in Z^2. Two points in X are said to be 4-adjacent points (8-adjacent points) if they are 4-neighbors

[6] The proofs of this and other lemmas are given in the appendix.

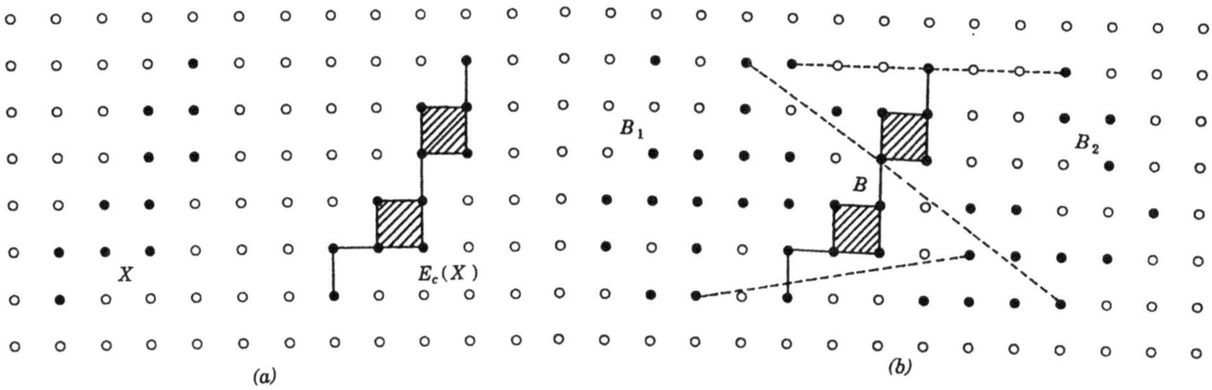

Fig. 1. (a) The continuous extension $E_c(X)$ of a set X. (b) Two bounded sets B_1 and B_2 separated by the bounded set B.

(8-neighbors). The open line segment

$$\alpha_{x_1x_2} = \{x \in R^2 : x = \lambda x_1 + (1-\lambda)x_2, 0 < \lambda < 1\},$$

which joins two 4-adjacent points (8-adjacent points) x_1, $x_2 \in X$, is called a 4-adjacency (8-adjacency) in X. A 4-path (8-path) in X, from point x_1 to point x_n, is a collection $\pi_{x_1x_n} = \{x_1, x_2, \ldots, x_n\}$ of n points in X such that x_i is 4-adjacent (8-adjacent) to x_{i+1} for all $i = 1, 2, \ldots, n-1$. A set $X \in P$ is said to be 4-connected (8-connected) if it cannot be partitioned into two subsets that are not 4-adjacent (8-adjacent) to each other. It is straightforward to show that a set $X \in P$ is 4-connected (8-connected) if and only if for every pair (x_1, x_2) of points in X there exists a 4-path (8-path) in X from point x_1 to point x_2. Clearly, if a set is 4-connected, then it is also 8-connected.

If A_X^4 denotes the collection of all 4-adjacencies in X and if $\sum_X^4$ denotes the collection of all open unit squares whose sides are contained in A_X^4, then the union $X \cup A_X^4 \cup \sum_X^4$ is called the continuous extension $E_c(X)$ of the set X. This concept is illustrated in figure 1(a). We now have the following definition.

DEFINITION 2. Two bounded sets B_1, $B_2 \in S$ are said to be *separated* by a bounded set $B \in S$ if $L_{b_1b_2} \cap E_c(X) \neq \emptyset$ for every pair (b_1, b_2) of points such that $b_1 \in B_1$ and $b_2 \in B_2$.

Figure 1(b) depicts an example of two bounded sets B_1 and B_2 that are separated by the bounded set B. By convention we shall assume that if $B_1 = \emptyset$ or $B_2 = \emptyset$, then B_1 and

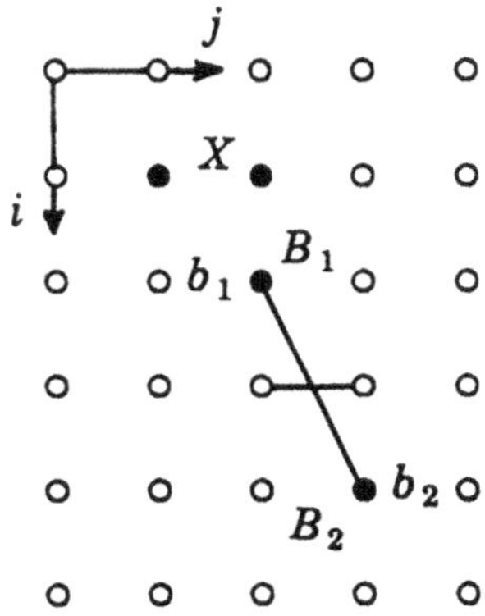

Fig. 2. A nonconnected convex set that violates Lemma 2.

B_2 are separated by any bounded set $B \in S$, whereas B_1 and B_2 are separated by $B = \emptyset$ only when $B_1 = \emptyset$ or $B_2 = \emptyset$. It is intuitive to believe that if a convex-8-connected set X hits two bounded sets B_1 and B_2 that are separated by a bounded set B, then it should also hit B. This statement is formalized in Proposition 13, which is a direct consequence of the following lemma.

LEMMA 2. Let $B_1, B_2 \in S$ be two bounded sets separated by a bounded set $B \in S$, let X be a convex-8-connected set, and let $L_{b_1b_2}$ be the line segment that joins any two points $b_1 \in X \cap B_1$ and $b_2 \in X \cap B_2$. Then (a) if $L_{b_1b_2}$ passes through a point b in B, then $b \in X \cap B$ and (b) if $L_{b_1b_2}$ hits a 4-adjacency $\alpha_{bb'}^4$ in B that joins two 4-adjacent points b, $b' \in B$, then $\{b, b'\} \cap X \neq \emptyset$.

For Lemma 2 to be true, X must be not only

convex but also 8-connected. For example (see figure 2), the set $X = \{(1,1),(1,2),(2,2),(4,3)\}$ is a nonconnected convex set. Take $B_1 = \{(2,2)\}$ and $B_2 = \{(4,3)\}$. In this case B_1 and B_2 are two bounded sets separated by the bounded set $B = \{(3,2),(3,3)\}$. Take $b_1 = (2,2)$ and $b_2 = (4,3)$. If $\alpha^4_{(3,2),(3,3)}$ is the 4-adjacency between points (3,2) and (3,3) and if $L_{(2,2),(4,3)}$ is the line segment joining points (2,2) and (4,3), then $L_{(2,2),(4,3)} \cap \alpha^4_{(3,2),(3,3)} \neq \emptyset$ but neither of the points (2,3) or (3,3) is in X.

PROPOSITION 13. Let B_1, $B_2 \in S$ be two bounded sets separated by a bounded set $B \in S$. For every convex-8-connected set $X \in S$, if $X \cap B_1 \neq \emptyset$ and $X \cap B_2 \neq \emptyset$, then $X \cap B \neq \emptyset$.

Proof. Assume that $X \cap B_1 \neq \emptyset$ and $X \cap B_2 \neq \emptyset$. Take points b_1 and b_2 in Z^2 such that $b_1 \in X \cap B_1$ and $b_2 \in X \cap B_2$. According to Definition 2, the line segment $L_{b_1 b_2}$ that joins the points b_1 and b_2 hits the continuous extension $E_c(B)$ of set B. In this case $L_{b_1 b_2}$ will pass through at least one point in B or will intersect at least one 4-adjacency in B or both. The proof is now a direct consequence of Lemma 2.

When X is assumed to be a bounded set, we have the following necessary and sufficient condition for convexity and 8-connectivity.

PROPOSITION 14. A bounded set X in S is a convex-8-connected set if and only if for every pair (B_1, B_2) of bounded, convex-4-connected sets that are separated by their intersection $B_1 \cap B_2$, we have that

$$\text{if } X \cap B_1 \neq \emptyset \text{ and } X \cap B_2 \neq \emptyset, \quad \text{then } X \cap (B_1 \cap B_2) \neq \emptyset. \tag{48}$$

Proof. Assume first that X is a convex-8-connected set. Equation (48) is a direct consequence of Proposition 13 with $B = B_1 \cap B_2$.

Assume now that (48) is satisfied for a bounded nonempty (we shall consider the empty set $\emptyset$ to be both convex and 8-connected by definition), nonconvex set X. In this case $H(X)$ will contain a point x_0 that does not belong to X and that is adjacent to at least one point in X. Indeed, if no such point can be found, then X will be convex or empty, which contradicts our assumptions. Assume, without loss of generality, that if $x_0 = (m,n)$, then x_0 is adjacent to a point $x_1 \in X$ such that $x_1 = (m-1,n)$. If there exists a point $x_2 = (m+M,n)$ for some $M \geq 1$ such that $x_2 \in X$ (see figure 3(a)), then the sets $H_1 = \{(m-1,n),(m,n)\}$ and $H_2 = \{(m,n),(m+1,n),\ldots,(m+M,n)\}$ are two bounded, convex-4-connected sets that are separated by their intersection $H = H_1 \cap H_2 = \{(m,n)\}$. In this case $X \cap H_1 \neq \emptyset$ and $X \cap H_2 \neq \emptyset$, but $X \cap H = \emptyset$, which contradicts our assumption (48). If there is no point $x_2 = (m+M,n)$ for some $M \geq 1$ such that $x_2 \in X$, since X is assumed to be bounded (and therefore $H(X)$ is bounded), then we can always find a point $x_2 \in H(X)$, $x_2 \notin X$ such that either (i) x_2 is on an edge of pol(X) or (ii) x_2 is in the interior of pol(X) and is 4-adjacent to a point $x_3 \in H^c(X)$. In case (i) we can always find an edge $L_{p_1 p_2}$ of the polytope pol(X) such that $p_1, p_2 \in X$ and $x_2 \in L_{p_1 p_2}$ (see figure 3(b)). In case (ii) we can always find an edge $L_{p_1 p_2}$ of the polytope pol(X) such that p_1, $p_2 \in X$ and $L_{p_1 p_2} \cap \alpha^4_{x_2 x_3} \neq \emptyset$ (see figure 3(c)). Assume that $p_1 = (m_1, n_1)$, $p_2 = (m_2, n_2)$, and $x_2 = (m_3, n_3)$, where, without loss of generality, $m_2 > m_1$ and $n_2 > n_1$. Consider $H_1 = H(\{(m_1,n_1),(m_1,n_3),(m_3,n_3)\})$ and $H_2 = H(\{(m_3,n_3),(m_1,n_3),(m_1,n_2),(m_2,n_2)\})$. Observe that H_1 and H_2 are two bounded convex-4-connected sets that are separated by their intersection $H = H_1 \cap H_2 = H(\{(m_1,n_3),(m_3,n_3)\}) \subseteq X^c$ with $X \cap H_1 \neq \emptyset$, $X \cap H_2 \neq \emptyset$, but $X \cap H = \emptyset$ which again contradicts our assumption (48). Therefore X should be convex.

Assume now that X is a convex, nonconnected set; then there will be at least a pair (x_1, x_2) of points in X such that there is no path $\pi_{x_1 x_2}$ in X from point x_1 to point x_2. These points cannot be horizontally or vertically collinear because a convex set of horizontally or vertically collinear points should be 4-connected. Consider the minimal rectangle $R_{x_1 x_2}$, with horizontal and vertical edges, that contains both points x_1 and x_2 (this rectangle has the minimal possible area of all rectangles with horizontal and

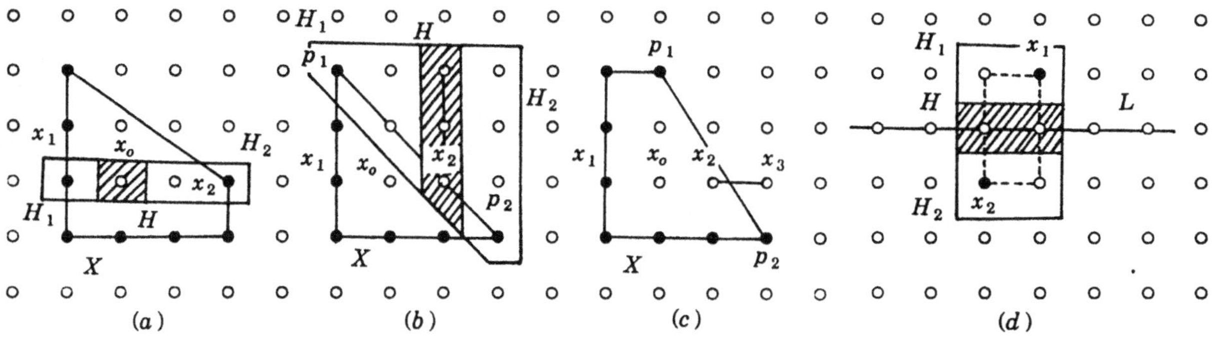

Fig. 3. Illustration related to the proof of Proposition 14.

vertical edges that contain a pair (x_1, x_2) of points in X for which there is no path $\pi_{x_1 x_2}$ in X from point x_1 to point x_2). In this case there will be a horizontal or vertical line L such that $(L \cap R_{x_1 x_2}) \cap Z^2 \subseteq X^c$ (see figure 3(d)); otherwise, $R_{x_1 x_2}$ will not be minimal, which is a contradiction. Line L divides the rectangle $R_{x_1 x_2}$ into two rectangles R_1 and R_2 such that $R_1 \cap R_2 = L \cap R_{x_1 x_2}$ and $R_1 \cup R_2 = R_{x_1 x_2}$. Consider sets $H_1 = R_1 \cap Z^2$ and $H_2 = R_2 \cap Z^2$ and a set H such that $H = L \cap R_{x_1 x_2} \cap Z^2$. It is clear that H_1 and H_2 are two bounded convex-4-connected sets that are separated by $H = H_1 \cap H_2$. Since $x_1 \in H_1$ and $x_2 \in H_2$, we have that $X \cap H_1 \neq \emptyset$, $X \cap H_2 \neq \emptyset$, and since $H = L \cap R_{x_1 x_2} \cap Z^2 \subseteq X^c$ we have that $X \cap H = \emptyset$, which contradicts (48); therefore X should be 8-connected. This completes the proof.

Proposition 14 suggests the following definition.

DEFINITION 3. An a.s. bounded DRS $\mathbf{X}$ (i.e., a DRS $\mathbf{X}$ such that $\Pr[\mathbf{X} \in S] = 1$) is an *a.s. convex-8-connected DRS*, if

$$\Pr[\mathbf{X} \cap (B_1 \cap B_2) = \emptyset, \mathbf{X} \cap B_1 \neq \emptyset, \mathbf{X} \cap B_2 \neq \emptyset] = 0, \tag{49}$$

for every pair (B_1, B_2) of bounded, convex-4-connected sets that are separated by their intersection.

From Definition 3 and (13b) it is now clear that a.s. convexity and connectedness are uniquely described by the functional $Q_{\mathbf{X},n}$ defined by (13), with $n = 2$. Indeed, from (13b), with $n = 2$, and from (49) we have that an a.s. bounded DRS $\mathbf{X}$ is a.s. convex-8-connected if and only if

$$Q_{\mathbf{X},2}(B_1 \cap B_2; B_1, B_2) = 0, \tag{50}$$

for every pair (B_1, B_2) of bounded, convex-4-connected sets that are separated by their intersection. From (18) and (50) we also have that an a.s. bounded DRS $\mathbf{X}$ is a.s. convex-8-connected if and only if

$$T_{\mathbf{X}}(B_1 \cup B_2) + T_{\mathbf{X}}(B_1 \cap B_2) = T_{\mathbf{X}}(B_1) + T_{\mathbf{X}}(B_2) \tag{51a}$$

or, equivalently (see (41) and (51a)),

$$R_{\mathbf{X}}(B_1, B_2) = T_{\mathbf{X}}(B_1 \cap B_2), \tag{51b}$$

for every pair (B_1, B_2) of bounded, convex-4-connected sets that are separated by their intersection (compare also (51a) with (17) and (51b) with (44b)). Therefore the autocorrelation functional $R_{\mathbf{X}}(B_1, B_2)$ can also serve as a measure of convexity for a DRS $\mathbf{X}$.

It is now instructive to briefly discuss the similarities and differences between discrete and continuous a.s. convexity. Condition (51a) is similar to that obtained by Matheron for the continuous case [28]. Our condition is associated with both convexity and connectivity, whereas Matheron's condition is associated only with convexity since in the continuous case connectivity is automatically guaranteed by convexity. The definition of separability of two sets by a third is instrumental in deriving the continuous result. However, it is not clear whether this definition is still valid in the discrete case.

Definition 2 introduces the discrete version of such a concept and allows us to proceed with the technicalities of Propositions 13 and 14 and, finally to obtain (51).

The fact that two different types of connectivity are required by our analysis is surprising. However, a careful consideration of the problem at hand provides us with a good explanation. Recall that our objective here is to derive conditions that guarantee a.s. convexity of the DRS **X**. Since connectivity is also a desirable property [2] and since in the discrete case connectivity is not automatically guaranteed by convexity, we have decided to seek conditions for both convexity and connectivity. However, we should keep in mind that any condition of interest should characterize the largest possible class of convex, connected, discrete sets and it should be in terms of the smallest possible collection of test sets B. Among the classes of convex-4-connected and convex-8-connected sets, the class of convex-8-connected sets is the largest and is assigned to **X**, whereas the class of convex-4-connected sets is the smallest and is assigned to the test set B.

Appendix

Proof of Lemma 1. Observe that $\text{pol}(X_1) \subseteq \text{pol}_1(X)$ since every edge of $\text{pol}(X_1)$ connects a pair of points $(x_1, x_2) \in X_1 \times X_1$, this edge being inside $\text{pol}_1(X)$. In this case $X_1 = \text{pol}_1(X) \cap Z^2 \supseteq \text{pol}(X_1) \cap Z^2 = H(X_1)$, or $X_1 \supseteq H(X_1)$. On the other hand, $X_1 \subseteq H(X_1)$, and therefore $X_1 = H(X_1)$, i.e., X_1 is a convex set in P. A similar proof leads to the convexity of X_2. Since $\text{pol}_1(X) \cap \text{pol}_2(X) = L \cap \text{pol}(X)$, we have that $\text{pol}_1(X) \cap \text{pol}_2(X) \cap Z^2 = L \cap \text{pol}(X) \cap Z^2$, or $(\text{pol}_1(X) \cap Z^2) \cap (\text{pol}_2(X) \cap Z^2) = L \cap \text{pol}(X) \cap Z^2$, or $X_1 \cap X_2 = L \cap H(X) = L \cap X$, since X is assumed to be convex. Also, since $\text{pol}_1(X) \cup \text{pol}_2(X) = \text{pol}(X)$, we have that $(\text{pol}_1(X) \cup \text{pol}_2(X)) \cap Z^2 = \text{pol}(X) \cap Z^2$, or $(\text{pol}_1(X) \cap Z^2) \cup (\text{pol}_2(X) \cap Z^2) = \text{pol}(X) \cap Z^2 = H(X)$, or $X_1 \cup X_2 = X$, since X is assumed to be convex. This completes the proof.

Proof of Lemma 2. (a) Assume first that $L_{b_1 b_2}$ passes through a point $b \in B$. Since $b_1 \in X \cap B_1$ and $b_2 \in X \cap B_2$, we have that $b_1, b_2 \in X$. Since the polytope $\text{pol}(X)$ is convex, we have that $L_{b_1 b_2} \subseteq \text{pol}(X)$, or $b \in L_{b_1 b_2} \cap Z^2 \subseteq \text{pol}(X) \cap Z^2 = H(X) = X$, because X is assumed to be convex; therefore $b \in X$ and $b \in X \cap B$. (b) Assume now that $L_{b_1 b_2}$ hits a 4-adjacency $\alpha^4_{bb'}$ in B that joins two 4-adjacent points b, $b' \in B$. Since $b_1, b_2 \in X$ and since X is convex, the line L that passes through points b_1 and b_2 divides X into two convex sets X_1 and X_2 such that $X_1 \cup X_2 = X$ (see Lemma 1). Since $b_1, b_2 \in X$ and since X is assumed to be 8-connected, there will be an 8-path $\pi_{b_1 b_2} = \{p_1 = b_1, p_2, \ldots, p_n = b_2\}$ in X from point b_1 to point b_2. If $\pi_{b_1 b_2} \cap \{b, b'\} \neq \emptyset$, then $\{b, b'\} \cap X \neq \emptyset$ since $\pi_{b_1 b_2} \subseteq X$. Assume now that $\pi_{b_1 b_2} \cap \{b, b'\} = \emptyset$. The bounded sets $\{b_1\}$ and $\{b_2\}$ are separated by the bounded set $\{b, b'\}$, and in this case $\{b\} \subseteq H_1$ and $\{b'\} \subseteq H_2$, where H_1 and H_2 are the two closed half-planes in R^2 such that $H_1 \cup H_2 = R^2$ and $H_1 \cap H_2 = L$. Assume without loss of generality, that $X_1 \subseteq H_1$ and $X_2 \subseteq H_2$. Consider the closed set Q of all points in R^2 that are between $C(\pi_{b_1 b_2})$ and $L_{b_1 b_2}$, where $C(\pi_{b_1 b_2}) = \pi_{b_1 b_2} \cup \{\alpha_{p_i p_{i+1}}, \textit{for } i = 1, 2, \ldots, n-1\}$. It is now clear that Q can be decomposed as $Q = Q_1 \cup Q_2$, where $Q_1 = Q \cap H_1$ and $Q_2 = Q \cap H_2$. Observe that $Q_1 \subseteq \text{pol}(X_1)$ and $Q_2 \subseteq \text{pol}(X_2)$. If $b \in Q_1$, then $b \in H(Q_1 \cap Z^2)$ since $\{b\} \subseteq Q_1 \cap Z^2 \subseteq H(Q_1 \cap Z^2)$. Since $Q_1 \subseteq \text{pol}(X_1)$ and since X_1 is convex, we have that $H(Q_1 \cap Z^2) \subseteq H(\text{pol}(X_1) \cap Z^2) = H(H(X_1)) = X_1$; therefore $\{b\} \subseteq X_1 \subseteq X$, from which we obtain that $\{b, b'\} \cap X \neq \emptyset$. If $b \notin Q_1$, then $b' \in Q_2$; otherwise, there should be an adjacency $\alpha_{p_i p_{i+1}}$ between points $p_i, p_{i+1} \in \pi_{b_1 b_2}$ such that $\alpha_{p_i p_{i+1}} \cap \alpha^4_{bb'} \neq \emptyset$, which is not possible. Since $b' \in Q_2$ and by following a similar argument as before, we have that $\{b'\} \subseteq X_2 \subseteq X$, from which we again obtain that $\{b, b'\} \cap X \neq \emptyset$. This completes the proof.

Acknowledgments

The author thanks Nikos Sidiropoulos for valuable suggestions and discussions of the connections between discrete random sets and binary random fields and thanks Chuanju Wen fo pointing out that both 4- and 8-connectivity are needed in section 5 and for correcting some of the proofs in the same section.

The support of the U.S. Office of Naval Research, Mathematical Sciences Division, under ONR grant N00014-90-J-1345 is gratefully acknowledged.

References

1. U. Grenander, *Lectures in Pattern Theory*, vols. 1–3, Springer-Verlag: New York, 1976, 1978, 1981.
2. J. Serra, *Image Analysis and Mathematical Morphology*, Academic Press: London, 1982.
3. A.K. Jain, *Fundamentals of Digital Image Processing*, Prentice-Hall: Englewood Cliffs, NJ, 1989.
4. B.D. Ripley, "Modelling spatial patterns (with discussion)," *J. Roy. Statist. Soc. Ser. B*, vol. 39, pp. 172–212, 1977.
5. D. Stoyan, "Applied stochastic geometry: a survey," *Biomed. J.*, vol. 21, pp. 693–715, 1979.
6. R.E. Miles, "A survey of geometrical probability in the plane, with emphasis on stochastic image modelling," *Comput. Graph. Image Process.*, vol. 12, pp. 1–24, 1980.
7. A. Baddeley, "Stochastic geometry: an introduction and reading-list," *Internat. Statist. Rev.*, vol. 50, pp. 179–193, 1982.
8. A.J. Baddeley, "Stochastic geometry and image analysis," *CWI Newsletter*, no. 4, pp. 2–20, 1984.
9. D. Stoyan, W.S. Kendall, and J. Mecke, *Stochastic Geometry and Its Applications*, John Wiley: Berlin, 1987.
10. D.L. Snyder and M.I. Miller, *Random Point Processes in Time and Space*, Springer-Verlag: New York, 1991.
11. N. Cressie, *Statistics of Spatial Data*, John Wiley: New York, 1991.
12. N. Ahuja and A. Rosenfeld, "Mosaic models for texture," *IEEE Trans. Patt. Anal. Mach. Intell.*, vol. 3, pp. 1–10, 1981.
13. N. Ahuja and B.J. Schachter, *Pattern Models*, John Wiley: New York, 1983.
14. J.W. Modestino, R.W. Fries, and A.L. Vickers, "Stochastic image models generated by random tessellations of the plane," *Comput. Graph. Image Process.*, vol. 12, pp. 74–98, 1980.
15. J. Woods, "Two-dimensional discrete Markovian fields," *IEEE Trans., Informat. Theory*, vol. 18, pp. 232–240, 1972.
16. J. Besag, "Spatial interaction and the statistical analysis of lattice systems (with discussion)," *J. Roy. Statist. Soc. Ser. B*, vol. 36, pp. 192–236, 1974.
17. R. Chellappa, "Two-dimensional discrete Gaussian Markov random field models for image processing," in *Progress in Pattern Recognition*, vol. 2, L. Kanal and A. Rosenfeld, eds., North Holland: New York, 1985, pp. 79–112.
18. D. Geman and S. Geman, "Stochastic relaxation, Gibbs distributions, and the Bayesian restoration of images," *IEEE Trans. Patt. Anal. Mach. Intell.*, vol. 6, pp. 721–741, 1984.
19. B.D. Ripley, "Statistics, images, and pattern recognition," *Canad. J. Statist.*, vol. 14, pp. 83–111, 1986.
20. B.D. Ripley, *Statistical Inference for Spatial Processes*. Cambridge University Press: Cambridge, England, 1988.
21. U. Grenander, "The 1985 Rietz Lecture: advances in pattern theory," *Ann. Statist.*, vol. 17, pp. 1–30, 1989.
22. R.C. Dubes and A.K. Jain, "Random field models in image analysis," *J. Appl. Statist.*, vol. 16, pp. 131–164, 1989.
23. Y. Amit, U. Grenander, and M. Piccioni, "Structural image restoration through deformable templates," *J. Am. Statist. Assoc.*, vol. 86, pp. 376–387, 1991.
24. U. Grenander, Y. Chow, and D.M. Keenan, *Hands: A Pattern Theoretic Study of Biological Shapes*, Springer-Verlag: New York, 1991.
25. D.G. Kendall, "A survey of the statistical theory of shape (with comments)," *Statist. Sci.*, vol. 4, pp. 87–120, 1989.
26. I.L. Dryden and K.V. Mardia, "Size and shape analysis of landmark data," *Biometrika*, vol. 79, pp. 57–68, 1992.
27. D.G. Kendall, "Foundations of a theory of random sets," *Stochastic Geometry*, E.F. Harding and D.G. Kendall, eds., John Wiley: London, 1974, pp. 322–376.
28. G. Matheron, *Random Sets and Integral Geometry*, John Wiley: New York, 1975.
29. P.J. Davy, "Aspects of random set theory," *Adv. Appl. Probab.*, vol. 10 (suppl.), pp. 28–35, 1978.
30. J. Serra, "The Boolean model and random sets, " *Comput. Graph. Image Process.*, vol. 12, pp. 99–126, 1980.
31. N. Cressie and G.M. Laslett, "Random set theory and problems of modeling," *SIAM Rev.*, vol. 29, pp. 557–574, 1987.
32. J. Serra (ed.), *Image Analysis and Mathematical Morphology, Vol. 2: Theoretical Advances*, Academic Press: New York, 1988.
33. R.M. Haralick, S.R. Sternberg, and X. Zhuang, "Image analysis using mathematical morphology," *IEEE Trans. Patt. Anal. Mach. Intell.*, vol. 9, pp. 532–550, 1987.
34. P. Maragos and R.W. Schafer, "Morphological systems for multidimensional signal processing," *Proc. IEEE*, vol. 78, pp. 690–710, 1990.
35. E.R. Dougherty, *An Introduction to Morphological Image Processing*, SPIE Optical Engineering Press: Bellingham, WA, 1992.
36. J. Goutsias and C. Wen, "Discrete random set models for shape synthesis and analysis," *Proc. Soc. Photo-Opt. Instrum. Eng.*, vol. 1606, pp. 174–185, 1991.
37. A. Yuille, L. Vincent, and D. Geiger, "Statistical morphology," *Proc. Soc. Photo-Opt. Instrum. Eng.*, vol. 1568, pp. 271–282, 1991.
38. F. Chen and P.A. Kelly, "Algorithms for generating and segmenting morphologically smooth binary images," *Proc. 26th Conf. on Information Sciences and Systems*, Princeton, NJ, March 18–20, 1992, pp. 902–906.
39. A. Yuille, L. Vincent, and D. Geiger, "Statistical morphology and Bayesian reconstruction," *J. Math. Imag. Vis.*, vol. 1, No. 3, 1992.

40. F. Preteux, "Advanced mathematical morphology: from an algebraic to a stochastic point of view," *Proc. Soc. Photo-Opt. Instrum. Eng.*, vol. 1769, pp. 44–58, 1992.
41. G.S. Watson, "Mathematical morphology," in *A Survey of Statistical Design and Linear Models*, J.N. Srivastava, ed., North Holland: New York, 1975, pp. 547–553.
42. A. Rosenfeld, "The fuzzy geometry of image subsets," *Patt. Recog. Lett.*, vol. 2, pp. 311–317, 1984.
43. D. Sinha and E.R. Dougherty, "Fuzzy mathematical morphology," *J. Vis. Commun. Image Rep.*, vol. 3, pp. 286–302, 1992.
44. G. Choquet, "Theory of capacities," *Ann. Inst. Fourier*, vol. 5, pp. 131–295, 1953–1954.
45. G. Matheron, "Random sets theory and its applications to stereology," *J. Micros.*, vol. 95, pp. 15–23, 1972.
46. B.D. Ripley, *Spatial Statistics*, John Wiley: New York, 1981.
47. D. Stoyan, "Stereology and stochastic geometry," *Internat. Statist. Rev.*, vol. 58, pp. 227–242, 1990.
48. D. Schonfeld and J. Goutsias, "Optimal morphological pattern restoration from noisy binary images," *IEEE Trans. Patt. Anal. Mach. Intell.*, vol. 13, pp. 14–29, 1991.
49. D. Schonfeld and J. Goutsias, "On the morphological representation of binary images in a noisy environment," *J. Vis. Commun. Image Rep.*, vol. 2, pp. 17–30, 1991.
50. N.D. Sidiropoulos, "Statistical inference, filtering, and modeling of discrete random sets," Ph.D. thesis, Department of Electrical Engineering, University of Maryland, College Park, 1992.
51. G.S. Watson, "Texture analysis," *Geol. Soc. Am. Bull.*, vol. 142, pp. 367–391, 1975.
52. N. Cressie and F.L. Hulting, "A special statistical analysis of tumor growth," *J. Am. Statist. Assoc.*, vol. 87, pp. 279–283, 1992.
53. P.J. Diggle, "Binary mosaics and the spatial pattern of heather," *Biometrics*, vol. 37, pp. 531–539, 1981.
54. D. Jeulin, "Multi-component random models for the description of complex microstructures," *Mikroskopie*, vol. 37, pp. 130–137, 1980.
55. D. Jeulin, "Random structure analysis and modelling by mathematical morphology," *Proc. 5th Internat. Symp. Continuum Models of Discrete Systems*, Nottingham, July 14–20, 1985, pp. 217–226.
56. D. Jeulin, "Mathematical morphology and materials image analysis," *Scanning Micros.*, vol. 2 (suppl.), pp. 165–183, 1988.
57. J. Masounave, A.L. Rollin, and R. Denis, "Prediction of permeability of non-woven geotextiles from morphometry analysis," *J. Micros.*, vol. 121, pp. 99–110, 1981.
58. D. Jeulin, "Some aspects of mathematical morphology for physical applications," *Physica A*, vol. 157, pp. 13–20, 1989.
59. D. Jeulin and P. Jeulin, "Synthesis of rough surfaces by random morphological models," *Stereol. Iugosl.*, vol. 3 (suppl. 1), pp. 239–246, 1981.
60. J. Serra, "Boolean random functions," *J. Micros.*, vol. 156, pp. 41–63, 1988.
61. D. Jeulin, "Morphological modeling of images by sequential random functions," *Signal Process.*, vol. 16, pp. 403–431, 1989.
62. J. Serra, "Boolean random functions," in *Image Analysis and Mathematical Morphology, Vol. 2: Theoretical Advances*, J. Serra, ed., Academic Press: New York, 1988, pp. 317–342.
63. F. Preteux and M. Schmitt, "Boolean texture analysis and synthesis," *Image Analysis and Mathematical Morphology, Vol. 2: Theoretical Advances*, J Serra, ed., Academic Press: New York, 1988, pp. 377–400.
64. R.M. Haralick, X. Zhuang, C. Lin, and J.S.J. Lee, "The digital morphological sampling theorem," *IEEE Trans. Acoust., Speech, Signal Process.*, vol. 37, pp. 2067–2090, 1989.
65. G. Bernroider, "The foundation of computational geometry: theory and application of the point-lattice-concept within modern structure analysis," in *Lecture Notes in Biomathematics*, vol. 23, R.E. Miles and J. Serra, eds., Springer-Verlag: Berlin, 1978, pp. 153–170.
66. H.J.A.M. Heijmans and A. Toet, "Morphological sampling," *Comput. Vis. Graph. Image Process.*, Image Understanding, vol. 54, pp. 384–400, 1991.
67. B.D. Ripley, "Locally finite random sets: foundations for point process theory," *Ann. Probab.*, vol. 4, pp. 983–994, 1976.
68. B.D. Ripley, "The foundations of stochastic geometry," *Ann. Probab.*, vol. 4, pp. 995–998, 1976.
69. B.D. Ripley, "On stationarity and superposition of point processes," *Ann. Probab.*, vol. 4, pp. 999–1005, 1976.
70. M. Baudin, "Multidimensional point processes and random closed sets," *J. Appl. Probab.*, vol. 21, pp. 173–178, 1984.
71. R.J. Adler, *The Geometry of Random Fields*, John Wiley: New York, 1981.
72. P. Billingsley, *Probability and Measure*, 2nd ed., John Wiley: New York, 1986.
73. G. Birkhoff, *Lattice Theory*, 3rd ed. American Mathematical Society: Providence, RI, 1984.
74. H.J.A.M. Heijmans and C. Ronse, "The algebraic basis of mathematical morphology. I. Dilations and erosions," *Computer Vis. Graph. Image Process.*, vol. 50, pp. 245–295, 1990.
75. J. Goutsias, "Modeling random shapes: an introduction to random set theory," to appear in *Mathematical Morphology: Theory and Hardware*, R.M. Haralick, ed., Oxford University Press: New York, 1993.
76. J. Goutsias, "Mutually compatible Gibbs random fields," *IEEE Trans. Informat. Theory*, vol. 35, pp. 1233–1249, 1989.
77. J. Goutsias, "Unilateral approximation of Gibbs random field images," *Comput. Vis. Graph. Image Process.*, Graphical Models and Image Processing, vol. 53, pp. 240–257, 1991.
78. D.K. Pickard, "A curious binary lattice process," *J. Appl. Probab.*, vol. 14, pp. 717–731, 1977.
79. D.K. Pickard, "Unilateral Ising models," *Adv. Appl. Probab.*, vol. 10 (suppl.), pp. 58–64, 1978.
80. D.K. Pickard, "Unilateral Markov fields," *Adv. Appl. Probab.*, vol. 12, pp. 655–671, 1980.

81. C.J. Geyer and E.A. Thompson, "Constrained Monte Carlo maximum likelihood for dependent data (with discussion)," *J. Roy. Statist. Soc. Ser. B*, vol. 54, pp. 657–699, 1992.
82. M. Potamianos and J. Goutsias, "Partition function estimation of Gibbs random field images using Monte Carlo simulations," 1993, to appear, *IEEE Trans. Informat. Theory.*
83. K.S. Shanmugan and A.M. Breipohl, *Random Signals: Detection, Estimation, and Data Analysis*, John Wiley: New York, 1988.
84. R.L. Stevenson and G.R. Arce, "Morphological filters: statistics and further syntactic properties," *IEEE Trans. Circuits Syst.*, vol. 34, pp. 1292–1305, 1987.
85. D.L.B. Jupp, A.H. Strahler, and C.E. Woodcock, "Autocorrelation and regularization in digital images I. Basic theory," *IEEE Trans. Geosci. Remote Sensing*, vol. 26, pp. 463–473, 1988.
86. D.L.B. Jupp, A.H. Strahler, and C.E. Woodcock, "Autocorrelation and regularization in digital images II. Simple image models," *IEEE Trans. Geosci. Remote Sensing,* vol. 27, pp. 247–258, 1989.
87. T.Y. Kong and A. Rosenfeld, "Digital topology: introduction and survey," *Comput. Vis. Graph. Image Process.*, vol. 48, pp. 357–393, 1989.

John Goutsias received the Diploma in electrical engineering in 1981 from the National Technical University of Athens, Greece, and the M.S. and Ph.D. degrees in 1983 and 1986, respectively, in electrical engineering from the University of Southern California.

Dr. Goutsias is an Associate Professor with the Department of Electrical Engineering and Computer Engineering of the Johns Hopkins University at Baltimore, MD, where he has been since 1986. His current research interests include one-dimensional and multi-dimensional digital signal processing, and image modeling, processing, and analysis.

Dr. Goutsias is an associate Editor of the IEEE Transactions on Signal Processing, and the Journal of Visual Communication and Image Representation, a technical committee member of the 1992 and 1993 SPIE's Symposia on Visual Communications and Image Processing, a co-chairman of the 1993 SPIE Conference on Image Algebra and Morphological Image Processing, a member of the IEEE Signal Processing Society, the IEEE Information Theory Society, Technical Chamber of Greece, and Eta Kappa Nu, and a registered Professional Electronics Engineer in Greece.

Journal of Mathematical Imaging and Vision 2, 217–232 (1992).

Inverse Problems for Granulometries by Erosion

JULIETTE MATTIOLI AND MICHEL SCHMITT
Thomson-CSF, Laboratoire Central de Recherches, Domaine de Corbeville, 91404 Orsay Cedex, France

Abstract. Let us associate to any binary planar shape X the erosion curve Ψ_X defined by $\Psi_X : r \in \mathbb{R}^+ \to A(X \ominus rB)$, where $A(X)$ stands for the surface area of X and $X \ominus rB$ for the eroded set of X with respect to the ball rB of size r. Note the analogy to shape quantification by granulometry. This paper describes the relationship between sets X and Y verifying $\Psi_X = \Psi_Y$. Under some regularity conditions on X, Ψ_X is expressed as an integral on its skeleton of the quench function q_X (distance to the boundary of X). We first prove that a bending of arcs of the skeleton of X does not affect Ψ_X: Ψ_X quantifies *soft* shapes. We then prove, in the generic case, that the five possible cases of behavior of the second derivative Ψ''_X characterize five different situations on the skeleton $\mathrm{Sk}(X)$ and on the quench function q_X: simple points of $\mathrm{Sk}(X)$ where q_X is a local minimum, a local maximum, or neither, multiple points of $\mathrm{Sk}(X)$ where q_X is a local maximum or not. Finally, we give infinitesimal generators of the reconstruction process for the entire family of shapes Y verifying $\Psi_X = \Psi_Y$ for a given X.

Key words. mathematical morphology, erosion curve, skeleton, quench function, granulometry.

1 Introduction

Shape description is a very important problem in pattern analysis. It provides descriptions for objects according to their shape, which can be used for pattern recognition. The principle [1]–[3] is to synthesize the information contained in a shape into a curve, called the *erosion curve*, which is translation and rotation invariant and which gives a global information of the shape.

This paper deals with the following questions: (i) Having an erosion curve, what can we say about the original shape? How can one reconstruct a shape solely from the knowledge of the erosion curve? (ii) What information has been lost during the computation of the erosion curve?

The plan of the paper is as follows: After recalling some basic notions of mathematical morphology, we give properties of the skeleton of a compact planar shape and its links with erosions by disks. Then we study the erosion curve Ψ_X and show that its second derivative gives information about shape and characterizes classes of shape. Finally, we present a method for building a shape solely from knowledge of Ψ_X.

2 Notions of Mathematical Morphology

Morphological shape analysis uses the idea of a Boolean operator to compare an arbitrary reference shape called the *structuring element* and the image.

If we consider an isotropic structuring element, a ball of radius r centered at the origin, denoted by rB, the *eroded set* of X with respect to rB is given by $X \ominus rB = \{x, rB_x \subset X\}$, where B_x stands for the translation of B at point x. According to the usual duality principle with respect to the complementation, the *dilation* is expressed by $X \oplus rB = \{x, rB_x \cap X \neq \emptyset\}$, and we have $(X^c \ominus rB)^c = X \oplus rB$. The *opening* of X by B is the domain swept out by all translates of B that are included in X: $X_B = \cup\{B_x, B_x \subset X\}$, and the *closing* of X by rB, $X^{rB} = ((X^c)_{rB})^c$, is the set of points z such that any translate of rB containing z hits X.

For building the erosion curve (respectively, the opening curve), we compute for each $r > 0$ the area $\phi_r(X)$, where (ϕ_r) is a family of morphological transformations by structuring elements of increasing size. For example, if $\phi_r(X) = X \ominus rB$, we have $\Psi_X(r) = A(X \ominus rB)$

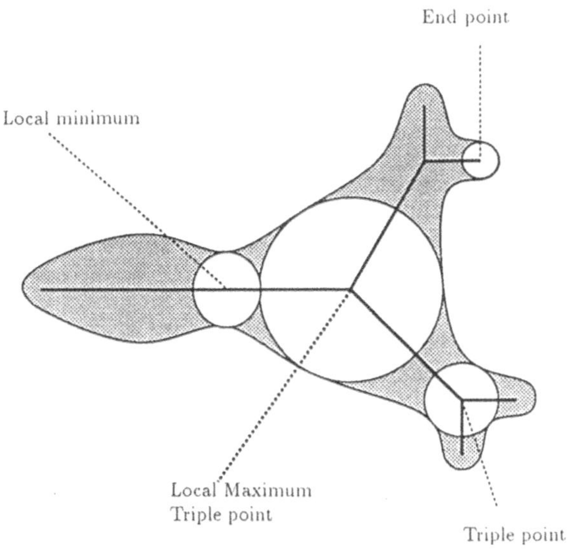

Fig. 1. Skeleton of a planar compact shape.

with $r \geq 0$, where $A(X)$ stands for the surface area of X. It is obvious that this curve is not sensitive to a translation and rotation of the shape.

3 Properties of the Skeleton

Because the properties of the function $\Psi_X : r \rightarrow A(X \ominus rB)$, called the *erosion curve*, are based on the skeleton of X, we first recall its definitions and properties and its links with erosions by disks.

The *skeleton* Sk(X) of a compact planar set X is the locus of the maximal inscribable closed balls in X [4] (see figure 1).

If we denote by $s_r(X)$ the set of centers of the maximal balls of radius $r > 0$, then $\mathrm{Sk}(X) = \cup_{r>0} s_r(X)$ and the object reconstruction is given by

$$X = \cup_{r>0} s_r(X) \oplus rB. \tag{1}$$

The datum of the set X is equivalent to that of its skeleton together with the maximal radius r associated with each point of Sk(X) [5], [6]. This maximal-radius function, called the *quench function* and denoted by q_X, is defined on the skeleton by $\forall x \in \mathrm{Sk}(X)$, $q_X(x) = d(x, X^c)$, where d is the Euclidean distance and X^c is the complement of X.

The expression for the skeleton of $X \ominus r_0 B$ is

$$\begin{aligned} \mathrm{Sk}(X \ominus r_0 B) &= \cup_{r>r_0} s_r(X), \\ X \ominus r_0 B &= \cup_{r>r_0} s_r(X) \oplus (r - r_0)B, \\ q_{X \ominus r_0 B} &= q_X - r_0. \end{aligned} \tag{2}$$

In other words, the skeleton of the eroded set $X \ominus r_0 B$ is composed of points x of the skeleton of X where $q_X(x) > r_0$.

Note that, unfortunately, we do not have a similar formula for the skeleton of the opened set $X_{r_0 B}$, contrary to [5, p. 377], because $\mathrm{Sk}(X_{r_0 B}) \not\subset \mathrm{Sk}(X)$ in general. We have only the reconstruction formula

$$X_{r_0 B} = \cup_{r>r_0} s_r(X) \oplus rB. \tag{3}$$

The skeleton of X is not necessarily a finite graph [6], [7]. Nevertheless, if the boundary ∂X of a compact planar set X is a finite union of C^3 arc loops, then Sk(X) is a connected finite graph with simple arcs [8], [9].

Thus if the boundary of X is a finite union of C^3 arc loops, the skeleton Sk(X) of X has a finite number of end points and multiple points. We consider each edge of Sk(X) as a curve $\gamma(s)$, where s is a curvilinear abscissa. We finally define on each edge the function q by $q(s) = q_X(\gamma(s)) = d(\gamma(s), X^c)$.

We define the *downstream* (respectively, *upstream*) of a point $x \in X$, denoted by Av(x) (respectively, Am(x)), to be the set of $y \in X$ that satisfy the relation [6] $d(y, X^c) = d(x, X^c) - d(x, y)$ (respectively, $d(y, X^c) = d(x, X^c) + d(x, y)$).

If $x \in \mathrm{Sk}(X)$, then its upstream is reduced to the point x itself and, conversely,

$$x \in \mathrm{Sk}(X) \Leftrightarrow \mathrm{Am}(x) = \{x\}.$$

We define the *edge* of x, denoted by Ar(x), to be the union of the upstream and the downstream of x : $\mathrm{Ar}(x) = \mathrm{Am}(x) \cup \mathrm{Av}(x)$.

Remark 3.1. Two nonidentical edges are either disjoint or cross at one point $x \in \mathrm{Sk}(X)$, which must be their common upstream extremity.

4 Study of the Function $\Psi_X : r \in \mathbf{R}^+ \to A(X \ominus rB)$

Let Γ be a finite union of C^3 arc loops. Let X be the connected compact component defined by $X = \Gamma_{\text{int}} \cup \Gamma$ where Γ_{int} is the interior of Γ (Jordan's theorem [10]). By hypothesis, X has no holes. The boundary of X is $\partial X = \Gamma$. We say that a point m of Γ is a *critical point of the curvature* if there exists an opened neighborhood $\mathcal{V}(m)$ of m such that the curvature is constant and strictly positive on $\mathcal{V}(m) \cap \Gamma$.

The right framework for our theorem is the following: We suppose that $\forall r \in \mathbb{R}^+$ the number of connected components of $\mathrm{Sk}(X \ominus rB)$ is finite and that the boundary ∂X of X is a finite union of C^3 arcs loop without a critical point. This class of shapes is a very wide one that includes polygons and avoids pathological cases. Under this assumption ∂X is parametrizable by g and each point $p \in \partial X$ is a contact point between a unique maximal inscribable circle of center m. The mapping σ that associates at each point $p \in \partial X$ the point m is a retract of X, and $\sigma \circ g$ is a parametrization of the skeleton. There exists a diffeomorphism λ such that $f = \lambda \circ \sigma \circ g$ is a curvilinear parametrization of $\mathrm{Sk}(X)$ such that an end point (respectively, a triple point, ...) has one (respectively, three, ...) curvilinear abscissa, and a point that is neither an end point nor a multiple point has only two curvilinear abscissas. In fact, the curvilinear abscissa goes over the skeleton by following the boundary parametrization (see figure 2).

We define by $\mathrm{Ar}(s)$ the upstream of p, i.e. $\mathrm{Ar}(s) = [m, p]$.

PROPOSITION 4.1. For all s such that s is a curvilinear abscissa of a point m in the skeleton $\mathrm{Sk}(X)$ that is not an end point or a multiple point, the function $q : s \to d(m, X^c)$ is differentiable and we have

$$\frac{dq}{ds}(s) = -\sin\theta(s), \tag{4}$$

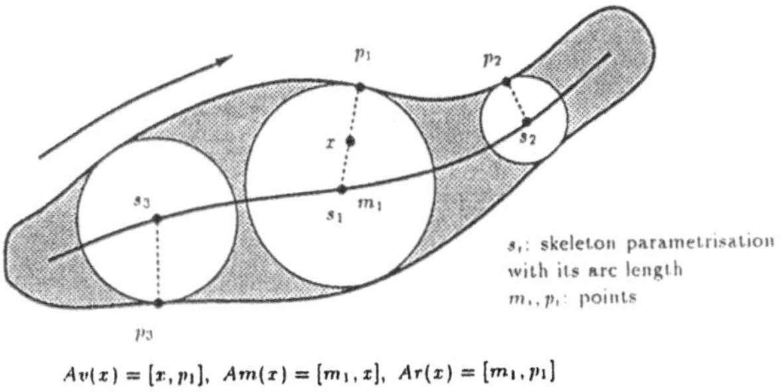

Fig. 2. The curvilinear abscissa over the skeleton by following the boundary parametrization.

where $\theta(s)$ is the angle between the normal on s at $\mathrm{Sk}(X)$ and $\mathrm{Ar}(s)$.

The proof of this proposition and other propositions and theorems are given in the appendix.

PROPOSITION 4.2. If s is the curvilinear abscissa of a point $m \in \mathrm{Sk}(X)$ that is an end point (but not critical), then the function q is right and left differentiable and

$$\left(\frac{dq}{ds}\right)_r(s) = -\left(\frac{dq}{ds}\right)_l(s) = -1, \tag{5}$$

with

$$\left(\frac{dq}{ds}\right)_r(s) = \lim_{\varepsilon \to 0^+} \frac{q(s+\varepsilon) - q(s)}{\varepsilon},$$
$$\left(\frac{dq}{ds}\right)_l(s) = \lim_{\varepsilon \to 0^-} \frac{q(s+\varepsilon) - q(s)}{\varepsilon},$$
$$\theta^+(s) = -\theta^-(s) = \frac{\pi}{2}.$$

If $m \in \mathrm{Sk}(X)$ is a multiple point, let $(s_i)_{i=1}^n$ be the curvilinear abscissas of m; then the function q is right and left differentiable at each s_i and

$$\left(\frac{dq}{ds}(s_i)\right)_r = -\sin\theta^+(s_i),$$
$$\left(\frac{dq}{ds}(s_i)\right)_l = -\sin\theta^-(s_i), \tag{6}$$

with $\sum_{i=1}^n (\theta^+(s_i) - \theta^-(s_i)) = \pi$.

In fact, q is Lipschitz and twice differentiable except for s such that $\gamma(s)$ is an end point or a multiple point of $\mathrm{Sk}(X)$ (see figure 3).

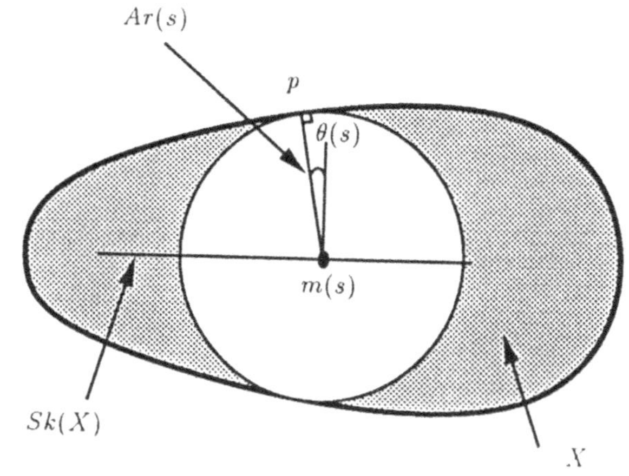

Fig. 3. The quench function q is differentiable at s.

Using Remark 3.1, Proposition 4.1, and the parametrization of the skeleton $\mathrm{Sk}(X)$ and of the boundary ∂X of X, we are able to express the perimeter and the surface area of X and of $X \ominus rB$ for all $r < R$, where $R = \max\{r, rB \subset X\}$ is the size of the greatest disk contained in X, as an integral on $\mathrm{Sk}(X)$ or on $\mathrm{Sk}(X \ominus rB)$ of a function of q and of its first and second derivative.

PROPOSITION 4.3. Under the same assumptions, expressions of the perimeter and the surface area of X and of $X \ominus rB$ are given by

$$P(X) = \int_{\mathrm{Sk}(X)} \frac{1 - q(s)q''(s) - q'^2(s)}{\sqrt{1 - q'^2(s)}} ds, \tag{7}$$

$$P(X \ominus rB) = \int_{\mathrm{Sk}(X\ominus rB)} \frac{1 - q(s)q''(s) - q'^2(s) + rq''(s)}{\sqrt{1 - q'^2(s)}} ds, \tag{8}$$

$$A(X) = \int_{\mathrm{Sk}(X)} \frac{q(s)}{2\cos\theta(s)} \left[2\cos^2\theta(s) - q(s)q''(s)\right] ds, \tag{9}$$

$$A(X \ominus rB) = \int_{\mathrm{Sk}(X\ominus rB)} \frac{q(s) - r}{2\cos\theta(s)} \left[2\cos^2\theta(s) - q(s)q''(s) + rq''(s)\right] ds. \tag{10}$$

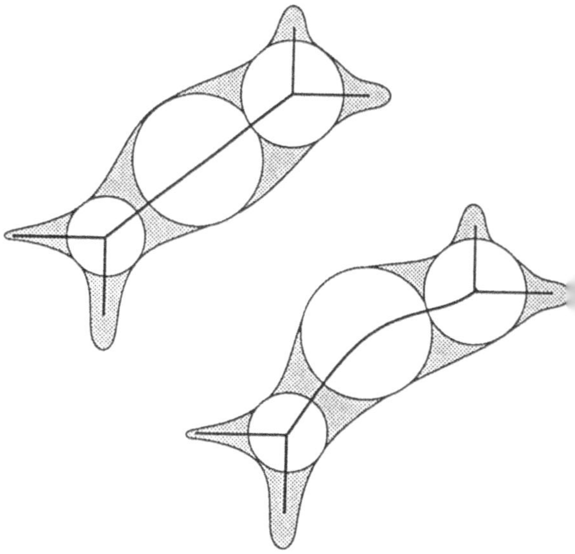

Fig. 4. Two shapes having the same erosion curve.

The first result is that all these integrals depend only on the curvilinear abscissa s of the skeleton and not on the actual skeleton shape.

PROPOSITION 4.4. Two shapes having the same topology of skeleton, the same edge length, and the same q function have the same erosion curve (see figure 4).

This result may be expressed in as follows: The erosion curve quantifies *soft* shapes because small bending of the skeleton does not change the erosion curve. But note that two shapes having the same erosion curve do not necessarily have the same topology of the skeleton.

Let us study the erosion curve in more detail.

THEOREM 4.5. The function $\Psi_X : r \in \mathbb{R}^+ \to A(X \ominus rB)$ is continuous, decreasing, and differentiable for all $r \in]0, R[$ with $R = \max\{r, rB \subset X\}$, and we have

$$\Psi'_X(r) = -P(X \ominus rB), \tag{11}$$

where $P(\cdot)$ represents the perimeter measure [11].

We recall that X is defined by $\Gamma_{\text{int}} \cup \Gamma$, where Γ is a finite union of C^3 arc loops without a critical point. This hypothesis means that each point of $\text{Sk}(X)$ is the center of a maximal inscribable circle with a finite number of contact boundary points (i.e., an end point (respectively, a triple point, ...) is the center of a maximal inscribable circle with one (respectively, three, ...) contact boundary points). If we assume only that Γ is a finite union of C^3 arc loops, then equation (11) becomes

$$\lim_{\varepsilon \to 0^+} \frac{\Psi_X(r+\varepsilon) - \Psi_X(r)}{\varepsilon} = -P(\overline{\overset{\circ}{X}} \ominus (r+\varepsilon)B),$$
$$\lim_{\varepsilon \to 0^-} \frac{\Psi_X(r+\varepsilon) - \Psi_X(r)}{\varepsilon} = -P(X \ominus (r+\varepsilon)B).$$

Remark 4.6. In the convex case we have Miles's formulas [12]. For all compact convex sets $X \subset \mathbb{R}^2$ such that $X = X_B$ and for all $r \in [0,1]$

$$\begin{cases} P(X \ominus rB) = P(X) - 2\pi r, \\ A(X \ominus rB) = A(X) - P(X)r + \pi r^2. \end{cases} \quad (12)$$

We examine what happens during an infinitesimal erosion of $X \ominus rB$ by εB for $\varepsilon > 0$. Five cases are possible, and they are summarized in Theorem 4.7 below. If they occur at different sizes of erosions, these five cases can be distinguished on the erosion curve.

The erosion curve's second derivative is not continuous everywhere; the right and the left second derivatives are

$$\begin{cases} \Psi_l''(r) = \lim_{\varepsilon \to 0^+} \frac{1}{\varepsilon}[\Psi'(r+\varepsilon) - \Psi'(r)], \\ \Psi_r''(r) = \lim_{\varepsilon \to 0^-} \frac{1}{\varepsilon}[\Psi'(r+\varepsilon) - \Psi'(r)]. \end{cases}$$

THEOREM 4.7.

(i) If $\Psi''_{X_r}(r) = \Psi''_{X_l}(r) < +\infty$, the erosion by rB is simple.
(ii) If $\Psi''_{X_l}(r) < \Psi''_{X_r}(r) = +\infty$, there is a disconnection of $X \ominus rB$ after an infinitesimal erosion.
(iii) If $\Psi''_{X_r}(r) < \Psi''_{X_l}(r) < +\infty$, there is a vanishing of a connected component of $X \ominus rB$ containing a multiple point of $\text{Sk}(X)$ that is a local maximum of the quench function q_X after an infinitesimal erosion.
(iv) If $\Psi_{X_r''}(r) < \Psi_{X_l''}(r) = +\infty$, there is a vanishing of a connected component of $X \ominus rB$ not containing a multiple point of $\text{Sk}(X)$ after an infinitesimal erosion.
(v) If $\Psi_{X_l''}(r) < \Psi_{X_r''}(r) < +\infty$, there is a simple vanishing of a multiple point of $\text{Sk}(X)$ after an infinitesimal erosion.

These results are summarized in table 1.

When $\Psi''_X(r)$ exits, let n be the number of end points $(m_i)_{i=1}^n$ of $\text{Sk}(X \ominus rB)$ and let $(\alpha_i)_{i=1}^n$ be the curvilinear abscissa of m_i; then we have

$$\Psi''_X(r) = 2\sum_{i=1}^{n} \cot\theta(\alpha_i) - \int_{\text{Sk}(X\ominus rB)} \frac{q''(s)}{\cos\theta(s)} ds. \quad (13)$$

If X is a polygon, then $\int_{\text{Sk}(X\ominus rB)} q''(s) / \cos\theta(s) ds = 0$ and n is the number of vertices of X.

Note that if two cases occur at the same r, nothing can be said, and the equivalences are only implications.

Note that the order of multiplicity of the multiple points cannot be computed. For example, in the convex-polygon case [11], [13] it can be shown that the number of edges cannot be computed from the erosion curve. This is stated more precisely by the following proposition.

PROPOSITION 4.8. Given an erosion curve Ψ_X of a convex polygon inscribed in a circle, there exists $n_0 > 0$ such that for each $n > n_0$ a convex polygon with n sides can be constructed that has Ψ_X as erosion curve.

In fact, the only information contained in the granulometric curve is $\sum_{i=1}^n \tan\delta_i/2$, where δ_i is the angle of the polygon at vertex i. This proposition is only a particular case of what we will see in the following.

5 Principle of Reconstruction

We now tackle the problem of constructing shapes from the knowledge of the erosion curves Ψ_X. We will keep the same conditions as above:

Fundamental theorem			
Infinitesimal erosion	**Erosion curve**	$X\ominus(r-\varepsilon)B \Leftrightarrow X\ominus(r+\varepsilon)B$	**Axes of freedom for reconstruction**
1 The erosion is simple: The number of end points, multiple points and of connected components of $Sk(X \ominus rB)$ and of $Sk(X \ominus (r+\epsilon)B)$ are equal.	$\Psi_{X_r}''(r) = \Psi_{X_l}''(r) < +\infty$		Lengths of arcs of the skeleton
2 There is a disconnection: The number of connected components of $X \ominus rB$ is greater than the $X \ominus (r+\epsilon)B$ one.	$\Psi_{X_l}''(r) < \Psi_{X_r}''(r) = +\infty$		The choice of arbitrary connected components
3 There is a vanishing of a connected component of $X \ominus rB$ containing a multiple point of $Sk(X)$ which is a local maximum of the quench function q_X.	$\Psi_{X_r}''(r) < \Psi_{X_l}''(r) < +\infty$	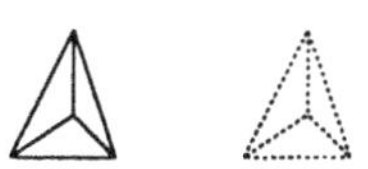	The order and the geometry of the multiple point
4 There is a vanishing of a connected component of $X \ominus rB$ not containing a multiple point of $Sk(X)$.	$\Psi_{X_r}''(r) < \Psi_{X_l}''(r) = +\infty$		The choice of edges of the skeleton which will be connected
5 There is a simple vanishing of a multiple point of $Sk(X)$	$\Psi_{X_l}''(r) < \Psi_{X_r}''(r) < +\infty$	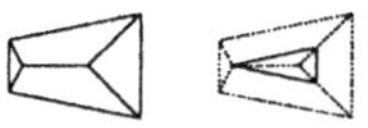	The order and the geographical place of the multiple point

Table 1. Summary of Theorem 4.7.

the skeleton of X will be a finite union of line segments whose multiple points are only triple points. We will read the curve with decreasing r from R to 0. At each step we will point out the degrees of freedom in the reconstruction process. Note that the reconstruction process gives us all shapes having the same erosion curve.

Suppose that for a given $r \in]0, R]$ the set $X \ominus rB$ has already been constructed. The skeleton $\mathrm{Sk}(X \ominus rB)$ and the function $q_r : s \in \mathrm{Sk}(X \ominus rB) \to q(s) - r$ are also known. We propose a method for building a set Y, having the same erosion curve, such that there exists $\varepsilon > 0$ independent of r, $Y \ominus \varepsilon B = X \ominus rB$. ($\varepsilon$ is smaller than the difference of two successive values of r where Ψ_X'' discontinuous.)

- If r is not a discontinuity of Ψ_X'' (case (i)), then at each extremity α_i of $\mathrm{Sk}(X \ominus rB)$ we extend the line segment of length equal to $\eta_i = \varepsilon / q_r'(\alpha_i)$. The values of $q_{r-\varepsilon}'(\alpha_i + \eta_i)$ are constrained only by one equation involving $\Psi_X''(r-\varepsilon)$ (equation (13)). This infinity of solutions gives the first degrees of freedom in the reconstruction process. We first draw parallel exterior arcs of the boundary of $X \ominus rB$ at distance ε and then add tips at every extremity such that the total surface area of these tips is equal to a suitable constant introduced by $\Psi_X(r)$ (see figure 5).
- If r is a discontinuity of Ψ_X'', then we proceed as follows:
  - If $\Psi_{X_r}''(r) - \Psi_{X_l}''(r) > 0$ (case (v)), we arbitrarily choose an end point s_0 of $\mathrm{Sk}(X \ominus rB)$ that we transform into a triple point (see figure 6). If δ_0, δ_1 are the angles between the new line segments at s_0 and the old

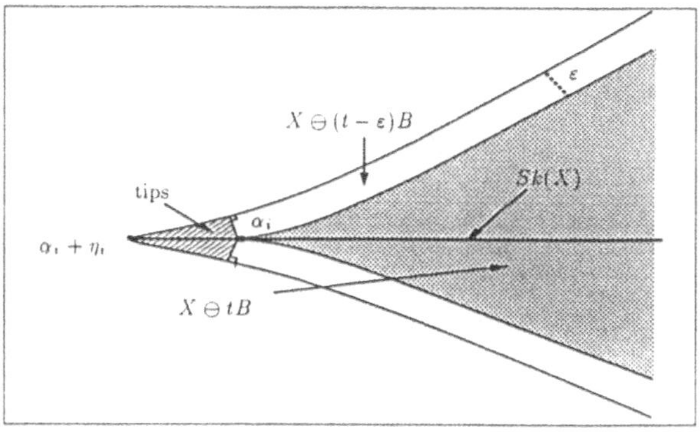

Fig. 5. Reconstruction of $X \ominus (t-\varepsilon)B$ from $X \ominus rB$ when r is not a discontinuity of Ψ_X''.

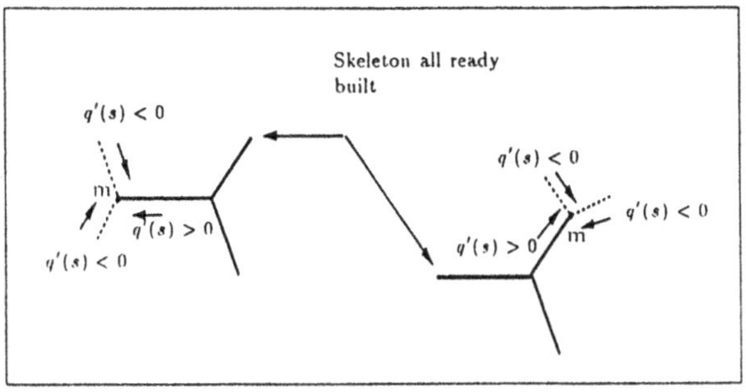

Fig. 6. Arbitrarily chosen end-point s_0 of $\mathrm{Sk}(X \ominus rB)$ transformed into a triple point.

line segment, we have

$$\begin{cases} \tan(\delta_0+\delta_1) - \tan\delta_0 - \tan\delta_1 \\ \qquad = -\dfrac{1}{2}[\Psi_{X_r''}(r) - \Psi_{X_l''}(r)], \\ \delta_0 + \delta_1 \\ \qquad = -\arcsin(q'(s_0)). \end{cases} \quad (14)$$

We extend the skeleton of $\eta_i = \varepsilon / q_r'(\alpha_i)$ for all other extremities $\alpha_i \neq s_0$ and of, respectively, $\varepsilon/\sin\delta_0$ and $\varepsilon/\sin\delta_1$ for the new branch of the skeleton. In general, system (14) has two symmetrical solutions in δ_0, δ_1. We implicitly infer $q_{r-\varepsilon}'(\alpha_i + \eta_i)$ of the value $\Psi_X''(r-\varepsilon)$ and draw in the same way as we drew the boundary of $X \ominus (r-\varepsilon)B$.

If $\Psi_{X_r}''(r) - \Psi_{X_l}''(r) < 0$ (case (iii)), a new connected component is created, the skeleton of which has a triple point. This triple point is a local maximum of q_X, and if δ_0, δ_1, δ_2 are angles between each pair of the three edges of this new piece of $\mathrm{Sk}(X)$, we have

$$\sum_{i=0}^{2} \tan\delta_i = \frac{1}{2}[\Psi_{X_r''}(r) - \Psi_{X_l''}(r)]$$

$$\text{with } \sum_{i=0}^{2} \delta_i = 2\pi. \quad (15)$$

This equation has, in general, an infinity of solutions with one degree of freedom, for example, δ_0.

- If $\Psi_{X_r}''(r) = +\infty$ (case (ii)), we have a narrow part on X. We reconnect two arbitrary different components of $\mathrm{Sk}(X \ominus rB)$, i.e., the union of two end points belonging to two different components.
- If $\Psi_{X_l}''(r) = +\infty$ (case (iv)), we add a new connected component, the skeleton of which is an edge with a local maximum of q_X on it.

Note that at each step the curvature of the skeleton is not imposed from the erosion curve.

6 Conclusion

The study of $\Psi_X : r \in \mathbb{R}^+ \to A(X \ominus rB)$ has shown Ψ_X is continuous and differentiable and that the discontinuities of its second derivative give us characteristic shape information. There are three types of discontinuities, each characteristic of an unambiguous event. For a given r

(a) If $\Psi_{X_r}''(r) = +\infty$, then the shape X under study has a narrow part (case (ii)).
(b) If $\Psi_{X_l}''(r) = +\infty$, then X has a swell (case (iv)).
(c) If $\Psi_{X_l}''(r)$ and $\Psi_{X_r}''(r)$ exist but are different, then there is a multiple point on the skeleton of X (Cases (iii) and (v)).

Nevertheless, it is impossible to know the order of this multiple point.

If we suppose that the five cases of table 1 could not happen at the same time, then the number of narrow parts and swells on the shape

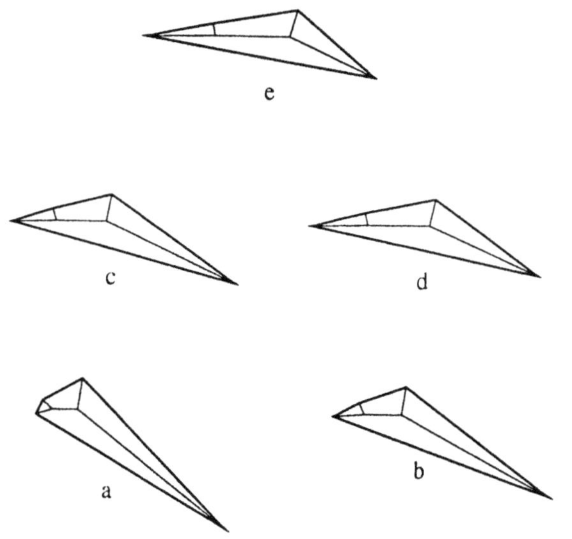

Fig. 7. Examples of polygons having the same function Ψ_X. We have then only one degree of freedom in the reconstruction process, the distance between the two triple points PQ. (a) $PQ = 2$, (b)$PQ = 3$, (c)$PQ = 4$, (d)$PQ = 5$, (e)$PQ = 6$.

and the number of multiple points in the skeleton are contained in the erosion curve, or else we have only a lower bound of these numbers. Besides, the polygonal feature is not measurable from the erosion curve because bending the skeleton does not change the erosion curve, and if we are sure that the shape is a polygon, we have only a lower bound of the number of its vertices (see figure 7).

We have solved a nontrivial inverse problem of mathematical morphology: to reconstruct shapes solely from the knowledge of the erosion curve.

Concerning the opening curve (i.e., $\Phi : r \rightarrow A(X_{rB})$), the first results are very similar, but the study of the opening curve together with the closing curve, i.e.,

$$\Omega : r \rightarrow \begin{cases} A(X^{-rB}), r < 0, \\ A(X_{rB}), \ \ r \geq 0, \end{cases}$$

where $(X^{rB})^c = X^c_{rB}$ and X^c represents the complementation of X, will give us more shape information such as the convexity.

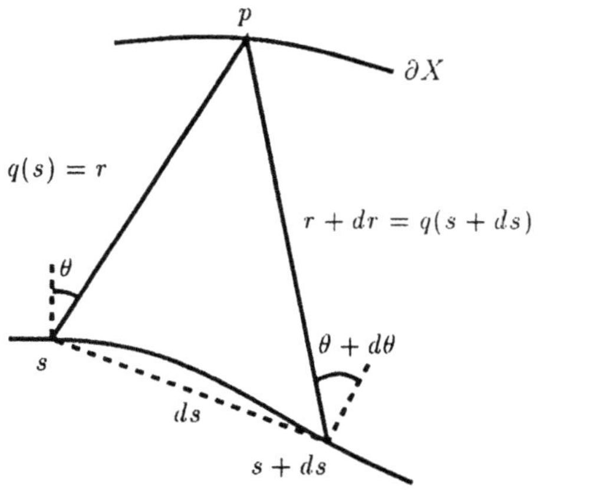

Fig. 8. Notation.

Appendix

Proof of Proposition 4.1. q_X is a Lipschitz function because of the restriction of the distance function $m \rightarrow d(m, X^c)$ to the skeleton. $q = q_X \circ f$, where f is a curvilinear parametrization of Sk(X), i.e., $\forall m \in$ Sk(X), $\exists s$ such that $f(s) = m$ and $||f'(s)|| = 1$, with $\int ||f'(s)||ds = 2\times$ length of the finite graph Sk(X) is also a Lipschitz function. Then q is almost everywhere differentiable [7].

Let p be a boundary point of X, i.e., $p \in \partial X$ (see figure 8). Let m be the center of the maximal inscribable circle C_p at p, then the line segment $[m, p]$ is the upstream of p : Ar(s) = $[m, p]$. If m is not an end point or a terminal point of the skeleton, the equation of the circle C_p is given by $\overrightarrow{mp} \cdot \overrightarrow{mp} = q^2(s) \Leftrightarrow -2\ \overrightarrow{mp}\ (\overrightarrow{dm}\ /ds) = 2q(s)q'(s)$. But $q(s) = d(m,p)$ and $(\overrightarrow{dm}\ /ds) = \overrightarrow{f'(s)} = \vec{t}$ is the tangent vector on m at Sk(X). Then $q'(s) = -(\overrightarrow{mp}\ /||\ \overrightarrow{mp}\ ||)\cdot \vec{t} = -\sin\theta(s)$, where $\theta(s)$ is the angle between the normal on s at Sk(X) and Ar(s).

Proof of Proposition 4.2.

- Let m be an end point of Sk(X) (see figure 9). There exist two tangent vectors $\vec{t_1}$ and $\vec{t_2}$ (such that $f(s) = m$, $\vec{t_1} = f'(s^-)$, $\vec{t_2} = f'(s^+)$) and two normal vectors $\vec{n_1}$ and $\vec{n_2}$ on m at Sk(X). Let p be the end point of the downstream Av(m) of m. p exists and is unique because

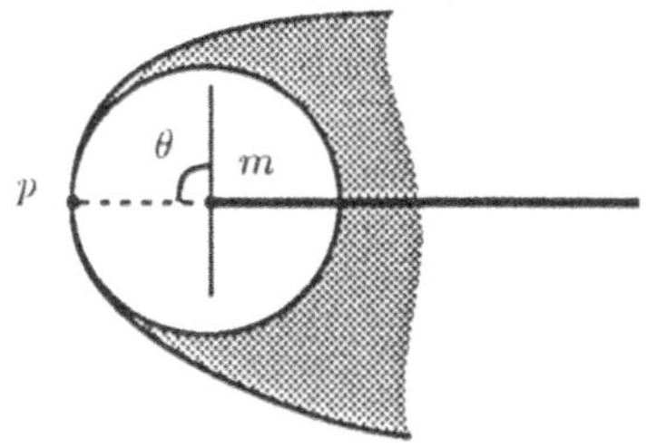

Fig. 9.

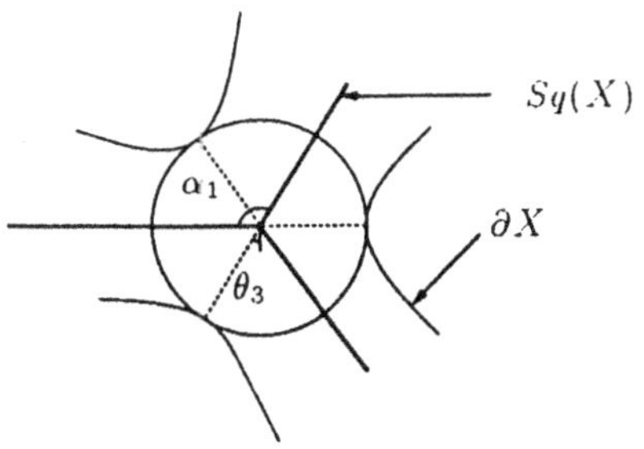

Fig. 10.

the maximal inscribable circle of center m is an osculator circle of ∂X and $p \in \partial X$. Then

$$\left(\frac{dq}{ds}\right)_r (s) = -\left(\frac{dq}{ds}\right)_l (s) = -1$$

because $\theta_i = (\vec{n_i}, \overrightarrow{mp})$ and $\theta_i \equiv \frac{\pi}{2} \pmod{\pi}$ for $i = 1, 2$.

- Let m be a multiple point of Sk(X). For simplicity, we assume that m is a triple point as in figure 10. Let α_i be the angle between two consecutive branches of Sk(X) ($i = 0, 1, 2$). $\theta^+(s_i)$ (respectively, $\theta^-(s_i)$) is defined by the α_i, $i = 0, 1, 2$. The skeleton geometry gives us the value of $(dq/ds)_r$ and $(dq/ds)_l$. For $(\alpha_i)_{i=0}^2$, the angle between $[f(s_i - \varepsilon), m]$ and $[m, f(s_i + \varepsilon)]$ with $\varepsilon > 0$, we have

$$\begin{aligned}
\theta^-(s_0) &= \alpha_1 - \frac{\pi}{2},\\
\theta^-(s_1) &= \alpha_2 - \frac{\pi}{2},\\
\theta^-(s_2) &= \alpha_0 - \frac{\pi}{2},\\
\theta^+(s_0) &= -\alpha_2 + \frac{\pi}{2},\\
\theta^+(s_1) &= -\alpha_0 + \frac{\pi}{2},\\
\theta^+(s_2) &= -\alpha_1 + \frac{\pi}{2},\\
q_r'(s_i) &= -\sin\theta^+(s_i) \quad (i \bmod 2),\\
q_l'(s_i) &= -\sin\theta^-(s_i) \quad (i \bmod 2).
\end{aligned}$$

Proof of Proposition 4.3.

- Let p be a boundary point, and let m be the center of C_p, the maximal inscribable circle at p. Let s (respectively, t) be the curvilinear abscissa of m (respectively, of p). If m is neither an end point nor a multiple point, then

$$\begin{cases} \overrightarrow{mp} \cdot \overrightarrow{mp} = q^2(s) \\ -2\, \overrightarrow{mp}\, \dfrac{\overrightarrow{dm}}{ds} = 2q(s)q'(s) \end{cases}$$

$$\begin{aligned}
\iff -q(s)q''(s) - q'^2(s)
&= \overrightarrow{mp} \cdot \vec{f''}(s) + \frac{\overrightarrow{dp}}{ds} \cdot \vec{f'}(s)\\
&\quad - \vec{f'}(s) \cdot \vec{f'}(s), \qquad (16)
\end{aligned}$$

with $\overrightarrow{mp} \cdot \vec{f''}(s) = \kappa^*(s)q(s)\cos\theta(s)$, $\vec{f'}(s) \cdot \vec{f'}(s) = 1$, and $\frac{dp}{ds} \cdot \vec{f'}(s) = \frac{dt}{ds}\cos\theta(s)$. Equations (16) $\Leftrightarrow$ $1 - q(s)q''(s) - q'^2(s) - \kappa^*(s)q(s)\cos\theta(s) - (dt/ds)\cos\theta(s) = 0$, where $\kappa^*(s)$ represents the curvature of the skeleton at $m = f(s)$.

$$\iff dt = \left(\frac{1 - q(s)q''(s) - q'^2(s)}{\cos\theta(s)} - \kappa^*(s)q(s)\right) ds.$$

But $1 - q'^2(s) = \cos^2\theta(s)$. Finally,

$$\iff dt = \left(\frac{1 - q(s)q''(s) - q'^2(s)}{\sqrt{1 - q'^2(s)}} - \kappa^*(s)q(s)\right) ds. \qquad (17)$$

- If m is not an end point of Sk(X), then

$$\frac{dA}{ds}(s) = \frac{1}{2}\frac{q(s)}{\cos\theta(s)}(2\cos^2\theta(s) - q(s)q''(s) - \kappa^*(s)q(s)\cos\theta(s))$$

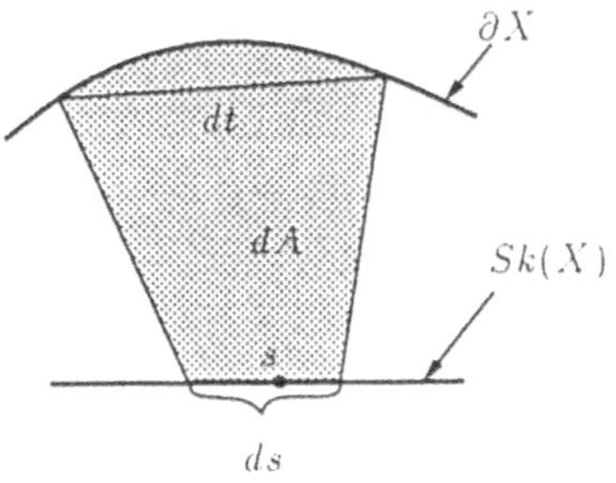

Fig. 11.

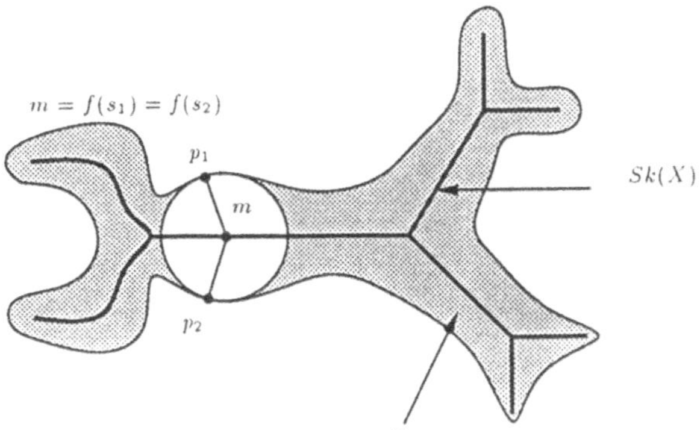

Fig. 12.

because in the neighborhood of s we have (see figure 11)

$$dA = \frac{1}{2}q(s)(dt + ds\cos\theta(s)).$$

Finally, we find

$$(17) \iff dA = \frac{1}{2}q(s)[\cos^2\theta(s) - q(s)q''(s) - \kappa^*(s)q(s)\cos\theta(s) + \cos^2\theta(s)]ds,$$

$$\Leftrightarrow dA = \frac{1}{2}\frac{q(s)}{\cos\theta(s)}(2\cos^2\theta(s) - q(s)q''(s) - \kappa^*(s)q(s) \cos\theta(s))ds.$$

- $\forall m \in \text{Sk}(X)$ that are neither end points nor multiple points (see figure 12), $\exists s_1, s_2$ such that

$$\begin{cases} \kappa^*(s_1) = -\kappa^*(s_2), \\ q(s_1) = q(s_2) = d(m, X^c), \\ q'(s_1) = -q'(s_2), \\ q''(s_1) = q''(s_2). \end{cases}$$

For this reason we have $\int_{\text{Sk}(X)} \kappa^*(s)q^2(s)ds = 0$. (The number of end points and multiple points is assumed to be finite.) We express the surface area of X as

$$A(X) = \int_{\text{Sk}(X)} \frac{1}{2}\frac{q(s)}{\cos\theta(s)} (2\cos^2\theta(s) - q(s)q''(s))ds \iff (9)$$

- $P(X) = \int_{\partial X} dt$; then

$$(17) \iff P(X) = \int_{\text{Sk}(X)} \frac{dt}{ds}ds = \int_{\text{Sk}(X)} \frac{1 - q(s)q''(s) - q'^2(s)}{\sqrt{1 - q'^2(s)}}ds \iff (7).$$

- By using (2), (7), and (9), we deduce (8) and (10).

Proof of Theorem 4.5. Equation (2) allows us to build a sequence of intervals $([\alpha_k^r, \beta_k^r])_{k=1}^{n(r)}$ such that $\text{Sk}(X \ominus rB) = \cup_{k=1}^n f([\alpha_k^r, \beta_k^r])$, i.e., $\forall k, \forall s \in [\alpha_k^r, \beta_k^r]$, we identify $\text{Sk}(X \ominus rB)$ with $\cup_{k=1}^n \{s \in [\alpha_k^r, \beta_k^r]\}$. In fact, we have $\forall s \in \text{Sk}(X \ominus rB), q(s) \geq r$ and $q(\alpha_k^r) = q(\beta_k^r) = r$.

- Case 1: $r < \sup_{s\in\text{Sk}(X)} q(s)$, $(dq/ds)(\alpha_i^r) \neq 0$, $(dq/ds)(\beta_i^r) \neq 0$, $\forall i$. To simplify the notation, let

$$\begin{cases} \alpha_i^r = \alpha_i \text{ and } \beta_i^r = \beta_i, \\ k(s) = q(s) - r, \\ g(s) = \dfrac{k(s)}{2\cos\theta(s)}[2\cos^2\theta(s) - k(s)q''(s)]. \end{cases}$$

Then $\forall i$, $k(\alpha_i) = k(\beta_i) = 0$ and $g(\alpha_i) = g(\beta_i) = 0$.

$$\Psi_X(r + \varepsilon) = \sum_{i=1}^{n} \int_{\alpha_i'}^{\beta_i'} \frac{k(s) - \varepsilon}{2\cos\theta(s)} [2\cos^2\theta(s) - k(s)q''(s) + \varepsilon q''(s)]ds,$$

where

$$\begin{cases} \alpha_i' = \alpha_i + \eta_i \text{with } \eta_i = \dfrac{\varepsilon}{q'(\alpha_i)}, \\ \beta_i' = \beta_i + \mu_i \text{with } \mu_i = \dfrac{\varepsilon}{q'(\beta_i)} \end{cases}$$

We have

$$\begin{cases} \Psi_X(r) & = \sum_{i=1}^{n} \int_{\alpha_i}^{\beta_i} g(s)ds \\ \Psi_X(r+\varepsilon) & = \sum_{i=1}^{n} \int_{\alpha_i'}^{\beta_i'} \Big[g(s) \\ & \quad -\varepsilon \left(\dfrac{q''(s)k(s) - \cos^2\theta(s)}{\cos\theta(s)}\right) \\ & \quad +\varepsilon^2 \dfrac{q''(s)}{2\cos\theta(s)}\Big] ds. \end{cases} \tag{18}$$

– First step:

$$\frac{1}{\varepsilon}\left[\sum_{i=1}^{n} \int_{\alpha_i'}^{\beta_i'} g(s)ds - \sum_{i=1}^{n} \int_{\alpha_i}^{\beta_i} g(s)ds\right]$$
$$= \sum_{i=1}^{n} \frac{-1}{q'(\alpha_i)\eta_i} \int_{\alpha_i}^{\alpha_i+\eta_i} g(s)ds + \sum_{i=1}^{n} \frac{1}{q'(\beta_i)\mu_i} \int_{\beta_i}^{\beta_i+\mu_i} g(s)ds.$$

But $\forall i$

$$\lim_{\eta_i\to 0^+} \frac{-1}{q'(\alpha_i)\eta_i} \int_{\alpha_i}^{\alpha_i+\eta_i} g(s)ds = \lim_{\mu_i\to 0^+} \frac{1}{q'(\beta_i)\mu_i} \int_{\beta_i}^{\beta_i+\mu_i} g(s)ds = \frac{g(\alpha_i)}{q'(\alpha_i)} = 0$$

because $g(\alpha_i) = g(\beta_i) = 0$.

$$\Rightarrow \lim_{\varepsilon\to 0^+} \frac{1}{\varepsilon}\left[\sum_{i=1}^{n} \int_{\alpha_i'}^{\beta_i'} g(s)ds - \sum_{i=1}^{n} \int_{\alpha_i}^{\beta_i} g(s)ds\right] = 0. \tag{19}$$

– Second step:

$$((18) \text{ and } (19)) \Rightarrow \lim_{\varepsilon\to 0^+} \frac{\Psi_X(r+\varepsilon) - \Psi_X(r)}{\varepsilon}$$
$$= \lim_{\varepsilon\to 0^+} -\sum_{i=1}^{n} \int_{\alpha_i'}^{\beta_i'} \left(\frac{q''(s)k(s) - \cos^2\theta(s)}{\cos\theta(s)}\right) ds$$
$$= -\sum_{i=1}^{n} \int_{\alpha_i'}^{\beta_i'} \frac{1 - q'^2(s) - q''(s)q(s) + rq''(s)}{\cos\theta(s)} ds$$
$$= -P(X \ominus rB).$$

We compute

$$\lim_{\varepsilon\to 0^-} \frac{\Psi_X(r+\varepsilon) - \Psi_X(r)}{\varepsilon}$$

in the same way.

- Case of $r < \sup_{s\in \mathrm{Sk}(X)} q(s)$ and $\exists j$ such that $(dq/ds)(\alpha_j^r) = 0$:
  - If $\alpha_j = \beta_j$, then α_j is a local maximum of q and $[\alpha_j, \beta_j] = \{\alpha_j\}$. The computations are the same as those above because $\int_{\alpha_j}^{\beta_j} g(s)ds = 0$ and for $r + \varepsilon(\varepsilon > 0)$ this term does not exist.
  - If $\alpha_j = \beta_{j-1}$, then α_j is a local minimum of q:

$$\lim_{\varepsilon\to 0^-} \frac{\Psi_X(r+\varepsilon) - \Psi_X(r)}{\varepsilon} = -P(X \ominus rB)$$

(same computations as above) because numbers of connected components of $X \ominus rB$ and of $X \ominus (r+\varepsilon)B$ for $\varepsilon < 0$ are equal. $q(\alpha_j) = q(\beta_{j-1}) = r$ and $q'(\alpha_j) = 0$, with $q(s) \geq r$ on $[\alpha_{j-1}, \beta_{j-1}] \cup [\alpha_j, \beta_j]$, and $q(s) \leq r$ on $[\beta_{j-1}, \alpha_j]$. Assume that $q''(\alpha_j) \neq 0$ (unless we use a Taylor development in an upper order). Then

$$q(\alpha_j + \xi) = q(\alpha_j) + \xi q'(\alpha_j) + \frac{\xi^2}{2} q''(\alpha_j)$$

and

$$\varepsilon = q(\alpha_j + \xi) - q(\alpha_j) = \frac{\xi^2}{2} q''(\alpha_j).$$

We can suppose the uniqueness of j (unless we do the same calculus for the

$j_1, j_2, \ldots, j_p$, which are finite in number). Then

$$\Psi_X(r) = \sum_{\substack{i=1\\ i\neq j\\ j-1}}^{n} \int_{\alpha_i}^{\beta_i} g(s)ds + \int_{\alpha_{j-1}}^{\beta_{j-1}} g(s)ds + \int_{\alpha_j}^{\beta_j} g(s)ds$$
$$= \sum_{\substack{i=1\\ i\neq j\\ j-1}}^{n} \int_{\alpha_i}^{\beta_i} g(s)ds + \int_{\alpha_{j-1}}^{\beta_{\beta_j}} g(s)ds$$

because $\alpha_j = \beta_{j-1}$. For simplification, let

$$\lambda(s,\varepsilon) = g(s) - \varepsilon\left(\frac{q''(s)k(s) - \cos^2\theta(s)}{\cos\theta(s)}\right) + \varepsilon^2 \frac{q''(s)}{2\cos\theta(s)},$$

and we have

$$\Psi_X(r+\varepsilon) = \sum_{\substack{i=1\\ i\neq j\\ j-1}}^{n} \int_{\alpha'_i}^{\beta'_i} \lambda(s,\varepsilon)ds + \int_{\alpha'_{j-1}}^{\beta'_{j-1}} \lambda(s,\varepsilon)ds + \int_{\alpha'_j}^{\beta'_j} \lambda(s,\varepsilon)ds. \quad (20)$$

with

$$\begin{cases} \alpha'_i = \alpha_i + \eta_i & \text{for } i \neq j \text{ and } \eta_i = \dfrac{\varepsilon}{q'(\alpha_i)}, \\ \beta'_i = \beta_i + \mu_i & \text{for } i \neq j-1 \text{ and } \mu_i = \dfrac{\varepsilon}{q'(\beta_i)}, \\ \alpha'_j = \alpha_j + \xi & \text{and } \xi^2 = \dfrac{2\varepsilon}{q''(\alpha_j)}, \\ \beta'_{j-1} = \beta_{j-1} + \xi \text{and} & \xi^2 = \dfrac{2\varepsilon}{q''(\beta_{j-1})}, \\ \alpha_j = \beta_{j-1}. \end{cases}$$

$$(20) \Leftrightarrow \Psi_X(r+\varepsilon) = \sum_{\substack{i=1\\ i\neq j\\ j-1}}^{n} \int_{\alpha_i+\eta_i}^{\beta_i+\mu_i} \lambda(s,\varepsilon)ds + \int_{\alpha_{j-1}+\eta_{j-1}}^{\beta_{j-1}+\xi} \lambda(s,\varepsilon)ds + \int_{\alpha_j+\xi}^{\beta_j+\mu_j} \lambda(s,\varepsilon)ds \quad (21)$$

and

$$\lim_{\varepsilon\to 0^+} \frac{1}{\varepsilon}[\Psi_X(r+\varepsilon) - \Psi_X(r)] = -P(X \ominus rB) + \lim_{\varepsilon\to 0^+} \frac{1}{\varepsilon}\left[\int_{\alpha_{j-1}+\eta_{j-1}}^{\beta_{j-1}+\xi} g(s)ds - \int_{\alpha_{j-1}}^{\beta_{j-1}} g(s)ds\right] + \lim_{\varepsilon\to 0^+} \frac{1}{\varepsilon}\left[\int_{\alpha_j+\xi}^{\beta_j+\mu_j} g(s)ds - \int_{\alpha_j}^{\beta_j} g(s)ds\right]$$

But

$$\lim_{\varepsilon\to 0^+} \frac{1}{\varepsilon}\left[\int_{\alpha_j+\xi}^{\beta_j+\mu_j} g(s)ds - \int_{\alpha_j}^{\beta_j} g(s)ds\right] = \lim_{\xi\to 0^+} \frac{-2}{\xi^2 q''(\alpha_j)} \int_{\alpha_j}^{\alpha_j+\xi} g(s)ds = \lim_{\xi\to 0^+} \frac{-2g(\alpha_j)}{\xi q''(\alpha_j)} = 0$$

and $g(\alpha_j) = \frac{\xi^2}{2} q''(\alpha_j)$. Then

$$\lim_{\xi\to 0^+} \frac{-2}{\xi^2 q''(\alpha_j)} \int_{\alpha_j}^{\alpha_j+\xi} g(s)ds = \lim_{\xi\to 0^+} +\xi = 0$$

and

$$\lim_{\varepsilon\to 0^+} \frac{1}{\varepsilon}[\Psi_X(r+\varepsilon) - \Psi_X(r)] = -P(X \ominus rB).$$

- Case 2: $r < \sup_{s\in \mathrm{Sk}(X)} q(s)$ and $\exists \alpha, \beta$ such that $\forall s \in [\alpha,\beta] < (dq/ds)(s) = 0$. We have only

$$\lim_{\varepsilon\to 0^+} \frac{\Psi_X(r+\varepsilon) - \Psi_X(r)}{\varepsilon} = -P(\overline{\mathring{X}} \ominus (r+\varepsilon)B),$$
$$\lim_{\varepsilon\to 0^-} \frac{\Psi_X(r+\varepsilon) - \Psi_X(r)}{\varepsilon} = -P(X \ominus (r+\varepsilon)B).$$

Proof of Theorem 4.7. To simplify, let

$$\phi_X(r) = P(X \ominus rB) = \sum_{i=1}^{n} \int_{\alpha_i}^{\beta_i} \left[\frac{\cos^2\theta(s) - q''(s)k(s)}{\cos\theta(s)}\right] ds.$$

- Case (i):

$$\phi_X(r+\varepsilon) = \sum_{i=1}^{n} \int_{\alpha'_i}^{\beta'_i} \left[\frac{\cos^2\theta(s) - q''(s)k(s) + \varepsilon q''(s)}{\cos\theta(s)}\right] ds.$$

$$\lim_{\varepsilon\to 0^+} \frac{1}{\varepsilon}(\phi_X(r+\varepsilon) - \phi_X(r))$$
$$= \lim_{\varepsilon\to 0^+} \frac{1}{\varepsilon}\sum_{i=1}^{n} \int_{\alpha'_i}^{\beta'_i} \left[\frac{\cos^2\theta(s) - q''(s)k(s)}{\cos\theta(s)}\right] ds - \sum_{i=1}^{n} \int_{\alpha_i}^{\beta_i} \left[\frac{\cos^2\theta(s) - q''(s)k(s)}{\cos\theta(s)}\right] ds + \sum_{i=1}^{n} \int_{\alpha'_i}^{\beta'_i} \frac{q''(s)}{\cos\theta(s)} ds$$
$$= -\sum_{i=1}^{n} \left\{ \frac{1}{q'(\alpha_i)} \left[\frac{\cos^2\theta(\alpha_i) - q''(\alpha_i)k(\alpha_i)}{\cos\theta(\alpha_i)}\right] - \frac{1}{q'(\beta_i)} \left[\frac{\cos^2\theta(\beta_i) - q''(\beta_i)k(\beta_i)}{\cos\theta(\beta_i)}\right]\right\} + \sum_{i=1}^{n} \int_{\alpha_i}^{\beta_i} \frac{q''(s)}{\cos\theta(s)} ds.$$

But $q'(\alpha_i) = -\sin\theta(\alpha_i)$ and $q'(\beta_i) = -\sin\theta(\beta_i)$ and $k(\alpha_i) = k(\beta_i) = 0$.

$$\frac{1}{q'(\alpha_i)} \left[\frac{\cos^2\theta(\alpha_i) - q''(\alpha_i)k(\alpha_i)}{\cos\theta(\alpha_i)}\right] = \frac{\cos^2\theta(\alpha_i)}{-\sin\theta(\alpha_i)\cos\theta(\alpha_i)} = -\cot(\alpha_i)$$

$$\Leftrightarrow \lim_{\varepsilon\to 0^+} \frac{1}{\varepsilon}(\phi_X(r+\varepsilon) - \phi_X(r)) = \sum_{i=1}^{n} \left(\frac{\cos\theta(\alpha_i)}{\sin\theta(\alpha_i)} - \frac{\cos\theta(\beta_i)}{\sin\theta(\beta_i)}\right) + \sum_{i=1}^{n} \int_{\alpha_i}^{\beta_i} \frac{q''(s)}{\cos\theta(s)} ds$$

$$\Leftrightarrow \Psi''_X(r) = \sum_{i=1}^{n} \left\{ \cot\theta(\beta_i) - \cot\theta(\alpha_i) - \int_{\alpha_i}^{\beta_i} \frac{q''(s)}{\cos\theta(s)} ds \right\}.$$

- Case (ii): $\exists j$ such that $q'(\alpha_j) = 0$ and α_j is a local minimum of q, let

$$\begin{cases} \alpha'_i = \alpha_i + \eta_i & \text{for } i \neq j \text{ and } \eta_i = \dfrac{\varepsilon}{q'(\alpha_i)}, \\ \beta'_i = \beta_i + \mu_i & \text{for } i \neq j-1 \text{ and } \mu_i = \dfrac{\varepsilon}{q'(\beta_i)}, \\ \alpha'_j = \alpha_j + \xi & \text{and } \xi^2 = \dfrac{2\varepsilon}{q''(\alpha_j)}, \\ \beta'_{j-1} = \beta_{j-1} + \xi & \text{and } \xi^2 = \dfrac{2\varepsilon}{q''(\beta_{j-1})}, \\ \alpha_j = \beta_{j-1} & \text{and } q''(\alpha_j) > 0. \end{cases}$$

$$\phi_X(r+\varepsilon) = \sum_{i\neq j-1,j} \int_{\alpha'_i}^{\beta'_i} \left[\frac{\cos^2\theta(s) - q''(s)k(s) + \varepsilon q''(s)}{\cos\theta(s)}\right] ds + \int_{\alpha'_{j-1}}^{\beta'_{j-1}} \left[\frac{\cos^2\theta(s) - q''(s)k(s) + \varepsilon q''(s)}{\cos\theta(s)}\right] ds + \int_{\alpha'_j}^{\beta'_j} \left[\frac{\cos^2\theta(s) - q''(s)k(s) + \varepsilon q''(s)}{\cos\theta(s)}\right] ds.$$

Then

$$\lim_{\varepsilon\to 0^+} \frac{1}{\varepsilon}[\phi_X(r+\varepsilon) - \phi_X(r)] = \sum_{i\neq j-1,j} (\cot\theta(\alpha_i) - \cot\theta(\beta_i)) + \sum_{i=1}^{n} \int_{\alpha_i}^{\beta_i} \frac{q''(s)}{\cos\theta(s)} ds + \lim_{\varepsilon\to 0^+} \frac{1}{\varepsilon} \int_{\alpha_{j-1}+\eta_{j-1}}^{\beta_{j-1}+\xi}$$

$$\left[\frac{\cos^2\theta(s) - q''(s)k(s)}{\cos\theta(s)}\right] ds - \int_{\alpha_{j-1}}^{\beta_{j-1}} \left[\frac{\cos^2\theta(s) - q''(s)k(s)}{\cos\theta(s)}\right] ds + \lim_{\varepsilon\to 0^+} \frac{1}{\varepsilon} \int_{\alpha_j+\xi}^{\beta_j+\mu_j} \left[\frac{\cos^2\theta(s) - q''(s)k(s)}{\cos\theta(s)}\right] ds - \int_{\alpha_j}^{\beta_j} \left[\frac{\cos^2\theta(s) - q''(s)k(s)}{\cos\theta(s)}\right] ds.$$

Let

$$h(s) = \left[\frac{\cos^2\theta(s) - q''(s)k(s)}{\cos\theta(s)}\right].$$

$$\lim_{\varepsilon\to 0^+} \frac{1}{\varepsilon}\left[\int_{\alpha_j+\xi}^{\beta_j+\mu_j} h(s)ds - \int_{\alpha_j}^{\beta_j} h(s)ds\right] = \lim_{\xi\to 0^+}\left[\frac{-2h(\alpha_j)}{\xi q''(\alpha_j)}\right] + \frac{-2h(\beta_j)}{q'(\beta_j)} = -\infty$$

because $h(\alpha_j) = \cos\theta(\alpha_j)$ and $h(\beta_j) = \cos\theta(\beta_j)$. In the same way,

$$\lim_{\varepsilon\to 0^+} \frac{1}{\varepsilon}\left[\int_{\alpha_{j-1}+\eta_{j-1}}^{\beta_{j-1}+\xi} h(s)ds - \int_{\alpha_{j-1}}^{\beta_{j-1}} h(s)ds\right] = -\infty.$$

- Case (iii): $\exists j$ such that $q'(\alpha_j) = 0$ and α_j is a local maximum of q, let

$$\begin{cases} \alpha'_i = \alpha_i - \eta_i \text{ for } i \neq j \text{ and } \eta_i = \dfrac{\varepsilon}{q'(\alpha_i)}, \\ \beta'_i = \beta_i + \mu_i \text{ for } i \neq j \text{ and } \mu_i = \dfrac{\varepsilon}{q'(\beta_i)}, \\ \alpha'_j = \alpha_j + \xi \text{ and } \xi^2 = \dfrac{-2\varepsilon}{q''(\alpha_j)}, \xi < 0, \\ \beta'_j = \beta_j + \xi \text{ and } \xi^2 = \dfrac{-2\varepsilon}{q''(\beta_j)}, \xi < 0, \\ \alpha_j = \beta_j \quad \text{and } q''(\alpha_j) > 0. \end{cases}$$

$$\phi_X(r+\varepsilon) = \sum_{i\neq j} \int_{\alpha'_i}^{\beta'_i} \left[\frac{\cos^2\theta(s) - q''(s)k(s) + \varepsilon q''(s)}{\cos\theta(s)}\right] ds + \int_{\alpha'_j}^{\beta'_j} \left[\frac{\cos^2\theta(s) - q''(s)k(s) + \varepsilon q''(s)}{\cos\theta(s)}\right] ds$$

$$\Rightarrow \lim_{\varepsilon\to 0^-} \frac{1}{\varepsilon}[\phi_X(r+\varepsilon) - \phi_X(r)] = \sum_{i\neq j}(\cot\theta(\alpha_i) - \cot\theta(\beta_i)) + \sum_{i=1}^{n} \int_{\alpha_i}^{\beta_i} \frac{q''(s)}{\cos\theta(s)} ds + \lim_{\varepsilon\to 0^-} \frac{1}{\varepsilon} \int_{\alpha_j+\xi}^{\beta_j+\mu_j} \left[\frac{\cos^2\theta(s) - q''(s)k(s)}{\cos\theta(s)}\right] ds - \int_{\alpha_j}^{\beta_j} \left[\frac{\cos^2\theta(s) - q''(s)k(s)}{\cos\theta(s)}\right] ds.$$

Let

$$h(s) = \left[\frac{\cos^2\theta(s) - q''(s)k(s)}{\cos\theta(s)}\right],$$

and we have

$$\lim_{\varepsilon\to 0^-} \frac{1}{\varepsilon}\left[\int_{\alpha_j+\xi}^{\beta_j+\mu_j} h(s)ds - \int_{\alpha_j}^{\beta_j} h(s)ds\right] = \lim_{\varepsilon\to 0^-} \frac{1}{\varepsilon}\left[-\int_{\alpha_j}^{\alpha_j+\xi} h(s)ds - \int_{\beta_j+\mu_j}^{\beta_j} h(s)ds\right] = \lim_{\xi\to 0^-}\left[\frac{-2h(\alpha_j)}{\xi q''(\alpha_j)} + \frac{2h(\beta_j)}{\xi q''(\beta_j)}\right] = -\infty$$

because $h(\alpha_j) = \cos\theta(\alpha_j)$ and $h(\beta_j) = \cos\theta(\beta_j)$. We have proved that

$$\lim_{\varepsilon\to 0^-} \frac{1}{\varepsilon}[\Psi'_X(r+\varepsilon) - \Psi'_X(r)] = +\infty.$$

- Case (iv): Assume for simplification that the multiple point m is a triple point. Let δ_0, δ_1, and δ_2 be the angles between two consecutive branches of the skeleton at m, and let s_0, s_1, s_2

be the curvilinear abscissa of m. Let γ_i be the angle between $[s_{i-1} - \varepsilon, s_i - \varepsilon]$ and $[s_i - \varepsilon, s_{i+1} - \varepsilon]$ (i mod 3). Then we have

$$\begin{cases} 0 < \gamma_i < \pi, \\ \sum_{i=0}^{2} \gamma_i = 2\pi, \\ \delta_i = \pi - \frac{\gamma_i}{2}. \end{cases}$$

The first two equations become a triangle that is a convex polygon. If there is a vanishing connected component containing a multiple point, then there exist j_0, j_1, j_2 such that

$$\begin{cases} \theta(\alpha_{j0}) = \frac{\pi}{2} + \frac{\gamma_1}{2}, \theta(\beta_{j0}) = \frac{\pi}{2} - \frac{\gamma_2}{2}, \\ \theta(\alpha_{j1}) = \frac{\pi}{2} + \frac{\gamma_2}{2}, \theta(\beta_{j1}) = \frac{\pi}{2} - \frac{\gamma_0}{2}, \\ \theta(\alpha_{j2}) = \frac{\pi}{2} + \frac{\gamma_0}{2}, \theta(\beta_{j2}) = \frac{\pi}{2} - \frac{\gamma_1}{2}, \end{cases}$$

$$\begin{aligned} \Rightarrow \lim_{\varepsilon \to 0^+} & \frac{1}{\varepsilon}[\phi_X(r+\varepsilon) - \phi_X(r)] \\ & - \lim_{\varepsilon \to 0^-} \frac{1}{\varepsilon}[\phi_X(r+\varepsilon) - \phi_X(r)] \\ & = 2\sum_{k=0}^{2} \tan\left(\frac{\gamma_k}{2}\right) \end{aligned}$$

$$\Rightarrow \Psi''_{X_r}(r) - \Psi''_{X_e}(r) = -2\sum_{k=0}^{2} \tan\left(\frac{\gamma_k}{2}\right) < 0.$$

- Case (v): Assume for simplification that the multiple point m is a triple point. Then

$$\begin{aligned} & \lim_{\varepsilon \to 0^+} \frac{1}{\varepsilon}[\phi_X(r+\varepsilon) - \phi_X(r)] \\ & - \lim_{\varepsilon \to 0^-} \frac{1}{\varepsilon}[\phi_X(r+\varepsilon) - \phi_X(r)] \\ & 2\left[\tan\left(\frac{\gamma_0 + \gamma_1}{2}\right) + \tan\left(\frac{\gamma_0}{2}\right) \right. \\ & \left. + \tan\left(\frac{\gamma_1}{2}\right)\right] > 0 \\ & \Rightarrow \Psi''_{X_r}(r) - \Psi''_{X_e}(r) < 0. \end{aligned}$$

References

1. G. Matheron, *Random Sets and Integral Geometry*, John Wiley: New York, 1975.
2. M. Coster and J.L. Chermant, "*Précis d'analyse d'images,*" *CNRS Etud.*, 1985.
3. D. Stoyan, W.S. Kendall, and J. Mecke, *Stochastic Geometry and Its Applications*, John Wiley: New York, 1987.
4. L. Calabi and J.A. Riley, "The skeletons of stable plane sets," Parke Math. Lab., Inc., Carlisle, MA, Tech. Rep. AF 19, 1967.
5. J. Serra, *Image Analysis and Mathematical Morphology*, Academic Press: London, 1982.
6. G. Matheron, "Examples of topological properties of skeletons," in *Image Analysis and Mathematical Morphology, Vol. 2: Theoretical Advances*, J. Serra, ed., Academic Press: London, 1988.
7. G. Matheron, "On the negligibility of the skeleton and the absolute continuity of erosions," in *Image Analysis and Mathematical Morphology, Vol. 2: Theoretical Advances*, J. Serra, ed., Academic Press: London, 1988.
8. J. Riley and L. Calabi, "Certain properties of circles inscribed in simple closed curves," Park Math. Lab., Inc., Carlisle, MA, Tech. Rep. 59281, 1964.
9. J. Riley, "Plane graphs and their skeletons," Park Math. Lab., Inc., Carlisle, MA, Tech. Rep. 60429, 1965.
10. J. Dieudonné, *Eléments d'analyse*, vol. I, Gauthier-Villars: Paris, 1969.
11. J. Mattioli, "Squelette, érosion et fonction spectrale par érosion d'une forme binaire planaire," Rapport Interne, ASRF-91-8, 1991.
12. G. Matheron, La formule de Steiner pour les érosions, Centre de Géostatistique, Ecole des Mines, Paris, Tech. Rep. 496, 1977.
13. M. Schmitt, "On two inverse problems in mathematical morphology," in *Mathematical Morphology in Image Processing*, E.R. Dougherty, ed., Marcel Dekker: New York, 1991.

Juliette Mattioli received her Maîtrise de Mathématiques Pures at the University of Paris VII in 1989 and her DEA Artificial Intelligence and Image Analysis at Paris Dauphine University in 1990. During 1982–1990 she was a mathematics teacher. In 1990 she began her Ph.D. thesis at the Signal Analysis and Pattern Recognition Laboratory at the Central Research Laboratory of Thomson-CSF. Her research interests include shape recognition, mathematical morphology, and set-valued analysis.

Michel Schmitt received his engineer degree from the Ecole Polytechnique in 1982 and his engineer of the Corps des Mines degree in 1985. He joined the Mathematical Morphological Laboratory of J. Serra at the Ecole des Mines (1985), where he began a thesis on mathematical morphology: theoretical aspects, algorithmics and artificial intelligence. He joined the team of O.D. Faugeras at the Institut National de Recherche en Informatique et en Automatique Domaine de Volucean in 1985, carried on his research work, and obtained his Docteur en Morphologie Mathématique degree from the Ecole des Mines in 1989 and his Habilitation à diriger des recherches degree in 1991. Since October 1989 he has been on leave from the Corps des Mines at the Central Research Laboratory of Thomson-CSF and has been head of the Signal Analysis and Pattern Recognition Laboratory since 1991. He is also a professor at the ISIA in Sophia-Antipolis and an assistant professor at the Paris-Dauphine university. His current interests involve theoretical aspects of mathematical morphology and neural networks.

Journal of Mathematical Imaging and Vision 2, 233–250 (1992).

Design of a Multitask Neurovision Processor

GEORGE K. KNOPF AND MADAN M. GUPTA
Intelligent Systems Research Laboratory and Centre of Excellence on Neuro-Vision Research (IRIS), College of Engineering, University of Saskatchewan, Saskatoon, Saskatchewan, S7N0W0, Canada

Abstract. The architecture of a biologically motivated visual-information processor that can perform a variety of tasks associated with the early stages of machine vision is described. The computational operations performed by the processor emulate the spatiotemporal information-processing capabilities of certain neural-activity fields found along the human visual pathway. The state-space model of the neurovision processor is a two-dimensional nural network of densely interconnected nonlinear processing elements PE's. An individual PE represents the dynamic activity exhibited by a spatially localized population of excitatory and inhibitory nerve cells. Each PE may receive inputs from an external signal space as well as from the neighboring PE's within the network. The information embedded within the signal space is extracted by the feedforward subnet. The feedback subnet of the neurovision processor generates useful steady-state and temporal-response characteristics that can be used for spatiotemporal filtering, short-term visual memory, spatiotemporal stabilization, competitive feedback interaction, and content-addressable memory. To illustrate the versatility of the multitask processor design for machine-vision applications, a computer simulation of a simplified vision system for filtering, storing, and classifying noisy gray-level images is presented.

Key words. neural-activity field, neuro-vision processor, spatio-temporal processing, short-term visual memory, spatio-temporal stabilization, content addressable memory

1 Introduction

The human visual system employs billions of very slow and largely unreliable information-processing elements [1], called *neurons*, that perform an immense variety of complex computational operations in real time. Although each neuron along the visual pathway requires nearly 2 ms to transmit a response to a second neuron, the entire process of interpreting a time-varying scene occurs in only 70 to 200 [2]. In sharp contrast, the digital processor is over a million times faster than a biological neuron, but a computer-based vision system requires several minutes to hours to recognize the objects in a single stationary image frame. The incredible speed and overall versatility of our own visual experience is, therefore, the necessary proof that robust forms of visual-information processing are possible and can provide the computational framework whereupon comparable vision systems can be developed for engineering applications. In terms of system design, it would be desirable to replicate certain aspects of the computational architecture of human vision in order to design and develop an effective neurovision system. The term *neurovision* is used to refer to any artificial or machine vision system that embodies the basic computational principles of biological neural circuits.

Biological systems are able to perform complex visual-information-processing operations in real time because they use a computational architecture that is vastly different from that of a digital computer. The human visual pathway may be envisioned as a series of anatomical regions, such as the retina, lateral geniculate nucleus (LGN), primary visual cortex, and higher cortical regions of the brain. Each successive region along the pathway consists of one or more functionally distinct layers of nervous tissue that perform specific sets of operations on the neural information. The various nervous-tissue layers found within a common anatomical region operate in parallel.

All aspects of neural-information processing that occur within a nervous-tissue layer are assumed to exist as the spatiotemporal patterns of activity represented by a two-dimensional neural-activity field. The spatiotemporal field activity is the result of spatially localized assemblies of densely interconnected excitatory and inhibitory nerve cells. A state-space model of the functional dynamics of a generalized neural-activity field is described in this paper. This mathematical model is the basis of a multitask neurovision processor for machine-vision applications.

The neurovision processor is able to efficiently process spatiotemporal information because it uses a parallel structure containing numerous processing elements (PE's) that individually perform computations based solely on local inputs. The incorporation of a parallel distributed structure makes most of the computational power required by the early stages of machine vision unnecessary. For example, a standard image from a digital video camera is a two-dimensional array containing approximately 0.75 million pixels that change at a rate of 30 frames per second. If the vision algorithm were to perform only one instruction per pixel, it would be necessary for the computer system to perform approximately 23 million instructions per second [3]. However, any practical vision process requires far more than one instruction per pixel to be performed. Even for a high-speed supercomputer this requirement is impossible to achieve for any image of reasonable size. Thus the real-time constraints imposed on the machine-vision process makes the parallel structure of the neurovision processor necessary for exploiting the intrinsic parallelism within the time-varying two-dimensional video signals.

A state-space model for the computational organization of a neural-activity field is presented in section 2. The basic structure of the generalized neural-activity field, as represented by the state-space model, is extended to a multitask neurovision processor. A summary of the various visual-information-processing tasks that can be performed by the feedforward and feedback subnets of the state-space model are given in section 3. A computer simulation of a vision system used to filter, store, and classify noisy gray-level images is described in section 4. Finally, the basic concepts presented in this paper are summarized in section 5.

2 Mathematical Model of a Generalized Neural-Activity Field

All aspects of neural-information processing that occur within a nervous-tissue layer are assumed to exist as spatiotemporal patterns of activity represented by a two-dimensional neural-activity field $\mathbf{X}(k)$ [1], [4]–[9]. The spatiotemporal-field activity is a result of spatially localized assemblies of densely interconnected nerve cells called *neural populations*. Each neural population contains a mixture of excitatory and inhibitory nerve cells [7] that lie in close spatial proximity. The individual neurons within the populations that generate $\mathbf{X}(k)$ receive stimuli from other neurons in the nervous-tissue layer and from a common signal space $\mathbf{S}(k)$ external to the tissue layer, see figure 1.

The state activity exhibited by the field can be mathematically represented as a spatially discrete two-dimensional function. The amplitude of the function at some spatial coordinate is the state of the neural-population activity observed at that location. Thus the state activity of the neural field at time k can be represented by the two-dimensional matrix

$$\mathbf{X}(k) = \begin{bmatrix} x_{1,1}(k) & x_{1,2}(k) & \cdots & x_{1,J}(k) \\ x_{2,1}(k) & x_{2,2}(k) & \cdots & x_{2,J}(k) \\ \vdots & & \ddots & \\ x_{I,1}(k) & x_{I,2}(k) & \cdots & x_{I,J}(k) \end{bmatrix}, \quad (1)$$

where (i, j) is the spatial coordinate of the population in the (IJ)-dimensional array. The state variable $x_{i,j}(k)$ corresponds to the proportion of excitatory and inhibitory nerve cells active at time k within the neural population centered at location (i, j) of the field $\mathbf{X}(k)$. A neuron is assumed to be active if it is firing a sequence of action potentials and the measured axon potential at time k is greater than the threshold potential [8].

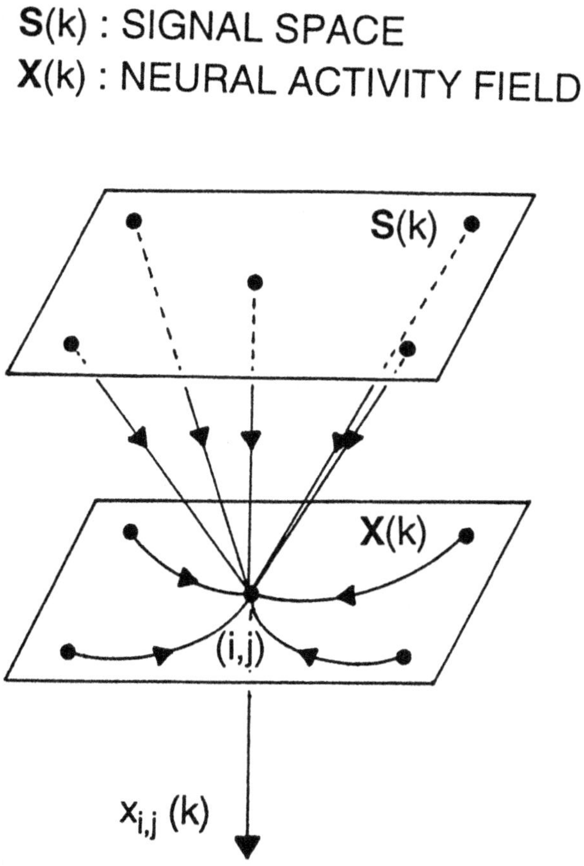

Fig. 1. Schematic diagram of a neural-activity field $\mathbf{X}(k)$ that represents the functional dynamics of a nervous-tissue layer. Connections are shown only to the neural population situated at location (i,j).

Also, the state activity of the signal space is $\mathbf{S}(k)$ at time k is mathematically represented by the matrix

$$\mathbf{S}(k) = \begin{bmatrix} s_{1,1}(k) & s_{1,2}(k) & \cdots & s_{1,Q}(k) \\ s_{2,1}(k) & s_{2,2}(k) & \cdots & s_{2,Q}(k) \\ \vdots & & \ddots & \\ s_{P,1}(k) & s_{P,2}(k) & \cdots & s_{P,Q}(k) \end{bmatrix}, \quad (2)$$

where (p,q) is the spatial coordinate of the external input arising within the (PQ)-dimensional signal space. The state variable $s_{p,q}(k)$ represents the signal data originating from a sensor or another neural-activity field. Since the field and signal space may originate at different anatomical regions along the visual pathway, they may have different spatial dimensions, that is, $(P \times Q)$ may not be equal to $(I \times J)$.

Two types of inputs are received by the constituent populations in the field. One type arises from the external inputs originating in the signal space $\mathbf{S}(k)$, and the second type originates from the lateral and recurrent connections [10], [11], between the populations within a common field $\mathbf{X}(k)$. The total applied input $\mathbf{U}(k)$ to the neural-activity field is, therefore, assumed to be given by

$$\mathbf{U}(k) = \mathbf{AS}(k) + \mathbf{BX}(k). \quad (3)$$

The spatial transformation of the external input signal is given by the matrix $\mathbf{A}$ of the feedforward subnet in the field, and the strength of the lateral and recurrent connections between the individual populations in the field is given by the matrix $\mathbf{B}$ of the feedback subnet.

The system matrices $\mathbf{A}$ and $\mathbf{B}$ are of dimensions $(IJ) \times (PQ)$ and $(IJ) \times (IJ)$, respectively. Matrix $\mathbf{A}$ of the feedforward subnet consists of $(I \times J)$ partitions ordered according to

$$\mathbf{A} = \begin{bmatrix} \mathbf{A}_{1,1} & \mathbf{A}_{1,2} & \cdots & \mathbf{A}_{1,J} \\ \mathbf{A}_{2,1} & \mathbf{A}_{2,2} & \cdots & \mathbf{A}_{2,J} \\ \vdots & & \ddots & \\ \mathbf{A}_{I,1} & \mathbf{A}_{I,2} & \cdots & \mathbf{A}_{I,J} \end{bmatrix}. \quad (4)$$

Each partition $A_{i,j}$ of (4) is constructed from the (i,j)th row of the function $a_{i,j,p,q}$ as follows:

$$\mathbf{A}_{i,j} = \begin{bmatrix} a_{ij,11} & a_{ij,12} & \cdots & a_{ij,1Q} \\ a_{ij,21} & a_{ij,22} & \cdots & a_{ij,2Q} \\ \vdots & & \ddots & \\ a_{ij,P1} & a_{ij,P2} & \cdots & a_{ij,PQ} \end{bmatrix}, \quad (5)$$

where $p = 1,2,\ldots,P$ and $q = 1,2,\ldots,Q$.

Similarly, the square matrix $\mathbf{B}$ of the feedback subnet consists of $(I \times J)$ partitions ordered as

$$\mathbf{B} = \begin{bmatrix} \mathbf{B}_{1,1} & \mathbf{B}_{1,2} & \cdots & \mathbf{B}_{1,J} \\ \mathbf{B}_{2,1} & \mathbf{B}_{2,2} & \cdots & \mathbf{B}_{2,J} \\ \vdots & & \ddots & \\ \mathbf{B}_{I,1} & \mathbf{B}_{I,2} & \cdots & \mathbf{B}_{I,J} \end{bmatrix}, \quad (6)$$

with each partition $\mathbf{B}_{i,j}$ in (6) constructed from the (i,j)th row of the function $b_{i,j,s,r}$ as follows:

$$\mathbf{B}_{i,j} = \begin{bmatrix} b_{ij,11} & b_{ij,12} & \cdots & b_{ij,1J} \\ b_{ij,21} & b_{ij,22} & \cdots & b_{ij,2J} \\ \vdots & & \ddots & \\ b_{ij,I1} & b_{ij,I2} & \cdots & b_{ij,IJ} \end{bmatrix}, \qquad (7)$$

where $s = 1, 2, \ldots, I$ and $r = 1, 2, \ldots, J$.

Thus the total applied input $\mathbf{U}(k)$ is represented by an (IJ)-dimensional matrix as

$$\mathbf{U} = \begin{bmatrix} \mathbf{u}_{1,1}(k) & \mathbf{u}_{1,2}(k) & \cdots & \mathbf{u}_{1,J}(k) \\ \mathbf{u}_{2,1}(k) & \mathbf{u}_{2,2}(k) & \cdots & \mathbf{u}_{2,J}(k) \\ \vdots & & \ddots & \\ \mathbf{u}_{I,1}(k) & \mathbf{u}_{I,2}(k) & \cdots & \mathbf{u}_{I,J}(k) \end{bmatrix}. \qquad (8)$$

In terms of biology, the actual influence of the input given by (3) on the field activity is a function of the distribution of the neural thresholds within the constituent populations [1], [5], [6], [8], [9], [12]. That is, the expected proportion of neurons in the population receiving at least threshold excitation per unit time is a mathematical function $\Phi[\cdot]$ of the total applied input $u_{i,j}(k)$ originating from both inside the field and the external signal space. This function depends on the probability distribution $g[u_{i,j}(k)]$ of the neural thresholds within the population and is given by the integral

$$\Phi[u_{i,j}(k)] = \int_{-\infty}^{u_{ij}(k)} g[u_{i,j}(k)]du_{i,j}(k). \qquad (9)$$

If the probability of neural thresholds is assumed to be a unimodal distribution about a single mean value θ^1 as in figure2(a), then the nonlinear mapping operator given by (9) is a sigmoidal function, as shown in figure 2(b). An important assumption in deriving the sigmoidal mapping function is that all neurons in a population have the same threshold characteristics. Alternatively, if the individual neurons in a population form several groups with distinct threshold characteristics, then the nonlinear mapping operator $\Phi[u_{i,j}(k)]$ becomes more complex. In general, an M-modal probability distribution for neural thresholds in a population can be expressed as

$$g[u_{i,j}(k)] = \frac{1}{2M}\sum_{m=1}^{M} v^m \cdot \operatorname{sech}^2 \{v^m(u_{i,j}(k) - \theta^m)\}, \qquad (10)$$

and from (9) the corresponding nonlinear map is given by

$$\Phi[u_{i,j}(k)] = \frac{1}{2}\left[1 + \frac{1}{M}\sum_{m=1}^{M} \tanh \{v^m(u_{i,j}(k) - \theta^m)\}\right], \qquad (11)$$

where a separate slope parameter v^m exists for each inflection point $\theta^m, m = 1, \ldots, M$. For a sigmoidal map the parameter v^1 determines the value of the maximum slope of $\Phi[u_{i,j}(k)]$ at θ^1through the relation

$$g[u_{ij}(k) = \theta^1] = \frac{v^1}{2}. \qquad (12)$$

However, the maximum slope at each θ^m of a nonlinear map $\Phi[u_{i,j}(k)]$ with multiple inflection points $M > 1$ must be calculated directly from (10) because each local maximum is a function of all constituent θ^m. An example of a multimodal neural-threshold distribution resulting in a map $\Phi[\cdot]$ with four inflection points is shown in figure 3. The input $\mathbf{Y}(k)$ that directly influences state activity of a field is given by

$$\mathbf{Y}(k) = \Phi[\mathbf{U}(k)], \qquad (13)$$

where $\Phi[\cdot]$ is a nonlinear mapping operator (11) that transforms the total applied input $\mathbf{U}(k) \in [-\infty, \infty]$ to a bounded state over the unit interval $\mathbf{Y}(k) \in [0, 1]$. The bounded input $\mathbf{Y}(k)$ received by the various individual neural populations within the field is represented by an (IJ)-dimensional matrix with the order

$$\mathbf{Y}(k) = \begin{bmatrix} y_{1,1}(k) & y_{1,2}(k) & \cdots & y_{1,J}(k) \\ y_{2,1}(k) & y_{2,2}(k) & \cdots & y_{2,J}(k) \\ \vdots & & \ddots & \\ y_{I,1}(k) & y_{I,2}(k) & \cdots & y_{I,J}(k) \end{bmatrix}. \qquad (14)$$

Many information-processing applications require that the nonlinear map have predefined characteristics. For a sigmoidal map [figure 2b] the point of inflection θ^1 is often assumed to be given by

$$\theta^1 = \frac{\beta_2 - \beta_1}{2}, \qquad (15)$$

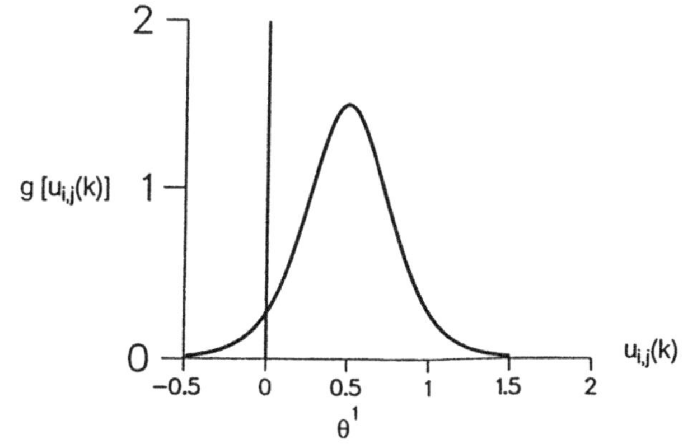

(a) Unimodal distribution of neural thresholds for $M = 1$.

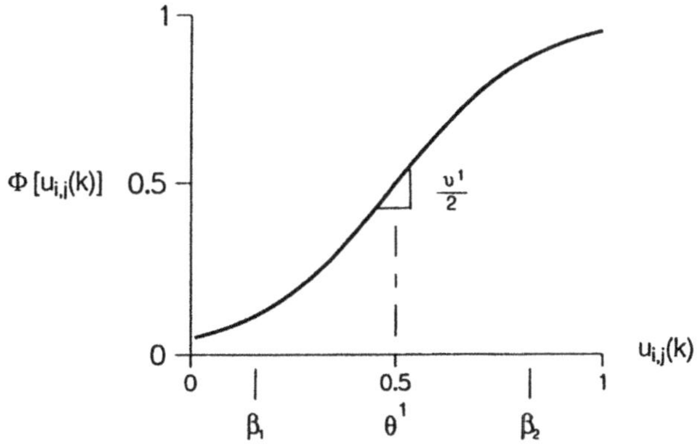

(b) Sigmoidal nonlinear mapping operator corresponding to (a).

Fig. 2. Unimodal neural threshold distribution $g[u_{i,j}(k)]$ and corresponding sigmoidal nonlinear mapping operator $\Phi[u_{i,j}(k)]$. The point of inflection is at $\theta^1 = 0.5$, and the corresponding slope parameter is $\upsilon^1 = 3.0$.

where the symmetrical parameters β_1 and β_2 are used to determine the maximum slope of the $\Phi[u_{i,j}(k)]$ about θ^1 by the relation

$$\frac{\upsilon^1}{2} = \frac{1.0}{(\beta_2 - \beta_1)}. \tag{16}$$

Note that as $\upsilon^1 \to \infty$ the map becomes binary and the applied inputs about θ^1 are limited to two possible states (0, 1). Alternatively, if the nonlinear mapping operator has multiple inflection points, i.e., if $M > 1$, as shown in figure 3, then it will quantize the applied inputs $\mathbf{U}(k)$ about $(M + 1)$ states.

The state evolution of the populations within the (IJ)-dimensional neural-activity field may be represented by the first-order difference equation [1], [4], [7], [9], [10], [12]

$$\mathbf{X}(k + 1) = \alpha\mathbf{X}(k) + (1 - \alpha)\mathbf{Y}(k), \tag{17}$$

where α is the rate of decay in the present state, $\mathbf{X}(k) \in [0, 1]$, and $(1 - \alpha)$is the rate of growth in the new state activity due to the inputs $\mathbf{Y}(k) \in [0, 1]$ received by the various populations. If $\alpha \to 0$, then the new state $\mathbf{X}(k + 1)$ is directly proportional to the present input state $\mathbf{Y}(k)$. Furthermore, by intercoupling more than

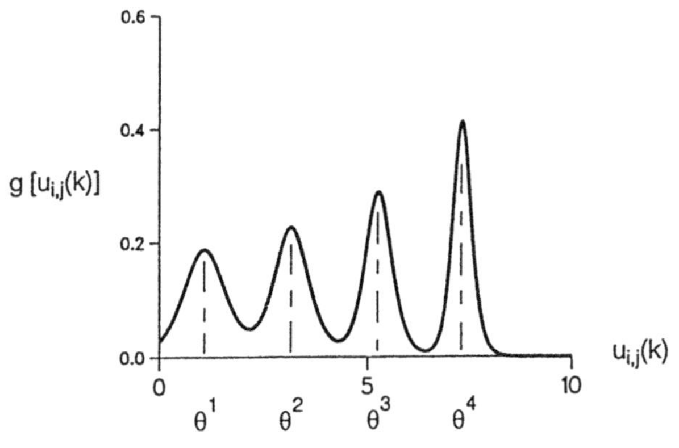

(a) Multimodal distribution of neural thresholds for $M = 4$.

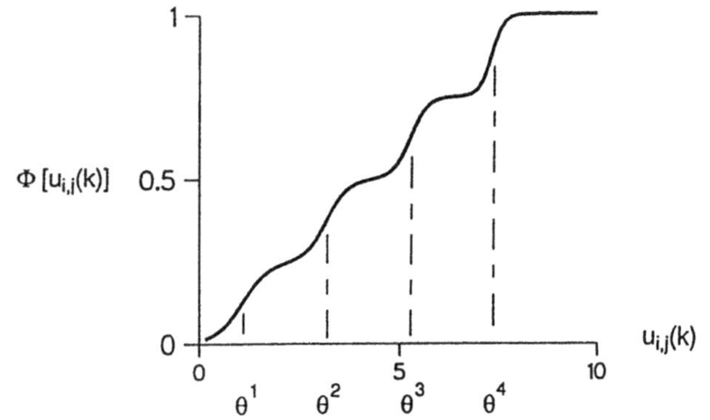

(b) Nonlinear mapping operator with multiple inflection points that correspond to (a).

Fig. 3. Multimodal neural-threshold distribution $g[u_{i,j}(k)]$ and corresponding nonlinear mapping operator $\Phi[u_{i,j}(k)]$: $v^1 = 1.5, v^2 = 1.8, v^3 = 2.3, v^4 = 3.28, \theta^1 = 1.075, \theta^2 = 3.175, \theta^3 = 5.275$, and $\theta^4 = 7.34$.

one field it is possible to generate higher-order dynamics [5], [8]. A diagram for the state-space model of the neural-activity field given by (1) to (17) is shown in figure 4. The state-space representation of a generalized neural-activity field, as shown in figure 4, is the theoretical basis of a multitask information processor for machine-vision applications. Fundamentally, the neurovision processor is a two-dimensional array of laterally distributed nonlinear PE's. Each PE within the processor represents the nonlinear first-order dynamic activity $x_{i,j}(k)$ exhibited by a spatially localized population of excitatory and inhibitory nerve cells.

3 Visual-Information Processing

A variety of different information-processing operations associated with the early stages of machine vision can be realized by this basic architecture. These diverse operations are achieved

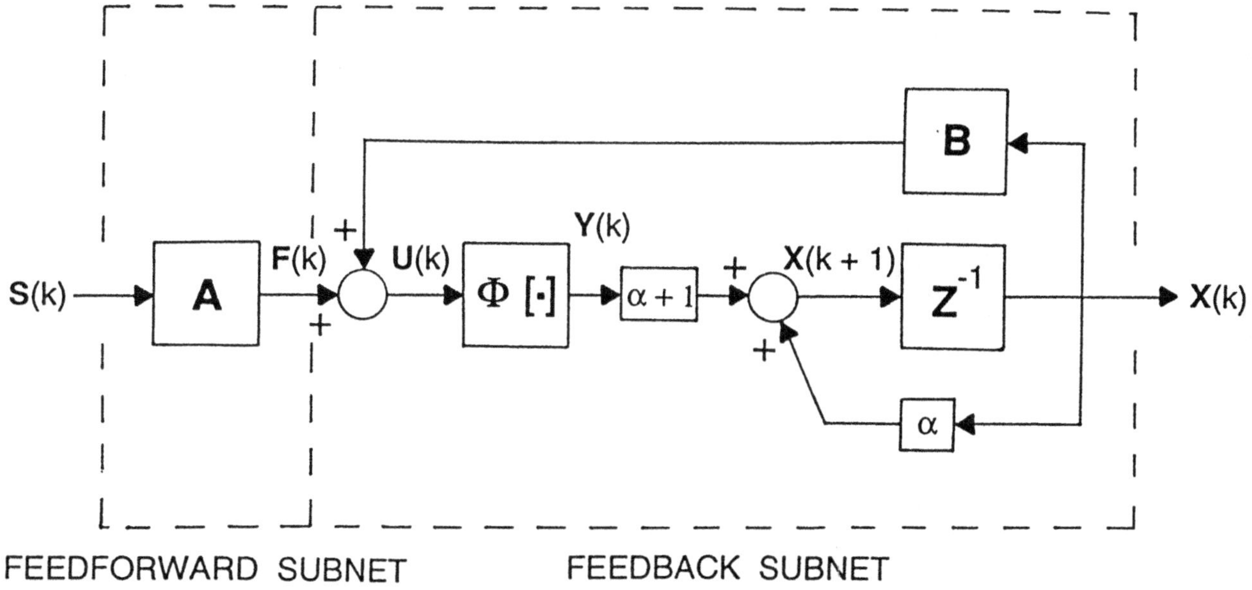

Fig. 4. State-space model for describing the information-processing capabilities of a generalized neural-activity field. This model is the basic architecture for the multitask neurovision processor design.

by selectively programming the system matrices **A** and **B** and the parameters v^m and θ^m of the nonlinear mapping operator $\Phi[\cdot]$. The coefficients of **A** in the feedforward subnet are made to act as a linear spatial filter [13], [14] that either smoothes or enhances features embedded within the external input $\mathbf{S}(k)$. The coefficients selected for **B** in the feedback subnet, is conjunction with properly chosen parameters for $\Phi[\cdot]$, determine the computational role of the neurovision processor in terms of spatiotemporal filtering, short-term visual memory, spatiotemporal stabilization, competitive feedback interaction, and content-addressable memory. The contribution of each constituent subnet to these visual-information-processing tasks will now be described.

3.1 Feedforward Subnet **A**: *Feature Extraction*

The overall feedforward input $\mathbf{F}(k)$ received by the neurovision processor at time k is given by the linear state equation

$$\mathbf{F}(k) = \mathbf{AS}(k), \tag{18}$$

where the matrix **A** transforms the input signal $\mathbf{S}(k)$ into a particular information attribute, or feature, represented by $\mathbf{F}(k)$. In other words, the coefficients of **A** represent the synaptic connections for the feedforward subnet that can be programmed to extract features from the signal space $\mathbf{S}(k)$.

Each coefficient of matrix **A** in the feedforward subnet is defined by a function $a_{i,j,p,q}$ that corresponds to the degree that the input signal at location (p,q), $s_{p,q}(k)$, influences the activity of the PE situated at location (i,j) of the processor $\mathbf{X}(k)$. For simplicity it is assumed that the relative magnitudes of the synaptic connections are isotropic and depend only on the lateral distance $(i-p, j-q)$ from the PE, that is,

$$a_{i,j,p,q} = a_{i,j,s,r} \tag{19}$$

for an input located at (p,q) and (s,r) in the signal space $\mathbf{S}(k)$ of equal distance from PE at (i,j), regardless of the direction.

The feedforward input $f_{i,j}(k)$ received by the PE at (i,j) is a discrete convolution of the signal component $s_{p,q}(k)$ with the transformation

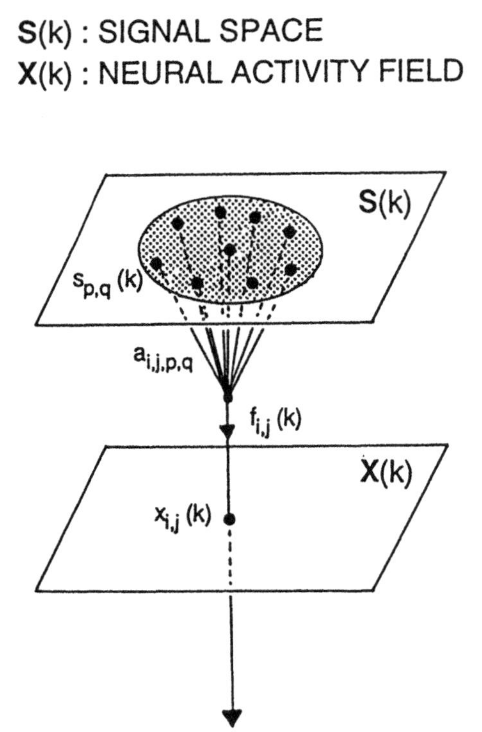

Fig. 5. Schematic diagram showing the mapping of information from a local region of the signal space $\mathbf{S}(k)$ to an individual PE at location (i, j) of the neurovision processor.

kernel centered at (i, j) given by

$$f_{i,j}(k) = \sum_{p=1}^{P} \sum_{q=1}^{Q} a_{i,j,p,q} \cdot s_{p,q}(k), \qquad (20)$$

where $a_{i,j,p,q}$ is the coefficient of the desired transformation kernel [13], [14]. Thus each PE in the neurovision processor extracts information $f_{i,j}(k)$ originating from a local region of the signal space $\mathbf{S}(k)$; see figure 5. A large number of functions can be used to define the transformation kernel given by $\mathbf{A}$. These functions may be used to extract nondirectional edges, directional edges, textures, and color information [2], [3], [13]–[15]. The two most common types of transformation kernels used to smooth and enhance two-dimensional signals are described below. These kernels are derived from Gaussian and Laplacian functions, respectively.

3.1.1 Signal smoothing. Often an external signal, such as a video image, is corrupted by excessive high-frequency noise that must be removed prior to scene analysis. In this context the transformation kernel $\mathbf{A}$ may be programmed to function as a low-pass filter in the spatial-frequency domain [13], [14]. An example of a kernel that smooths the intensities of noisy signals is a two-dimensional isotropic Gaussian function with individual coefficients given by

$$a_{i,j,p,q} = \frac{1}{2\pi\sigma^2} \exp\left[\frac{-\{(i-p)^2 + (j-q)^2\}}{2\sigma^2}\right], \qquad (21)$$

where σ is the spatial spread with respect to the lateral distance from the PE centered at (i, j). Increasing the size of σ will decrease the cutoff frequency in the spatial-frequency domain. A variety of other functions can be used to generate a smoothing kernel that performs low-pass filtering operations [13], [14].

3.1.2 Signal enhancement Alternatively, it may be necessary to sharpen information detail, such as edges, within the external input signal. In terms of the neurovision processor, this is achieved by using a gradient-transformation kernel with coefficients that function as a band-pass filter in the spatial-frequency domain. This gradient kernel may be approximated by a normalized isotropic Laplacian function [14] given by

$$a_{i,j,p,q} = \frac{1}{2}\left[2 - \frac{(i-p)^2 + (j-q)^2}{\sigma^2}\right] \exp\left[\frac{-\{(i-p)^2 + (j-q)^2\}}{2\sigma^2}\right], \qquad (22)$$

where σ is the spread of the central region of the function given over the PE at spatial location (i, j).

3.2 Feedback Subnet B *: Computational Role*

The computational role of the neurovision processor depends on the coefficients of $\mathbf{B}$ and the shape of the nonlinear mapping operator $\Phi[\cdot]$. The diagonal coefficients of the square matrix

B represent the recurrent connections that are fed back into the original PE's, and the non-diagonal coefficients of **B** represent the lateral connections between the spatially discrete PE's.

To separate the recurrent and lateral connections, let **D** be an $(IJ) \times (IJ)$ matrix containing only the diagonal coefficients, and let **L** be an $(IJ) \times (IJ)$ matrix containing the lateral (offdiagonal) coefficients of **B**. Thus the total input (3) is rewritten as

$$\mathbf{U}(k) = (\mathbf{D} + \mathbf{L})\mathbf{X}(k) + \mathbf{F}(k). \qquad (23)$$

Three broad classes of spatiotemporal-information processing are now examined. These classes involve the absence of feedback, $\mathbf{B} = \mathbf{0}$, recurrent feedback only, $\mathbf{D} \neq \mathbf{0}$ and $\mathbf{L} = \mathbf{0}$, and the presence of both lateral and recurrent feedback, $\mathbf{L} \neq \mathbf{0}$ and $\mathbf{D} \neq \mathbf{0}$. The matrix $\mathbf{0}$ is defined as the null matrix.

3.2.1 Absence of feedback, $\mathbf{B} = \mathbf{0}$ *: spatiotemporal filtering (STF).* The spatial and temporal changes that occur within the visual field are the most important and informative aspects of the sensory stimulus. Changes in the stimulus are continually present because of self-movement and the natural movement of objects in the field of view. Biological vision alleviates many of the distortions arising from fluctuations associated with sporadic movement by processing the visual information in terms of both present and past values. Temporal continuity is a critical component of visual-information processing because any physical object in a realistic environment is statistically unlikely to suddenly alter its present pattern of behavior [16]. As a result, robust forms of machine vision must account for the spatial and temporal components of visual information simultaneously and not treat them as mutually exclusive events.

In the absence of feedback connections among the laterally distributed PE's, $\mathbf{B} = \mathbf{0}$, the neurovision processor is able to perform only very localized STF operations. In terms of the neurovision processor, the spatial component of the STF operation arises from the predefined coefficients of **A**, (4) and (5), and the temporal component is a direct result of the first-order dynamics given by (17). Some support for modeling the spatial and temporal components of early biological vision in this fashion has been reported in the literature [17].

The nonlinear first-order response of a spatially isolated PE with a sigmoidal mapping operator (figure 2) acts as a low-pass temporal filter that eliminates high temporal frequencies usually associated with noise in the feedforward input (see figure 6). The steady-state response $x_{i,j}(k)$ of an isolated PE is a scaled version of the input, $f_{i,j}(k)$ that is,

$$x_{i,j}(k) = \Phi\left[f_{i,j}(k)\right] \qquad (24a)$$

$$= \Phi\left[a_{i,j,p,q} \cdot s_{p,q}(k)\right], \qquad (24b)$$

where $b_{i,j,i,j} = 0$. If the nonlinear mapping operator contains multiple inflection points, (figure 3), then the steady-state response would be a quantized value about one of the $(M + 1)$ possible states.

3.2.2 Recurrent feedback only, $\mathbf{D} \neq 0$ *and* $\mathbf{L} = 0$*: hysteresis phenomena.* The inclusion of the recurrent feedback, $\mathbf{D} \neq \mathbf{0}$, within the PE dynamics enable the neurovision processor to exhibit localized hysteresis behavior over a specific range of feedforward inputs $\mathbf{F}(k)$. One possible example of the multiple hysteresis loops formed by a single PE is shown in figure 7. An important property of localized hysteresis phenomena is that for the state activity to switch from rest to a higher state the feedforward inputs must be greater than some upper threshold value, i.e., $f_{i,j}(k) > \Omega_H^m$, where $m = 1, \ldots, M$. Similarly, the activity of a PE can return to its original rest only by a further feedforward input that is less than a lower threshold value, $f_{i,j}(k) < \Omega_L^m$. In this fashion the hysteresis loops exhibit important noise-immunity characteristics beneficial for preventing small perturbations within the feedforward inputs from drastically altering the state behavior of a PE. A neurovision processor with PE's that exhibit localized hysteresis phenomena can function as form of short-term visual memory (STVM) [1], [5], [6], [8], [18]. In this situation the state activity reverberates within the closed-loop circuit of the individual PE's such that the overall response of the neurovision processor $\mathbf{X}(k)$ remains relatively constant even after the input pattern is removed i.e,

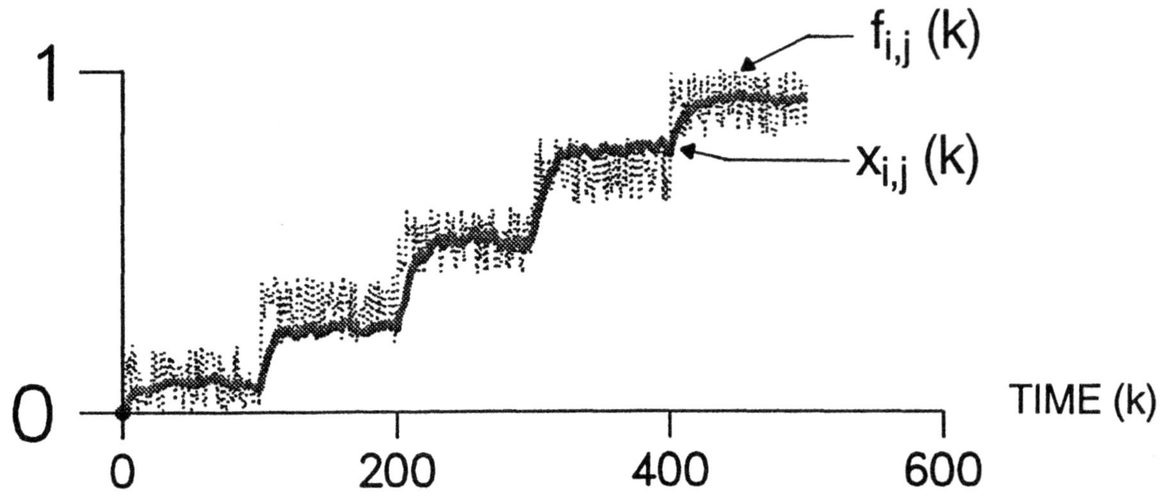

Fig. 6. Example of the temporal-filtering capabilities of a PE programmed for STF with $M = 1, \alpha = 0.9, \upsilon^1 = 3.0, \theta^1 = 0.5$ and $\mathbf{B} = \mathbf{0}/$. The standard deviation of the random noise superimposed on the signal is $\sigma = 0.2$.

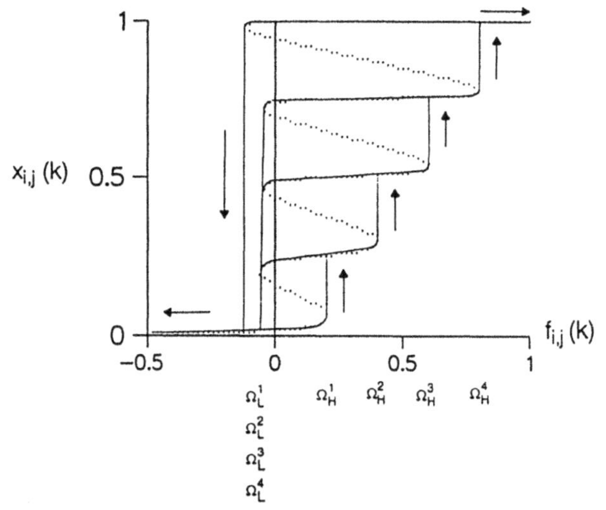

Fig. 7. The four hysteresis loops for short-term visual memory arising from the parameters $M = 4, \alpha = 0.9, \upsilon^1 = 1.5, \upsilon^2 = 1.8, \upsilon^3 = 2.3, \upsilon^4 = 3.28, \theta^1 = 1.075, \theta^2 = 3.175, \theta^3 = 5.275, \theta^4 = 7.34$, $\mathbf{D} = 8.0 \times \mathbf{I}$ and $\mathbf{L} = 0$.

$\mathbf{F}(k) = \mathbf{0}$. In this way any information retained by the PE's can be recalled while this reverberation continues. This state will continue until a strong inhibitory influence destroys the reverberating activity. STVM occurs when the lower thresholds for the hysteresis loops are all less than zero, i.e., $\Omega_L^m < 0$, where $m = 1, \ldots, M$.

An alternative application of this localized hysteresis phenomenon spatio temporal stabilization (STS). STS enables the PE's to stabilize the fluctuating feedforward inputs about a limited number of persistent states, but once the feedforward input is removed, the activity of each PE is allowed to return to its original rest state. This is achieved by selecting parameters for the map $\Phi[\cdot]$ that permit the lower thresholds for the hysteresis loops all to be greater than zero, i.e., $\Omega_L^m > 0$, where $m = 1, \cdots, M$.

3.2.3 Lateral and recurrent feedback, $\mathbf{L} \neq \mathbf{0}$ *and* $\mathbf{D} \neq \mathbf{0}$*: competitive feedback interactions.* Inclusion of the fixed lateral connections, $\mathbf{L} \neq \mathbf{0}$, increases the degree of sophistication and complexity in the processor dynamics because the state of each PE is now dependent on the state activity generated by all neighboring PE's. For example, the neurovision processor is able to exhibit feature-enhancement properties [1], [5], [7], [8], [11], [12], [19] if each PE is allowed to inhibit the state activity of its nearest neighbors. This is achieved by programming the diagonal matrix with positive coefficients for self-excitation, $\mathbf{D} > \mathbf{0}$ and the lateral matrix with negative coefficients for neighborhood inhibition, $\mathbf{L} < \mathbf{0}$. Lateral inhibition enables the processor to enhance the contrast or edges in the input signals $\mathbf{F}(k)$ [1], [5], [7], [8], [19]. In the human retina the amacrine and horizontal cells that lie in close spatial proximity are believed to exhibit such inhibitory interactions [14].

If the inhibitory connections of an individual PE are extended to include all other PE's within the processor array, then the neurovision processor will function as a global competitive network [10], [11], [19]–[22]. For competitive applications the feedforward inputs $\mathbf{F}(k)$ are taken as the initial states of the neurovision processor, that is, $\mathbf{X}(0) = \mathbf{F}(0)$ at $k = 0$. After the input pattern is removed, $\mathbf{F}(k) = \mathbf{0}$ for $k > 0$, the state activity of the processor will evolve toward a rest state. This basic competitive architecture enables the processor dynamics to select a single winning PE for a particular input. That is, the response behavior of the processor to an input pattern will gradually evolve to a state for which only the PE that best represents the input $\mathbf{F}(0)$ will remain active.

3.2.4 Lateral and recurrent feedback, $\mathbf{L} \neq \mathbf{0}$ *and* $\mathbf{D} \neq \mathbf{0}$*: Content addressable Memory.* For the applications described above, the coefficients for the matrix $\mathbf{B}$ are all predetermined and fixed. However, a variety of techniques have been developed to create different aspects of self-organization and CAM in neural networks with modifiable lateral and recurrent feedback [10], [11], [19]–[22]. One type of CAM that can be generated by the neurovision processor architecture is an autoassociative nearest-neighbor classifier that stores arbitrary input patterns $\mathbf{F}(k)$ in the coefficients of $\mathbf{B}$. Similar to the competitive feedback interactions described above, the feedforward inputs $\mathbf{F}(k)$ for CAM applications are also taken as the initial states of the neurovision processor, that is, $\mathbf{X}(0) = \mathbf{F}(0)$ at $k = 0$. Again, after the input pattern is removed, $\mathbf{F}(k) = \mathbf{0}$ the state activity of the processor will evolve toward a rest state. The resultant rest state is considered to be the recalled memory of the initial input generated by $\mathbf{F}(0)$

The neurovision processor is able to learn various analog spatial patterns if the coefficients of $\mathbf{B}$ are adjusted prior to recall by means of the Hebbian learning equation [11], [12], [19] written as

$$b_{i,j,s,r}(t+1) = \mu b_{i,j,s,r}(t) + (1-\mu)\mathbf{d}\left[f^t_{i,j}(0), f^t_{s,r}(0)\right], \qquad (25)$$

where $s = 1, 2, \ldots, I$ and $r = 1, 2, \ldots, J$. The variable t is a training pattern in the set $t = 1, 2, \ldots, T$, μ is the passive decay constant for learning, and $\mathbf{d}[\cdot]$ is the measure of the correlation between initial states $f^t_{i,j}(0)$ and $f^t_{s,r}(0)$ prior to recall and is given by

$$\mathbf{d}\left[f^t_{i,j}(0), f^t_{s,r}(0)\right] = f^t_{i,j}(0) \cdot f^t_{s,r}(0). \qquad (26)$$

The coefficients of $\mathbf{B}$ are assumed to be symmetric from the (i,j)th PE, to the (s,r)th PE i.e $b_{i,j,s,r} = b_{s,r,i,j}$. Also, the recurrent coefficients of $\mathbf{B}$ are self-excitating, $\mathbf{D} > \mathbf{0}$, and the lateral coefficients are neighbor inhibiting, $\mathbf{L} < \mathbf{0}$.

The recalled memory associated with each feedforward input $f_{i,j}(0)$ is provided by the neurovision processor dynamics, (1) to (17), such that the activity generated by the PE's that most closely resemble the input $\mathbf{F}(0)$ become maximally activated and those least resembling the input are nullified. During the recall process the activated PE's will saturate to 1 and the nullified PE's will saturate to 0. Only one PE will remain active after a long period of time if the activations are processed long enough. The spatial coordinate of the winning PE in the processor array is interpreted as the address of the recalled memory. Different feedforward inputs $\mathbf{F}(0)$ will activate different PE's in the neurovision processor. Thus the content of each input pattern $\mathbf{F}(0)$ will determine the recalled address in CAM.

4 Application to Machine-Vision Design

A brief description of the diverse computational operations that may be performed by the state-space model of a generalized neural-activity field, (figure 4) was given in section 3. Although this basic dynamic architecture can be programmed to generate a variety of information-processing operations, a single processor is able to perform only one task at a time. For any practical machine-vision application it is, therefore, necessary to construct an interactive multilevel system that uses numerous individually programmed visual-information processors. A computer simulation of a simplified vision architecture that filters, stores, and classifies noisy gray-level images is now presented. Note that the following simulation study is used to illustrate the versatility of the neurovision processor design and not to promote a new vision-system architecture. A block diagram of the three-level vision system used in this example is shown in figure 8. Each block within this diagram represents a single neurovision processor given by the state-space model shown in figure 4. The fixed parameters of these neurovision processors are given in table 1.

In this simulation study the sensor image $\mathbf{S}_1(k)$ is defined as a two-dimensional spatial pattern λ that varies with respect to the temporal function $\Gamma(k)$. The spatial pattern λ is a (50×50)-pixel array with individual gray levels distributed over the range [0, 1], and the temporal function $\Gamma(k)$ is a pulse wave given over the period $k \in [1, 150]$. In addition, time-varying random noise $\eta(k)$ bounded over the interval [0, 0.1] is superimposed on the sensor image.

Table 1. Fixed parameters assigned to the neurovision processors shown in figure 8 with $\alpha = 0.9$.[a]

	Machine-Vision Task		
Parameter	STF	STVM	CAM
Eq. (21)			
σ	2.0	0.4	0.4
Eq. (11)			
M	1	4	1
Eq. (11)			
v^1	3.0	1.5	3.0
v^2	–	1.8	–
v^3	–	2.3	–
v^4	–	3.25	–
Eq. (11)			
θ^1	0.5	1.075	0.5
θ^2	–	3.175	–
θ^3	–	5.275	–
θ^4	–	7.34	–
Eq. (23)			
D	0	$8.0 \times \mathbf{I}$	variable $\mathbf{D} > \mathbf{0}$, Eq. (25)
L	0	0	variable $\mathbf{L} < \mathbf{0}$, Eq. (25)

[a]**B** is the square matrix of the feedback subnet given by $\mathbf{B} = (\mathbf{D} + \mathbf{L})$, $\mathbf{0}$ is the null matrix, and $\mathbf{I}$ is the identity matrix.

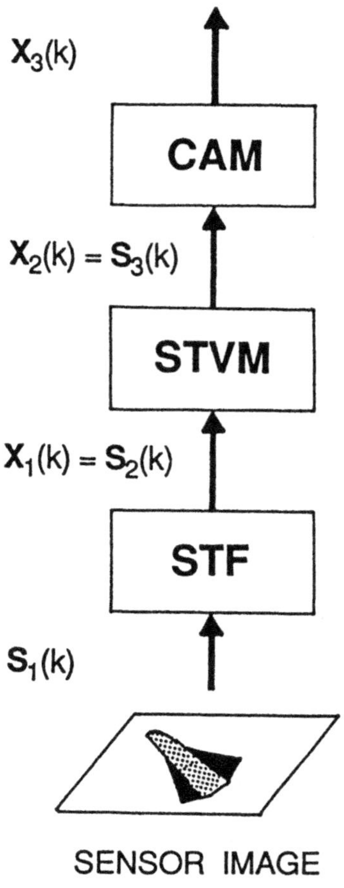

Fig. 8. Block diagram of a machine vision system with three neurovision processors that filter, store, and classify noisy gray-level images. Each block represents a state-space model as given in figure 4.

The neurovision processor at the first level performs a Gaussian or low-pass STF operation on the noisy gray-level image transmitted from the sensor. The filtered response $\mathbf{X}_1(k)$ generated by the STF processor is stored for a short time by the following STVM processor. The amount of time for which the image is stored in STVM depends on the temporal requirements of the CAM operation being performed by the neurovision processor located at the third level. Unlike the coefficients for the first two visual-information processors used in this machine-vision system, the coefficients of matrix **B** for the CAM processor are not fixed but are modified according to the learning al-

(a) CAM address:(43, 25)

(b) CAM address:(11, 37)

(c) CAM address:(10, 28)

(d) CAM address:(33, 21)

Fig. 9. Set of training images used to modify the coefficients of **B** for the CAM processor of the vision system shown in figure 8. For this example each image was presented to the system four times. The passive decay constant used in the learning algorithm (25) is $\mu = 0.9$.

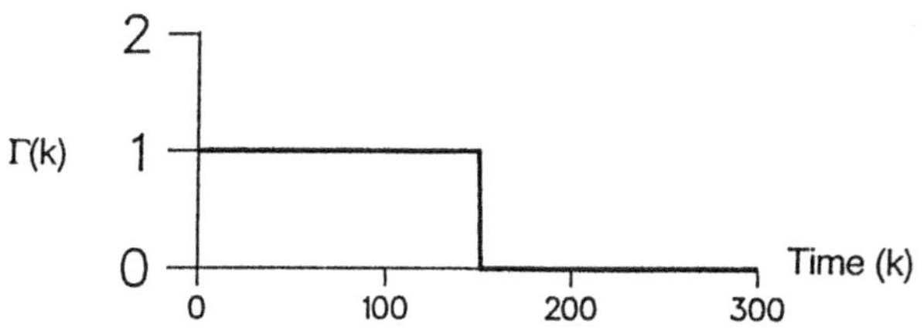

spatial pattern: λ

(a) Spatial pattern λ and temporal component $\Gamma(k)$ of the sensor image.

Fig. 10. State activity of the various neurovision processors in figure 8 of a noisy image of the Space Shuttle at times $k = 150$ and $k \gg 150$.

gorithm outlined in sub-section 3.2.4. In this example the steady-state activity generated by the STVM processor is taken as the intial state of the CAM processors that is, $\mathbf{X}_2(k) = \mathbf{X}_3(0)$. The activity exhibited by the CAM processor evolves toward a rest state, whereupon only one PE in the network remains active. The spatial coordinate or address of this PE corresponds to one of the training images shown in figure 9. Since the CAM processor is a single-layer network, it is sensitive to spatial distortions arising in the sensor image. A multilayered competitive network that is tolerant to deformations and shifts in object position, similar to the Neocognitron [19], [23], can be constructed from numerous neurovision processors that are programmed for CAM. This complex multilayered architecture is the focus of ongoing research and is, therefore, not included in the following discussion.

Figure 10 shows the state activities generated by the various neurovision processors in figure 8 of a noisy gray-level image of the Space Shuttle. The first two processors reach steady-state behavior at $k = 150$; see figure 10(b). At this time the spatial pattern in the sensor image is removed, $\mathbf{S}_1(k > 150) = \eta(k)$ and the memory recall begins within the CAM processor. As time progresses, the activity generated by the STF processor approaches rest state, i.e., $\mathbf{X}_1(k > 150) \rightarrow \mathbf{0}$. Meanwhile, the state activity exhibited by the STVM processor continues to reverberate within the network because of localized hysteresis behavior. This network state will continue until a stong negative input $\mathbf{S}_2(k) < \Omega_L^m$ is applied to the processor to erase the present contents of STVM. After a long time, $k \gg 150$, only one PE within the CAM processor remains active. As shown in figure 10(c), the CAM processor has correctly recalled the address associated with the training image given in figure 9(b).

5 Conclusions

In this paper a multitask neurovision processor based on the neural-information-processing capabilities of certain neural-activity fields found along the human visual pathway was proposed as

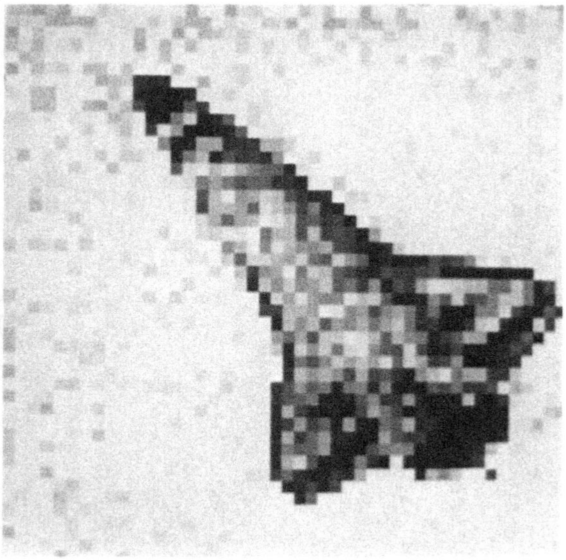

(a) $\mathbf{S}_1$ (150): Noisy sensor image

(b) $\mathbf{X}_1$ (150): STF activity

(c) $\mathbf{X}_2$ (150): STVM activity

(d) $\mathbf{X}_3$ (150): CAM activity

(b) State activity of the STF, STVM, and CAM neurovision processors at k = 150.

Fig. 10. (continued)

(a) S_1 ($\mathbf{k} \gg 150$): Noisy sensor image

(b) X_1 ($\mathbf{k} \gg 150$): STF activity

(c) X_2 ($\mathbf{k} \gg 150$): STVM activity

(d) X_3 ($\mathbf{k} \gg 150$): CAM address = (11, 37)

(c) State activity of the STF, STVM, and CAM processors at $k \gg 150$, after memory recall has been complete by the CAM processor.

Fig. 10. (continued)

a plausible computational circuit for constructing robust machine-vision systems. A state-space model for the functional architecture of a generalized neural-activity field was presented. This model formed the theoretical basis of the design of a multitask neurovision processor. In essence, this processor was represented as a two-dimensional neural network of densely interconnected nonlinear dynamic PE's. The parallel structure of the processor should reduce the computational power required by the early stages of machine vision. Furthermore, the use of dynamic PE's will ensure that the spatial and temporal components of the visual information are processed simultaneously. The interdependence of the processing of spatial and temporal information is a crucial attribute of any robust vision system.

The informative features embedded within a visual signal space that originated in a video camera or other neurovision processor were extracted by preprogramming the feedforward subnet. Modification of the feedback subnet permitted the processor to perform useful computational tasks, such as spatiotemporal filtering, spatiotemporal stabilization, short-term visual memory, competitive feedback interaction, and content-addressable memory. An example of an integrated-vision machine that used three functionally distinct neurovision processors to filter, store, and classify noisy gray-level images was described to illustrate the versatility of the neurovision processor design.

Acknowledgments

This work was supported in part by the Network of Centre of Excellence on Neuro-Vision Research (IRIS) and the Natural Sciences and Engineering Research Council of Canada.

References

1. H.R. Wilson and J.D. Cowan, "A mathematical theory of the functional dynamics of cortical and thalamic nervous tissue, " *Kybernetik*, vol. 13, pp. 55–80, 1973.
2. L. Uhr, "Psychological motivation and underlying concepts, " in *Structured Computer Vision*, S. Tanimoto and A. Klinger, (eds.), Academic Press: New York, 1980, pp. 1–30.
3. S.P. Levitan, C.C. Weems, A.R. Hanson, and E.H. Riseman, "The UMass image understanding architecture," in *Parallel Computer Vision*, L. Uhr,ed., Academic Press: New York pp. 215–248.1987
4. S. Amari, "Mathematical foundations of neurocomputing," *Proc. IEEE*, vol. 78, pp. 1443–1462, 1990.
5. M.M. Gupta and G.K. Knopf, "A multi-task visual information processor with a biologically motivated design," *J Vis. Commun. Image Rep.*, vol. 3, No. 3, pp. 230–246, 1992.
6. M.M. Gupta and G.K. Knopf, "A multi-task neuro-vision processor with extensive feedback and feedforward connections," in *Image Processing*, K.-H. Taou, ed., Proc. Soc. Photo-Opt. Instrum. Eng., vol 1606, pp. 482–495, 1991.
7. K. Kishimoto and S. Amari, "Existence and stability of local excitations in homogenous neural fields," *J. Math. Biol.*, vol. 7, pp. 303–318, 1979.
8. G.K. Knopf, "Theoretical studies of a dynamic neuro-vision processor with a biologically motivated design," Ph.D. dissertation, University of Saskatchewan, Canada, 1991.
9. H.R. Wilson and J.D. Cowan, "Excitatory and inhibitory interactions in localized populations of model neurons," *Biophys. J.*, vol. 12, pp. 1–24, 1972.
10. S. Grossberg, "Nonlinear neural networks: principles, mechanisms and architectures," *Neural Net.*, vol. 1, pp. 17–61, 1988.
11. P.K. Simpson, *Artificial Neural Systems*, Pergamon Press: New York, 1991.
12. D.S. Levine, "Neural population modeling and psychology: a review," *Math. Biosci.*, vol. 66, pp. 1–86, 1983.
13. R.C. Gonzalez and P. Wintz, *Digital Image Processing*, Addison-Wesley: Reading MA, 1977.
14. M.D. Levine, *Vision in Man and Machine*, McGraw-Hill: New York, 1985.
15. L. Uhr, "Highly parallel, hierarchical, recognition cone perceptual structures," in *Parallel Computer Vision*, L. Uhr, ed., Academic Press: New York: 1987, pp. 249–292.
16. H. Tunley, "Dynamic image segmentation and optic flow extraction," in *Proc. IEEE Int. Joint conf on Neural Networks*, Seattle WA, 1991, vol. 1, pp. 599–604.
17. H.R. Wilson, "Spatiotemporal characterization of a transient mechanism in the human visual system," *Vis. Res.*, vol. 20, pp. 443–452, 1980.
18. P.A. Anninos, B. Beek, T.J. Csermel, E.E. Harth, and G. Pertile, "Dynamics of neural structures," *J. Theor. Biol.*, vol. 26, pp. 121–148, 1970.
19. A.J. Maren, C.T. Harston, and R.M. Pap, *Handbook of Neural Computing Applications*, Academic Press: San Diego, CA, 1990.
20. J.J. Hopfield, "Neural networks and physical systems with emergent collective computational abilities," *Proc. Nat. Acad. Sci., USA*, vol. 79, pp. 2554–2558, 1982.
21. J.J. Hopfield and D.W. Tank, "Computing with neural circuits: a model," *Science*, vol. 233, pp. 625–633, 1986.
22. T. Kohonen, *Self-Organization and Associative Memory*, Springer-Verlag: Berlin, 1984.

23. K. Fukushima, S. Miyake, and T. Ito, "Neo-cognitron: a neural network model for a mechanism of visual pattern recognition," *IEEE Trans. Sys., Man, Cybern.*, vol. 13, pp. 826–834, 1983.

George Karl Knopf received a B.A. degree in the humanities and the B.E. degree in mechanical engineering from the University of Saskatchewan in 1984 and the M.Sc. and Ph.D. degrees in machine vision from the University of Saskatchewan in 1987 and 1991, respectively. Dr. Knopf is currently a research associate with the Centre of Excellence on Neuro-Vision Research (IRIS) at the University of Saskatchewan. He has coauthored numerous technical papers in the field of neurovision systems. His major research interests include machine-vision systems, neural networks, robotics, fuzzy approximate reasoning methods for ill-defined systems, and biological paradigms for engineering applications.

Madan M. Gupta is a professor and the Director of the Intelligent Research Laboratory and Centre of Excellence on Neuro-Vision Research at the University of Saskatchewan. He received the B. Eng. (with honors) in 1961 and the M.Sc. in 1962, both in electronics–communications engineering from Birla Engineering College (now the BITS), Pilani, India. He received the Ph.D. degree for his studies in adaptive control systems in 1967 from the University of Warwick, U.K. Dr. Gupta's fields of research have been adaptive control systems, noninvasive methods for the diagnosis of cardiovascular diseases, and fuzzy logic. His present research interests are neurovision, neurocontrol, neural–fuzzy computing, neuronal morphology of biological systems, intelligent systems, cognitive information, and new paradigms in information theory. In addition to publishing over 250 papers, Dr. Gupta has coauthored two books on fuzzy logic and has edited 12 volumes in the fields of adaptive control systems and fuzzy logic-computing and fuzzy neural networks. He was elected to IEEE Fellowship for his contributions to the theory of fuzzy sets and adaptive control systems and to the advancement of the diagnosis of cardiovascular disease. Recently, Dr. Gupta has been elected Visiting Professor and Special Advisor (in the area of high technology) to the European Center for Peace and Development (ECPD), University for Peace, established by the United Nations.

Journal of Mathematical Imaging and Vision 2, 251–259 (1992).

Wilson–Cowan Neural-Network Model in Image Processing

KARI MANTERE
Department of Applied Physics, University of Kuopio, P.O. Box 1627, SF-70211 Kuopio, Finland

JUSSI PARKKINEN
Department of Computer Science and Applied Mathematics, University of Kuopio, P.O. Box 1627, SF-70211 Kuopio, Finland

TIMO JAASKELAINEN
Department of Physics, University of Joensuu, P.O Box 111, SF-80101 Joensuu, Finland

MADAN M. GUPTA
Intelligent Systems Research Laboratory, College of Engineering, University of Saskatchewan, Saskatoon, Saskatchewan S7N 0W0, Canada

Abstract. The neural-network model based on the theory proposed by Wilson and Cowan has been simulated by using digitized real images. Mathematically, the model is based on coupled nonlinear differential equations that describe the functional dynamics of cortical nervous tissue, and the model can operate in different dynamical modes, depending on coupling strengths. The model is shown to store images in reduced form and to recognize edges of an object. Examples of how the network processes input images are shown.

1 Introduction

The vast majority of the research in image analysis is based on heuristic, application-oriented methods. However, there are different approaches to developing these methods. A basic approach is to think of the image as a two-dimensional function and to solve the problems mathematically. Another actively studied approach is based on (biologically motivated) neural networks. Here, the image is an input to a network of simple processing units, neurons. The topology and behavior of the network is often derived from knowledge about biological neural networks. These networks have such properties as memory and recognition of a specific pattern, and they can learn from examples.

The Wilson–Cowan neural network [1], [2] is based on the idea of two-layer (excitatory and inhibitory) network topology. The activity of each neuron is distributed to the neighboring neurons in both layers. By this distribution a property is embedded into the network: the acting unit is actually a set of neurons rather than an individual neuron. This gives a certain robustness to the behavior of the net. The Wilson–Cowan network can operate in different dynamical modes, depending on the state-transition scheme selected, which, in turn, depends on the mode parameters. In this paper we describe qualitive properties of a spatially stable steady-state network [2] by using digitized images.

Some theoretical analysis of Wilson–Cowan-type statistical neural networks has previously been performed. Therefore only basic outlines for understanding the mathematical background are given here, and the reader is referred to references [1], [2], and [3] for more detailed analysis.

2 Wilson–Cowan Model

The Wilson–Cowan model of neural networks is based on looking at the statistical behavior of a biological neural network. There are localized set of neurons that are understood to behave as a population instead of as individual neurons. These neuron populations are sometimes called netlets. This approach leads one to use expectation values over time in the development of the theory. Another characteristic of the Wilson–Cowan dynamic neural-network model is the two-layer network topology. The activity of each neuron is distributed to the neighboring neurons in both layers. One sublayer consists of excitatory neurons, and the other consists of inhibitory neurons. All possible interconnections between neurons and layers are permitted, and each connection is a function only of the distance between interconnection elements. Thus the layers are assumed to be isotropic and homogeneous.

Next we follow the guidelines of Wilson and Cowan in deriving the model. The following description refers to only the temporal dynamics of the model, but it may be generalized by taking into account the spatial interactions of neurons. A more detailed derivation can be found in the original papers [1], [2].

Let $X_P(t)(X_N(t))$ be the proportion of excitatory (inhibitory) neurons firing per unit time at the instant t. Each neuron is fired when it is in the sensitive state and it receives at least threshold excitation. Let us assume that each cell has refractory period of duration r. Then the proportions of sensitive excitatory and inhibitory cells are

$$1 - \int_{t-r}^{t} X_P(t')dt', \tag{1}$$

$$1 - \int_{t-r}^{t} X_N(t')dt'. \tag{2}$$

The response functions of neuronal activity due to the incoming synaptic activities are $f_P(z)$ and $f_N(z)$ for the excitatory and inhibitory neurons, respectively. These functions depend on the weighted sum of the neighbors in the excitatory and the inhibitory layers and on the external inputs $S_P(t)$ and $S_N(t)$, respectively. In the continuous case this can be expressed for the excitatory neurons as follows:

$$\int_{-\infty}^{t} \alpha(t-t')[w_1 X_P(t') - w_2 X_N(t') + S_P(t')]dt', \tag{3}$$

where w_1 and w_2 are positive connectivity coefficients and $\alpha(t)$ accounts for the effect of stimulation decay. A similar expression with different coefficients is applied to the inhibitory neurons. We obtain the firing of the neurons at the instant $t + \tau$ by multiplying the response functions of the neuronal activity by the number of sensitive neurons:

$$\begin{aligned} X_P(t+\tau) = {} & \left[1 - \int_{t-r}^{t} X_P(t)\right] \cdot \\ & f_P\Big\{\int_{-\infty}^{t} \alpha(t-t') \cdot \\ & [w_1 X_P(t') - w_2 X_N(t') \\ & + S_P(t')]dt'\Big\}, \end{aligned} \tag{4}$$

$$\begin{aligned} X_N(t+\tau) = {} & \left[1 - \int_{t-r}^{t} X_N(t)\right] \cdot \\ & f_N\Big\{\int_{-\infty}^{t} \alpha(t-t') \cdot \\ & [w_3 X_P(t') - w_4 X_N(t') \\ & + S_N(t')]dt'\Big\}. \end{aligned} \tag{5}$$

At this point an approximation called time coarse graining [1] is used:

$$\int_{t-r}^{t} X_P(t')dt' \approx r X_P(t), \tag{6}$$

$$\int_{-\infty}^{t} \alpha(t-t') X_P(t')dt' \approx k X_P(t). \tag{7}$$

By applying the coarse-graining approximation to (4) and (5) and by approximating the left-hand sides of (4) and (5) by two first terms of the Taylor expansion, we can derive the following equations:

$$\begin{aligned} \tau \frac{dX_P(t)}{dt} = {} & -X_P(t) + [k_P - r_P X_P(t)] \cdot \\ & f_P[w_1 X_P(t') - w_2 X_N(t') \\ & + S_P(t')], \end{aligned} \tag{8}$$

$$\begin{aligned} \tau \frac{dX_N(t)}{dt} = {} & -X_N(t) + [k_N - r_N X_N(t)] \cdot \\ & f_N[w_3 X_P(t') - w_4 X_N(t') \\ & + S_N(t')], \end{aligned} \tag{9}$$

where time constant τ has been assumed to be the same for both the excitatory and the inhibitory layers.

In this paper we test the network with images, and therefore spatial coordinates must also be added. Furthermore, because we use a discrete approach in time and space, the formulas must be converted to discrete form. In addition, the refractory periods r_P and r_N have been set equal to unity. Let $X_P(k,s,r)$ represent the state activity at time k for an excitatory neuron at spatial location (s,r), and, correspondingly, let the state activity of an inhibitory neuron be $X_N(k,s,r)$. The spatial–temporal behavior of the layers is defined by the discrete-space and the discrete-time equations [4]

$$\begin{aligned} X_P(k+1,s,r) &= (1-a)X_P(k,s,r) \\ &\quad +a[b-X_P(k,s,r)] \\ &\quad f_P[u_P(k,s,r)], \end{aligned} \tag{10}$$

$$\begin{aligned} X_N(k+1,s,r) &= (1-a)X_N(k,s,r) \\ &\quad +a[b-X_N(k,s,r)] \\ &\quad f_N[u_N(k,s,r)]. \end{aligned} \tag{11}$$

The parameter a governs the state-transition speed, and b is a scaling factor. The response functions $f_P(u_P(k,s,r))$ and $f_N(u_N(k,s,r))$ are typically monotonically increasing sigmoid functions and generally have a lower asymptote of 0 and an upper asymptote of 1. These functions prevent small activities from exciting any neuron elements and permit extremely large excitations by strong activities. The sigmoid function can be defined as

$$f(u(k,s,r)) = \frac{1}{1+\exp[-\nu(u(k,s,r)-\theta)]} - \frac{1}{1+\exp(\nu\theta)}, \tag{12}$$

where ν determines maximum slope and θ determines the position of the maximum slope (the last term in (12) only shifts the curve downward so that $f(0) = 0$). A piecewise-linear function and a step function have been also tested, where the piecewise linear function is defined as

$$f(u(k,s,r)) = \begin{cases} 0 & \text{if } u(k,s,r) < T_1, \\ qu(k,s,r)+p & \text{if } T_1 \le u(k,s,r) \le T_2, \\ 1 & \text{if } u(k,s,r) > T_2, \end{cases} \tag{13}$$

and the step function is defined as

$$f(u(k,s,r)) = \begin{cases} 0 & \text{if } u(k,s,r) < T_3, \\ 1 & \text{if } u(k,s,r) \ge T_3, \end{cases} \tag{14}$$

where T_1, T_2 and T_3 are some threshold values, q is a slope of the curve, and p is a constant. The functions $u_P(k,s,r)$ and $u_N(k,s,r)$ are the total inputs to each neuron element at time k and at location (s,r). They depend on connectivity weights w_{PP}, w_{NP}, w_{PN}, and w_{NN}, which determine the gain between different neuron units (excitatory–excitatory, excitatory–inhibitory, inhibitory–excitatory, and inhibitory–inhibitory, respectively). These correspond to w_1, w_2, w_3, and w_4 in the continuous case.

The inputs are also functions of synaptic-connectivity function $g(k,s,r)$, which determines the strength of a neuron's activity, depending on its connection distance on neighbors. Thus the total inputs for both layers are defined as

$$\begin{aligned} u_P(k,s,r) &= w_{PP}g_{PP}(k,s,r) \\ &\quad -w_{NP}g_{NP}(k,s,r) \\ &\quad +S_P(k,s,r), \end{aligned} \tag{15}$$

$$\begin{aligned} u_N(k,s,r) &= w_{PN}g_{PN}(k,s,r) \\ &\quad -w_{NN}g_{NN}(k,s,r) \\ &\quad +S_N(k,s,r), \end{aligned} \tag{16}$$

where $S_P(k,s,r)$ and $S_N(k,s,r)$ are external inputs to each layer. Functions $g_{PP}(k,s,r)$, $g_{PN}(k,s,r)$, $g_{NP}(k,s,r)$, and $g_{NN}(k,s,r)$ are

$$g_{PP}(k,s,r) = \sum_{i=0}^{I-1}\sum_{j=0}^{J-1} \varphi_{PP}(i-s,j-r)X_P(k,i,j), \tag{17}$$

$$g_{PN}(k,s,r) = \sum_{i=0}^{I-1}\sum_{j=0}^{J-1} \varphi_{PN}(i-s,j-r)X_P(k,i,j), \tag{18}$$

$$g_{NP}(k,s,r) = \sum_{i=0}^{I-1}\sum_{j=0}^{J-1} \varphi_{NP}(i-s,j-r)X_N(k,i,j), \tag{19}$$

$$g_{NN}(k,s,r) = \sum_{i=0}^{I-1}\sum_{j=0}^{J-1} \varphi_{NN}(i-s,j-r)X_N(k,i,j), \tag{20}$$

where φ_{PP}, φ_{PN}, φ_{NP} and φ_{NN} are normalized spatial-distribution functions for which we have used Gaussian functions. The variables I and J define the size of the area where the activity of a neuron is distributed.

The Wilson–Cowan network can operate in different dynamical states, depending on the parameters, and we have studied the qualitive properties of spatially stable steady states. Once the model reaches such a state, the neural activity is self-maintained without any further input [2]. Changes were made to the connectivity weights w_{PP}, w_{PN}, w_{NP}, and w_{NN}, to the range over which the connectivity functions g were affected, and to the distribution functions φ. The parameters were found experimentally by starting from the values published by Gupta and Knopf [4].

3 Computer Simulations

We have simulated the behavior of the Wilson–Cowan model by using as test objects real digitized images: a coffee cup (164 × 164 pixels) and a face image(164 × 164 pixels). For the tests we used a Sun Sparcstation II workstation, and the simulations were realized by using the MATLAB program. In all simulations the external input $S_N(k, s, r) = 0$ to the inhibitory layer and the output, which was obtained from the excitatory layer, were bounded by [0, 1].

Figure 1 shows the coffee-cup image with continuous input, and figure 2 shows the same image with the input turned off (set equal to zero) at the 20th iteration. In both figures the first picture is the original and the last is after 100 iterations. The other pictures are after 19, 30, 50, and 75 iterations. In both figures we have used the same parameter set.

With continuous input the image became sharpened, and by changing the connectivity weights and the spatial-distribution functions the degree of sharpness could be controlled. However, the output disappear if the input is applied for a sufficiently long time because all neurons will approach the upper asymptotic value of the response function. In figure 2, where the input was turned off at the 20th iteration, the network could store the input image, i.e., the network reached a stable state and the activity of the neurons was self-maintained. Furthermore, the edges of the image were enhanced. Figure 2 shows that the Wilson–Cowan network can store images and that it extracts certain features from input. The latter case becomes more obvious if we study figure 3 where the face image was used as input.

Figure 3(a) is the original image, and the other parts of the figure are for different parameter sets after 50 iterations when the input was removed after 20 iterations in each simulation. Figure 3(b) was made with the same parameter set as in figure 2, but because of the large amount of detail in the face image the output is confused. In figure 3(c) the parameters have changed so that even though the weights between layers are more distributed, they are smaller than those used in figure 3(b). Now the output of the system is a smoothed version of the input and the details have been removed, but the basic features of the stored image have remained clear. The output of the system will remain unchanged until it is erased by sufficiently strong inhibition of a new input image. Figure 3(d) represents the output when the image has changed into a binary one. Certain image areas that are above some threshold value have been moved to the excited state (dark areas), and the other areas are in the rest state (white areas). Thus by changing the mode parameters suitably it is possible to binarize images and to reduce unnecessary noise components. The same phenomenon appears in figure 3(c).

We have also made some simulations with a piecewise-linear function and with a step function. If the step function is used as the response function, the output is binarized because there are only two possible states for responses of the neuronal activity. With the piecewise-linear function the results are similar to those of the sigmoid function, depending on the slope of the linear response function. For both functions the threshold values (T_1 and T_3) must be set small enough so that neuronal activity will exceed them in reasonable time. With the sigmoid function the activity attains larger and larger values during iteration and choosing the

Fig. 1. Coffee-cup image with continuous input. Iteration steps from the upper-left corner to the lower-right corner rowwise: original, 19, 30, 50, 75, and 100 iterations.

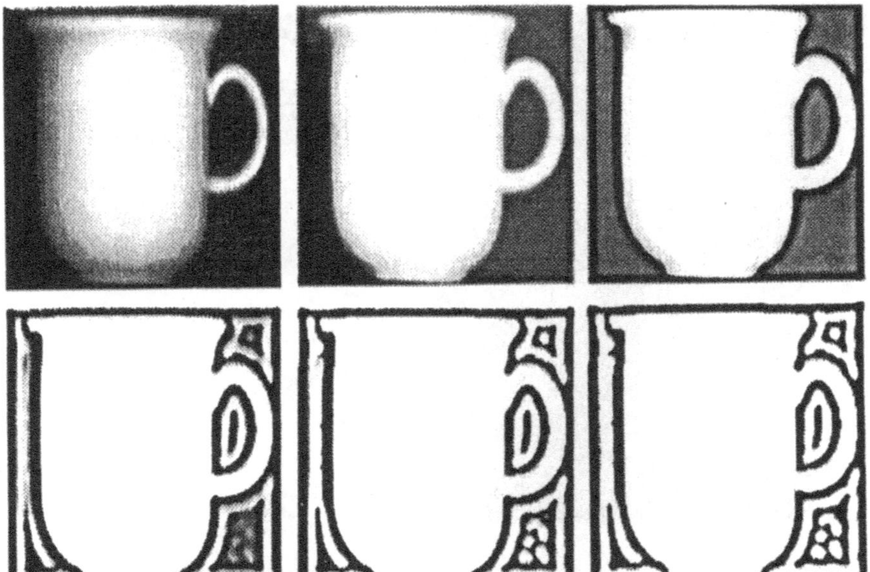

Fig. 2. Same as figure 1, but the input was turned off at the 20th iteration.

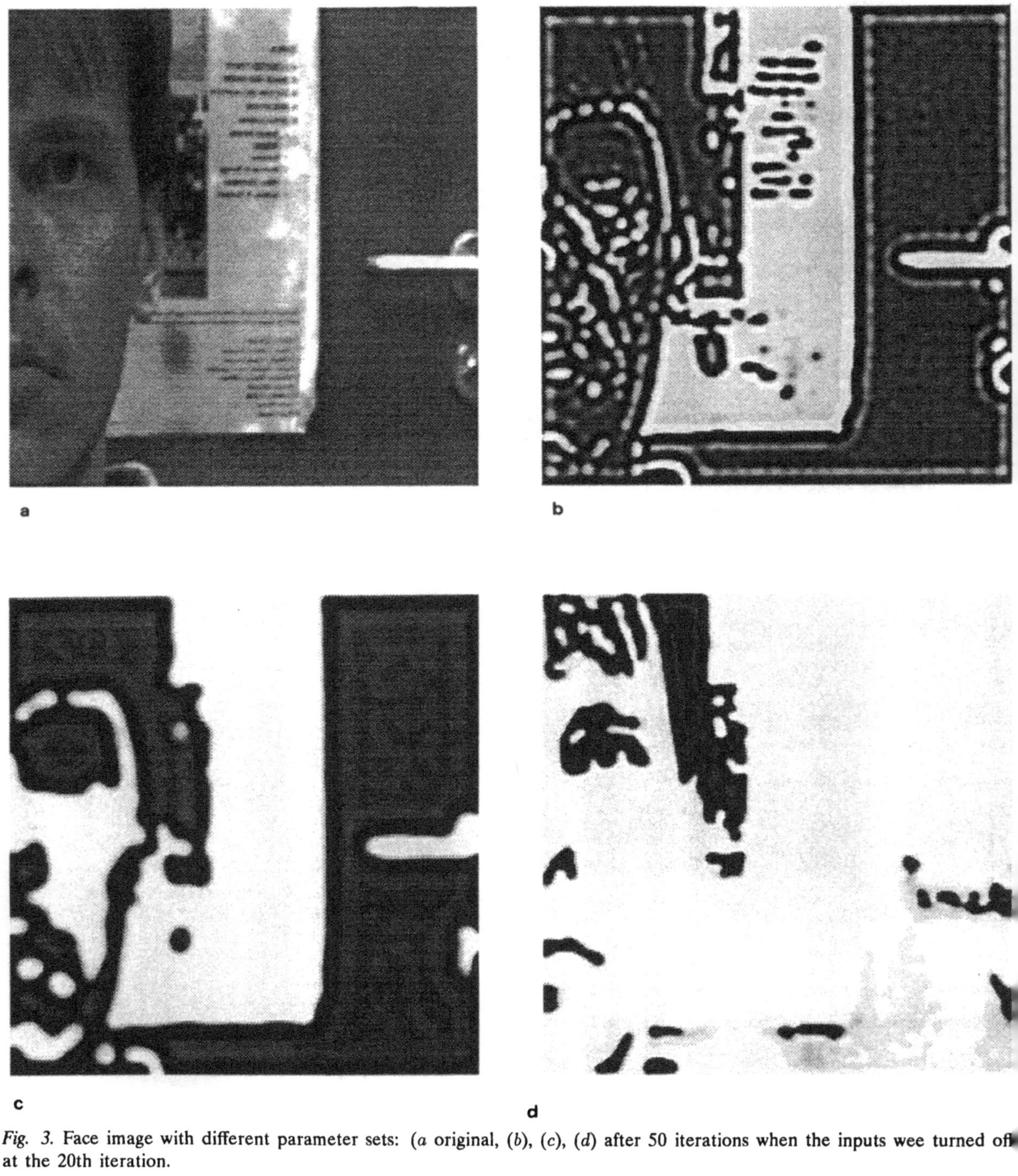

Fig. 3. Face image with different parameter sets: (*a* original, (*b*), (*c*), (*d*) after 50 iterations when the inputs wee turned off at the 20th iteration.

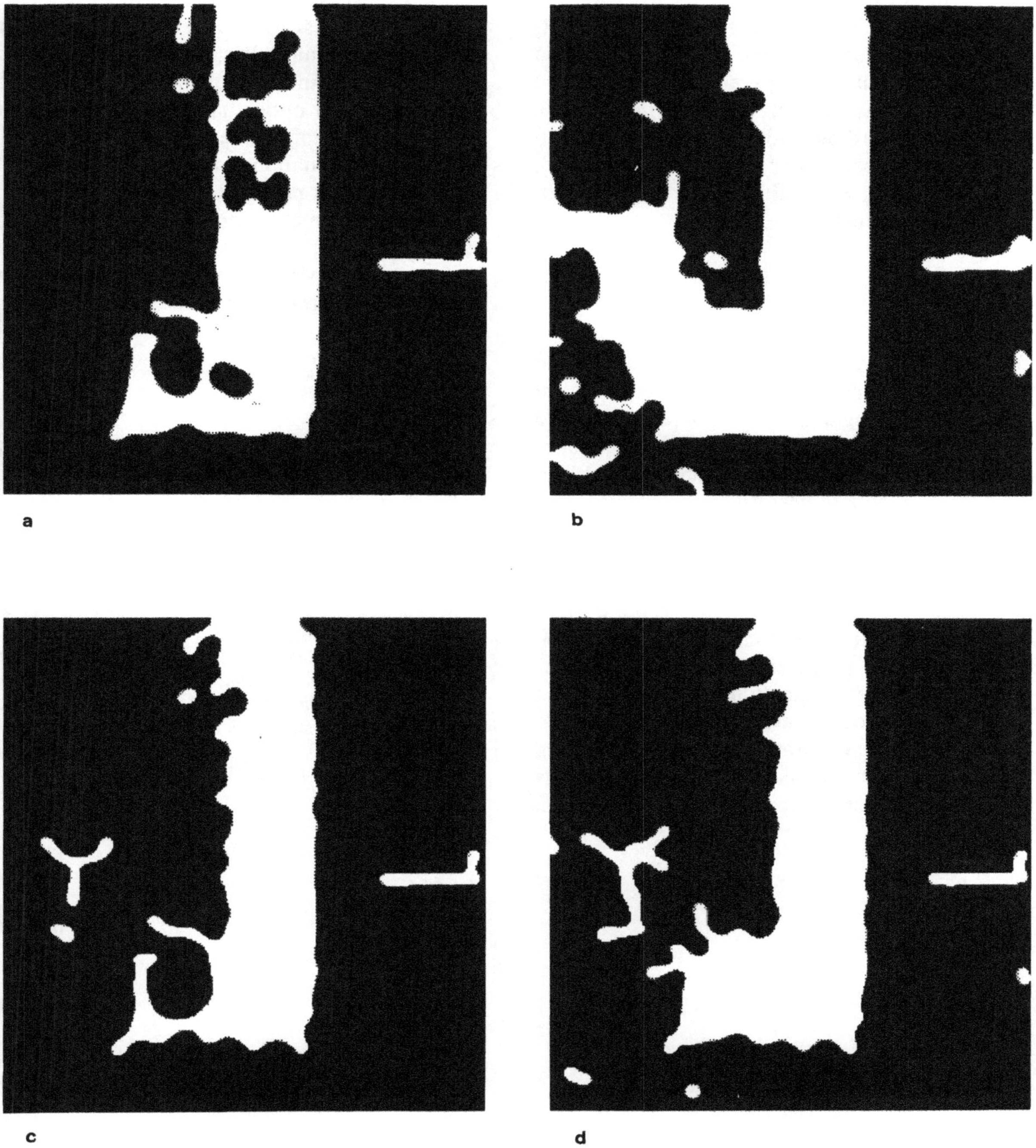

a b c d

Fig. 4. Examples with (*a*), (*b*) the piecewise-linear function and (*c*), (*d*) the step function.

position of the maximum slope is not crucial. Figure 4 shows two examples each of the use of the piecewise-linear and the step functions. These example show that the shift in the cut-off threshold regulates the saturated area on the image. All other parameters in figure 4 are the same as the parameters in figure 3(c). The cutoff thresholds for figure 4 were set to find outputs most similar to those of figure 3(c), thus illustrating the influence of the shape of the response function.

4 Discussion

We have studied the qualitive properties of the Wilson–Cowan neural-network model in image processing. Using digitized real images, we have shown that the network is able to store images and to reduce their information contents. The model is also able to perform edge enhancement of the object and to binarize it. We used the sigmoidal response function, and we showed that although the piecewise-linear and step functions are possible, they may not be applicable if one's only concern is the biological basis of the model.

The different properties of the network depend on parameter selection, and obviously the network can produce more properties than we have found. We did not use a specific stopping criterion in our simulations, and the outputs were obtained only at some specific iteration cycles. Iteration time, i.e., when the input was applied to the network, was also chosen experimentally. These issues need further research.

Parallel processing should be the preferable simulation technique because of the large number of neurons in the model. Optics have high parallelism, and optical systems can easily perform a number of mathematical operations [5]. The discrete equations of the model consist of pointwise matrix multiplication and addition operations, a correlation operation, and a thresholding operation. It is possible to perform these operations by using a known technique with a hybrid-optical processing unit.

References

1. H.R. Wilson and J.D. Cowan, "Excitatory and inhibitory interactions in localized populations of model neurons," *Biophys. J.*, vol. 12, pp. 1–24, 1972.
2. H.R. Wilson and J.D. Cowan, "A mathematical theory of the functional dynamics of cortical and thalamic nervous tissue," *Kybernetik*, vol. 13, pp. 55–80, 1973.
3. H. Sakaguchi, "Oscillatory and excitable behaviours in a population of model neurons," *Prog. Theor. Phys.*, vol. 79, pp. 1061–1068, 1988.
4. M.M. Gupta and G.K. Knopf, "A dynamic neural network for visual memory," in *Visual Communications and Image Processing '90: Fifth in a Series*, M. Kunt, ed., Proc. Soc., Photo-Opt. Instrum. Eng., vol. 1360, pp. 1044–1055, 1990.
5. J.L. Horner (ed.), *Optical Signal Processing*, Academic Press: San Diego, CA, 1987.

Kari Mantere received his M.Sc. in 1990 in physics from the University of Kuopio, Finland and he is presently a researcher at the Optical Information Processing research project of the Academy of Finland.

Timo Jaaskelainen received his Ph.D. in 1980 in physics from the University of Joensuu, Finland. He is an associate professor of physics and the Head of Department of Physics at the University of Joensuu. He has published over 60 papers in the areas of volume diffraction grating, optical information processing, and optical instrumentation in industry and medicine.

Jussi Parkkinen received his Ph.D. in 1989 in applied mathematics from the University of Kuopio, Finland. He is presently an associate professor of computer science at the Lappeenranta University of Technology. He has published about 30 papers in the areas of pattern recognition, digital image processing and optical information processing.

t the College of Engineering, University of Saskatchewan, Saskatoon, Canada. In addition to a number of papers, he has coauthored several books in the field of fuzzy sets. Please refer to another article in this issue.

Journal of Mathematical Imaging and Vision 2, 261–272 (1992).

Clustering Properties of Hierarchical Self-Organizing Maps

JOUKO LAMPINEN AND ERKKI OJA
Department of Information Technology, Lappeenranta University of Technology, P.O. Box 20, SF-53851 Lappeenranta, Finland

Abstract. A multilayer hierarchical self-organizing map (HSOM) is discussed as an unsupervised clustering method. The HSOM is shown to form arbitrarily complex clusters, in analogy with multilayer feedforward networks. In addition, the HSOM provides a natural measure for the distance of a point from a cluster that weighs all the points belonging to the cluster appropriately. In experiments with both artificial and real data it is demonstrated that the multilayer SOM forms clusters that match better to the desired classes than do direct SOM's, classical k-means, or Isodata algorithms.

Key words: cluster analysis, self-organizing maps, neural networks

1 Introduction

Most of the neural-network research in pattern recognition, image processing, and vision has been focused on supervised learning. Supervised neural networks, such as the multilayer perceptron (MLP), provide a highly efficient model-free method for designing an arbitrarily complex nonlinear classifier iteratively from examples. There are many sample cases showing the power of neural classifiers over classical methods; for a review see, e.g., [1].

A central theoretical result giving impetus to the increasing interest in neural networks is that an MLP with only one nonlinear hidden layer can approximate any continuous function on a compact domain to an arbitrary precision, or as a classifier it can form arbitrarily complex class boundaries [2], [3]. The close relationship between the outputs of the MLP and the optimal Bayes classifier has also been clarified [4].

However, there are some problem domains that cannot be solved with merely a powerful classifier. When the abstraction level of the classification task increases, the shapes of the regions associated together become increasingly complex, requiring impossibly large amounts of training data to form the class boundaries.

Perhaps the most important examples of such problems are in machine vision and image understanding. The essential tasks are locating and recognizing individual objects and compiling a useful interpretation from the objects and their relations. Both of these subtasks have proven to be extremely difficult. The classical approach of programming the *a priori* knowledge or model of the objects into the solution has severe limitations in handling all the natural variations in images. Also, the methods cannot easily adapt to unanticipated or changing situations.

To handle the large variability of natural scenes the image-analysis system must have a large number of free parameters in the early stages, and estimating the parameters requires a lot of data. Using any neural network trained by supervision for the entire image-analysis system would therefore require a huge network with a correspondingly huge number of manually classified samples, and collecting the samples would clearly be very expensive.

This dilemma can be solved by using unsupervised learning techniques in early stages to reduce the number of degrees of freedom in the data. Then the final supervised classifiers, giving semantic labels to objects or their primitives, can have a much smaller number of free parameters, thus requiring fewer preclassified training samples. This scheme is especially suitable for scene analysis, since it is fairly unexpensive to collect large amounts of image data and to train a neural network with them, as long as the images do not need manual analysis and classification.

The biological neural networks seem to have a similar basic structure, at least at the lowest levels. Although the very first stages in the sensory information pathways are genetically determined, the signals are thereafter fed to sensory maps. It has been shown that these maps can be formed by means of self-organization [5], and there is biological evidence from, e.g., deprivation experiments that the maps are indeed spanned by input data during the development of the network. Very profound conclusions about biological-image-analysis methods cannot be drawn at present. The role of expectations and guessing as means of creating bootstrap data for supervised learning is not known, but it is likely that the biological systems contain continuous hypothesis-generation and testing mechanisms and that feedback throughout the system is a very important factor in guiding the learning process.

In the next sections we discuss the use of a multilayer version of the self-organizing map (SOM) neural network, the hierarchical self-organizing map (HSOM), as a clustering preprocessor in an image-analysis system. The basic one-layer SOM, covered in section 2, divides the input space into convex regions in a fashion analogous to a one-layer feedforward network. The SOM is shown to have more desirable properties than do classical clustering methods: it provides a natural measure for the distance of a point from a cluster that is adaptive to the local statistics of the data. The SOM forms one complex shape following the data distributions in the space, so that regions of the map can be interpreted as clusters in the space, and the difference of the cluster indices is correlated to the weighted distance of all the points in the clusters. In the multilayer hierarchical SOM, discussed in section 3, the outputs of the first SOM are fed into another SOM as input, causing the SOM to divide into distinct cluster representations. The HSOM is shown to form arbitrarily complex clusters, in analogy with multilayer feedforward networks. Section 4 contains experimental results on synthetic and real data that confirm the desirable properties of the HSOM.

2 Self-Organizing Map

2.1 Basic SOM

The SOM, introduced by Kohonen [6], is one of the best-known unsupervised-learning neural networks. It belongs to the class of vector-coding algorithms. In vector coding the problem is to place a fixed number of vectors, the codewords, into the input space, which is usually a high-dimensional real space. Each codeword will correspond to and represent a part of the input space: the set of points in the space that are closer in distance to that codeword than to any other codeword. This produces a Voronoi tessellation into the space. The overall criterion in usual vector coding is to place the codewords in such a way that the average distances from the codewords to the input points belonging to their own Voronoi compartment are minimized.

One way to understand the SOM is to consider it to be a neural-network implementation of this basic idea: each codeword is the weight vector of a neural unit. However, there is an essential extra feature in the SOM. The neurons are arranged in a 1-, 2-, or multidimensional lattice such that each neuron has a set of neighbors, e.g., in two dimensions either four or eight neighbors can be chosen. The goal of learning is not only to find the most representative code vectors for the input space in mean square sense but also to realize a topological mapping from the input space to the grid of neurons. Mathematically, this can be defined as follows.

For any point x in the input space Ω, one or several of the codewords are closest to it. In the following, the distance is the Euclidean distance, but a generalization to other distance functions would be straightforward. Assume that m_b is the closest among all the codewords $m_1, \ldots, m_M$:

$$\| x - m_b \| = \min \| x - m_j \|, \quad j = 1, \ldots, M, \tag{1}$$

Where j is the usually multidimensional index giving the position of m_j in the lattice. To make the correspondence unique, assume that some tie-breaking rule is used if several codewords happen to be at exactly the same minimum distance from x. The unit b having the weight vector m_b is then called the *best-matching unit* for vector x, and index $b = b(x)$ can be considered to be the output of the map. Note that for fixed x equation (1) defines the index b of the best-matching unit and for fixed b equation (1) defines the Voronoi compartment of unit b as the set of points that satisfy (1). By the above relation the input space is mapped to the discrete set of neurons. If each neuron is taken to represent one cluster, then the clusters will have a convex polyhedral shape.

A topological mapping is defined as follows: if an arbitrary point $x \in \Omega$ is mapped to unit i, then all points in a neighborhood of x are mapped either to i itself or to one of the units in the neighborhood of i in the lattice. This implies that if i and j are two neighboring units, then their Voronoi compartments have a common boundary. Whether the topological property can hold for all units, however, depends on the dimensionalities of the input space and the neuron lattice. In some earlier works on topologically ordered neuron layers [7], such a mapping was made one-to-one by using a continuum instead of the discrete neuron lattice and by requiring that neighborhoods of points be mapped to neighborhoods. Because no genuine topological maps between two spaces of different dimensions can exist, a two-dimensional neural layer can only follow locally two dimensions of the multidimensional input space.

The Kohonen algorithm for self-organization of the code vectors is as follows [5]:

(i) Choose initial values randomly for the weight vectors m_i of the units i.
(ii) Repeat steps (iii) and (iv) until the algorithm has converged:
(iii) Draw a sample x from the probability distribution of the input samples, and find the best matching unit b according to equation (1).
(iv) Adjust the weight vectors of all units by

$$m_j := m_j + \gamma * h_{b,j} * (x - m_j), \qquad (2)$$

where γ is a gain factor and $h_{b,j}$ is the so-called neighborhood function; usually it is a function of the distance $b-j$ of units b and j measured along the lattice. (In the original version [6] the neighborhood function was equal to 1 for a certain neighborhood of b and was 0 elsewhere. The neighborhood and the gain γ should slowly decrease in time.)

The convergence and the mathematical properties of this algorithm have been considered by several authors, e.g., [6], [8], and [9].

2.2 SOM Optimization

The map algorithm is related to an energy function in [8] and [10]. Let V_b denote the set in the input space where (1) holds, i.e., the Voronoi compartment of unit b. Let $p(x)$ denote the probability density of the inputs x. Define the cost or energy function as

$$E(m_1, \ldots, m_M) = \sum_i \int_{V_i} \sum_k h_{i,k} \parallel x - m_k \parallel^2 p(x) d(x). \qquad (3)$$

The functional (3) is piecewise differentiable. Let us write it in the equivalent form

$$E = E(m_1, \ldots, m_M) = \int \sum_k h_{b,k} \parallel x - m_k \parallel^2 p(x) d(x), \qquad (4)$$

with b defined appropriately as the index of the best-matching unit. This moves the discontinuity of the V_i to the function

$$b = b(x, m_1, \ldots, m_M). \qquad (5)$$

The usual way to minimize a functional like E, in which the density $p(x)$ is unknown, is to resort to sample functions: for each x define

$$E_1(x, m_1, \ldots, m_M) = \sum_k h_{b,k} \parallel x - m_k \parallel^2 . \qquad (6)$$

Functional E is the mean value of this with respect to the density $p(x)$. Functional E_1 is well defined and unique (i.e., a function) almost everywhere in the space of its arguments, except the set of $x, m_1, \ldots, m_M$ defined by the condition that x has exactly the same distance from two or more points m_i:

$$S = \{x, m_1, \ldots, m_M | \parallel x - m_b \parallel = \parallel x - m_i \parallel \text{ for some } i \neq b\}. \quad (7)$$

In fact, for any fixed $m_1, \ldots, m_M$ the set S consists of all the borders of the Voronoi tessellation in the x-space. Denote the complement of S by $\overline{S}$. In S the index b, hence $h_{b,k}$, is not unique, but in $\overline{S}$ the index b is unique and is piecewise constant. It is not affected by any gradient with respect to x or one of the m_i. Note that S is a closed set and that $\overline{S}$ is open. This means that if $(x, m_1, \ldots, m_M) \in \overline{S}$, there is some ϵ-neighborhood that is also in $\overline{S}$. Any differential change in x or some m_i stays within this neighborhood. Because b cannot change its value within one connected region of S, it follows that b is constant over this neighborhood.

It now holds that E_1 is differentiable in all m_i as long as $(x, m_1, \ldots, m_M) \in \overline{S}$, and it holds that

$$\frac{dE_1}{dm_i} = -2h_{b,i}(x - m_i). \quad (8)$$

A steepest-descent minimization of E_1 leads directly to the usual SOM learning rule:

$$m_i(t+1) = m_i(t) - \frac{1}{2}\gamma\frac{dE_1}{dm_i(t)} \quad (9)$$

$$= m_i(t) + \gamma h_{b,i}(x(t) - m_i(t)) \quad (10)$$

$$i = 1, \ldots, n. \quad (11)$$

Thus the original SOM algorithm (with constant neighborhood function) is a gradient-descent method based on sample functions E_1.

It was shown by Kohonen [10] that when the goal is to minimize the original function E of (3), extra terms appear in the algorithm because of the discontinuities at the set $\overline{S}$.

The minimization of E becomes straightforward in a special case when there is no neighborhood,

$$h_{i,j} = \delta_{i,j}. \quad (12)$$

In this case the learning algorithm and the resulting behavior of the map become similar to vector quantization (VQ) according to the k-means algorithm [11]. This can be seen from the following expansion, which may have wider applicability in the analysis of the energy function E. It can be further expressed in a form that contains only the zeroth-, first-, and second-order moments over regions V_i but no other integrals. Let

$$\omega_i = \int_{V_i} p(x)dx,$$

$$c_i = \frac{1}{\omega_i}\int_{V_i} xp(x)dx,$$

$$\sigma_i = \int_{V_i} \parallel x - c_i \parallel^2 p(x)dx.$$

Then

$$E = \sum_i \sum_k h_{i,k} \int_{V_i} \parallel x - m_k \parallel^2 p(x)dx$$

$$= \sum_i \sum_k h_{i,k} \int_{V_i} [\parallel x - c_i \parallel^2 + \parallel c_i - m_k \parallel^2 - 2(x - c_i)^T(c_i - m_k)]p(x)dx$$

$$= \sum_i \sigma_i + \sum_i \sum_k h_{i,k}\omega_i \parallel c_i - m_k \parallel^2 .$$

It has been assumed that $\sum_k h_{i,k} = 1$ for all i.

If $h_{i,k} = \delta_{i,k}$, then we obtain

$$E = \sum_i \sigma_i + \sum_i \omega_i \parallel c_i - m_i \parallel^2 . \quad (13)$$

Now $m_i = c_i$ is at least a local minimum because at these points the gradient with respect to m_j is zero. At the same time, the sum $\sum_i \sigma_i$ is minimized. This is the basic VQ coding solution.

2.3 SOM and Clustering

Many proposed clustering algorithms have been based on minimum-spanning trees, graph theory, etc. [12] capable of forming arbitrarily complex clusters. The methods use different distance measures $D(x_i, C_k)$ for the points $x_1, \ldots, x_N$ to be clustered from the clusters $C_1, \ldots, C_K$, and they also use different iterative or one-pass algorithms by which all the points are allocated into clusters. A usual criterion is based on the distance of each point from its nearest cluster: for point x_i, let $D(x_i, C_{k(i)}) = \min_k D(x_i, C_k)$. Then the function to be minimized in clustering is

$$J(C_1, \ldots, C_K) = \sum_{i=1}^{N} D(x_i, C_{k(i)}). \quad (14)$$

If $D(x_i, C_k)$ is measured as the distance of x_i from the nearest point in the cluster C_k,

$$D(x_i, C_k) = \min_j \{d(x_i, y_j), y_j \in C_k\}, \quad (15)$$

the resulting clusters will be long chains in the space, which is desirable if the data from each class are known to have long irregular distributions. The single-linkage clustering method [12] uses this distance measure. However, the distance from the closest point does not take into account the cluster shape. A good measure would be an appropriately weighted distance from all the points in the cluster, assuming that it could be computed without large cost.

The SOM is now shown to use such a measure implicitly. The following analysis is based on the energy function E in (3). For comparisons with the standard clustering framework, we assume in the following that the input distribution is discrete uniform, i.e., there exists only a finite sample $x_1, \ldots, x_N$ of possible input vectors. We also assume that the neighborhood function $h_{b,j}$ is a function of the difference $j - b$ only and use the name $h(j - b)$ for it. The function will be assumed to be spherically symmetrical and monotonically nonincreasing in the sense that

$$h(k_1) = h(k_2) \quad \text{if} \parallel k_1 \parallel = \parallel k_2 \parallel, \quad (16)$$

$$h(k_1) \geq h(k_2) \quad \text{if} \parallel k_2 \parallel \geq \parallel k_1 \parallel, \quad (17)$$

where $\parallel k_1 \parallel$ is a norm of the discrete (multidimensional) index space.

Instead of the Voronoi compartments, it is now more appropriate to use the following Voronoi index sets:

$$I_b = \{i | \parallel x_i - m_b \parallel = \min \parallel x_i - m_j \parallel, j = 1, \ldots, M\}, \quad (18)$$

which gives the indices of all vectors x_i falling into the Voronoi compartment of m_b.

The energy function now becomes

$$E(m_1, \ldots, m_M) = \sum_i \sum_k h(i - k) \sum_{p \in I_i} \parallel x_p - m_k \parallel^2 . \quad (19)$$

The cost introduced by one data sample x_p is

$$E'(x_p) = \sum_k h(b(x_p) - k) \parallel x_p - m_k \parallel^2 . \quad (20)$$

The cost function $E'(x_p)$ can be interpreted as the distance from the point x_p to the cluster represented by the whole SOM network, and learning tries to minimize the total distance from points to the cluster.

When the SOM training has converged, the gradient of the cost function is zero for each unit, regardless of whether the state is a global or a local minimum:

$$\frac{dE}{dm_k} = -2 \sum_i h(i - k) \sum_{p \in I_i} (x_p - m_k) = 0. \quad (21)$$

Note that, because there are only a finite set of vectors x_i, a differential variation of m_k will not change the index sets I_i. Denote now the number of vectors x_p for $p \in I_i$ by N_i, and denote the mean of x_p for $p \in I_i$ by c_i, i.e., $c_i = (1/N_i) \sum_{p \in I_i} x_p$. We now make the approximations

$$N_1 = N_2 = \cdots = N_M, \quad (22)$$

and

$$\sum_i h(i - k) = 1 \quad \text{for all } k. \quad (23)$$

Equation (23) is no restriction because any constant value can be used instead of unity. Equation (22) can be motivated as follows: it is equivalent to the condition that, under the assumption of equal probabilities for all input vectors $x_1, \ldots, x_N$, each unit m_k has an equal probability of being the best-matching unit. In training with the Kohonen algorithm, for high-dimensional data the units will become equiprobable. Because of equation (22), equation (21) gives

$$\sum_i h(i - k)(c_i - m_k) = 0, \quad (24)$$

and, finally, because of (23),

$$m_k = \sum_i h(i-k)c_i. \tag{25}$$

This can be interpreted as a convolution of the sequence h with the sequence $c_1, \ldots, c_M$ to yield the corresponding vectors m_k. Now (25) can be substituted into the cost function (20) to yield

$$E'(x_p) = \sum_k h(b(x_p)-k) \cdot \left\| x_p - \sum_i h(i-k)\frac{1}{N_i}\sum_{j\in I_i} x_j \right\|^2. \tag{26}$$

In (26) the summation is first computed over all the units i and then over all the samples x_j mapped to the unit. By changing the summation to run over all the data samples we get

$$E'(x_p) = \sum_k h(b(x_p)-k) \cdot \left\| x_p - \sum_{r=1}^{N} h(b(x_r)-k)\frac{1}{N_{b(x_r)}} x_r \right\|^2. \tag{27}$$

To further simplify the expression we can approximate the unit locations m_j by the centers of their tessellations c_j since the neighborhood function is a low-pass filter (by (17) it is a non-increasing function), and then each unit will be in the weighted average of the tessellation centers of its neighbors. Clearly, the more curved the map is, the more the m_j move away from the c_j, since a curved map contains more high frequencies that the neighborhood filters out. For a locally linear map low-pass filtering has no effect and the m_j coincide with the c_j. Then Eq. (20) simplifies to

$$E'(x_p) = \sum_k h(b(x_p)-k) \cdot \left\| x_p - \frac{1}{N_k}\sum_{i\in I_k} x_i \right\|^2. \tag{28}$$

The cost introduced by x_p is then the weighted distance from all the other points in the training data. The weighting depends on how far away the points are mapped on the lattice. The virtue of the weighting is that the weighting always encompasses roughly the same amount of data samples and since the distances on the map reflect the distances in the input space, the weighting decreases as the distance of the data points increases.

3 Hierarchical SOM

The hierarchical SOM is here defined as a two-dimensional SOM whose operating principle is as follows:

(i) For each input vector x the best matching unit is chosen from the first-layer map and its index b is input to the second layer.
(ii) The best-matching unit for b is chosen from the second-layer map and its index is the output of the network.

One thing is immediately clear from the above: because each first layer map unit i has a convex polyhedral Voronoi region V_i defined by (1) and each second-layer unit j is the best-matching unit for a subset, say, $i_1, \ldots, i_K$, of the first-layer indices, the second-layer unit is in fact the best matching unit for any $x \in \cup_{k=1}^{K} V_{i_k}$. This region is an arbitrary union of nonoverlapping convex polyhedral regions. Any region in $\Re^n$ can be approximated by such a union to an arbitrary accuracy when the number of component regions V_{i_k} is arbitrarily large. Thus clusters of arbitrary shapes can be represented by the two-layer map.

By analogy to the basic approximation result of two-layer Perceptron networks [3], this is purely an existence result. There is no guarantee that a certain predetermined cluster shape could be learned by the map. However, in unsupervised learning this is an empty question because by definition no target clustering can exist.

In [9] and [13] the theory of hierarchical maps is derived from the principles of coding theory. It is shown that the hierarchical map minimizes the decoding squared error if the training neighborhood in SOM equals the probability distri-

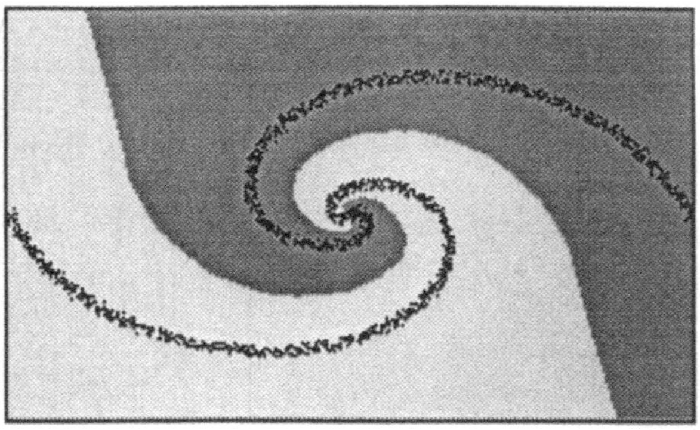

Fig. 1. Example of clusters forms of the HSOM for spiral data.

bution of errors in the codes. In hierarchical VQ the higher-order map quantization error is the source for the lower-order map code errors. In this context the hierarchical SOM is optimal in quantization.

In clustering terminology, whereas the first SOM layer forms one large cluster of all the data samples so that the total distance of the samples from the cluster is minimized, the second map in HSOM then splits the large cluster into equal-size parts. Since the distance relations of the data samples are preserved on the map, the cluster numbers or indices of the best-matching units can be used as a measure of distance of the original data samples. What is gained by the HSOM is that each high-dimensional data vector is mapped to a low-dimensional discrete value so that a comparison of the values implicitly contains a comparison of the original distances.

Figure 1 shows an example of decision regions of the HSOM. The black spiral-shaped stripes are the data points. The HSOM in the figure contained 100 units in the first layer and two units in the second layer. Regions mapped to the two clusters defined by the two second-layer units are shown as different gray levels.

The main advantage of HSOM clustering with respect to classical clustering methods, e.g., k-means, is the adaptive distance measure. In the k-means or Isodata family of methods [11] clusters that are too large are split into smaller ones and clusters that are too small are merged together until all the clusters are of the desired size. In practice it is very difficult to determine a suitable size for the clusters. Also, Isodata clustering algorithms can make only convex clusters because of the nearest-neighbor clustering rule.

Unlike simple linkage clustering, the HSOM offers a distance measure that takes into account all the points in the cluster. As can be seen from (20), the cost introduced by one data point contains the distance of the point from all the other clusters, weighted by the distance of each cluster along the lattice.

4 Experimental Results

In an experiment with artificial data, random Gaussian input classes with elongated shapes were generated (see figure 2) and the confusion matrices of clusterings, i.e., the distribution of each class among the clusters, were formed. The perfect clustering would be such that each cluster contains data points from only one class.

An appropriate measure for the goodness of any clustering ensemble is the mean columnwise entropy of the confusion matrix. It measures the width of the distribution of points from one class. For example, if we have 10 true classes and 40 clusters, the perfect clustering would map each class into a maximum of four clusters, corresponding to the entropy 2.0. An entropy value of 3, say, would indicate that each class is on average distributed into eight clusters.

In each experiment run we generated a new data ensemble of 10 Gaussian-distributed classes with random class centers, principal-axis directions, and variances. The variances of the classes were bounded so that the main-axis variance was randomly 1 to 10 times the minor-axis variance. The number of data samples in each class was either constant ($N = 300$) or random ($200 \leq N \leq 400$).

Clustering the data was tested with 1-dimensional maps of sizes 40, 20, 10, and 5 units for both SOM and HSOM. The first-layer map for the HSOM always contained 160 units.

The results averaged over several class ensembles are presented in figures 3 and 4. The figures show two entropy measures for both SOM and HSOM networks: clusters/class is the column-

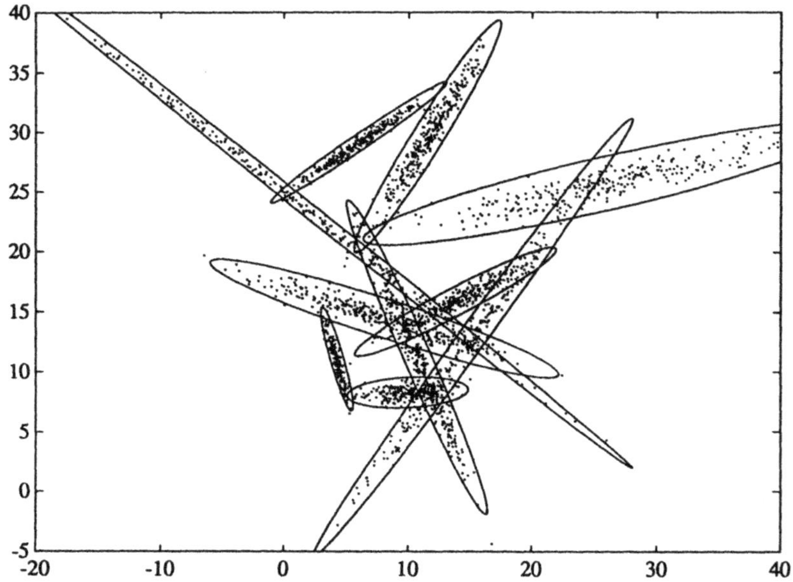

Fig. 2. Example test set of 10 clusters.

wise entropy, i.e., it tells how many clusters are used to cover one class, and classes/cluster is the rowwise entropy, i.e., it tells how many classes each cluster has collected samples from. The ordinate axes in the figures show the widths of the distribution compared to those for perfect clustering. For example, the value 2.0 for (10 classes)/(40 clusters) means that the classes are on average spread into eight clusters instead of four, which would be the optimum case.

In figure 3 all the classes contained the same number of data points ($N = 300$), which favors SOM-type clustering methods that try to make equal-size clusters. In figure 4 the class sizes were random ($200 \leq N \leq 400$), so that optimal VQ of the space is quite different from optimal clustering.

In every case the SOM has a better clusters/class measure, since the clusters of SOM are more compact and fewer clusters are needed to represent a class. The classes/cluster measure is better for HSOM, since the one-layer map cannot track the class boundaries as well as HSOM and each cluster collects points from nearby classes (cf. figure 1). The classification error depends on the classes/cluster measure, since the classifier cannot separate the classes once they have been mapped to the same cluster.

Direct classification errors were also measured for SOM and HSOM, and the results are given in table 1. Each unit was labeled into the class that gave the largest number of hits for the unit, and the classification errors were all the hits from any other classes. This corresponds to the *a posteriori* Bayes classifier, since the hit rates for each cluster measure the probability density of the classes in the cluster regions, and classification is done according to the largest probability.

Note that the class distributions in figure 2 would be rather easy to separate with any simple classifier and that the classification errors of unsupervised clustering cannot be compared with any result of direct-supervised classifiers. The purpose of the clustering network was, as explained in section 1, to reduce the complexity of the data when there were not enough preclassified samples to train a supervised classifier. For the same reason, the SOM network was not fine tuned by LVQ [5]. The classification error gives only an approximate lower limit for the number of errors if only the cluster identities are passed on to the classifier. For comparison, the Bayes classification error, estimated by the one-nearest-neighbor rule by using all the data samples, is about 5%, whereas the errors after

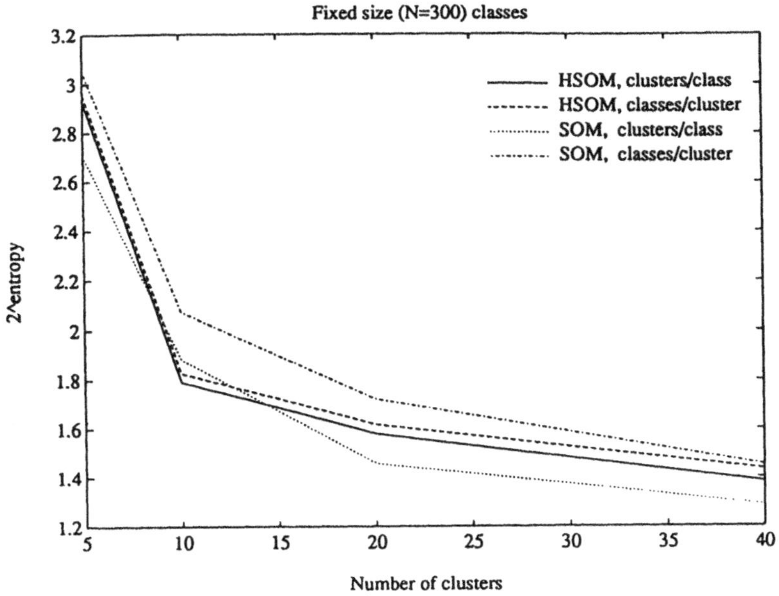

Fig. 3. Clustering entropies for fixed-size clusters.

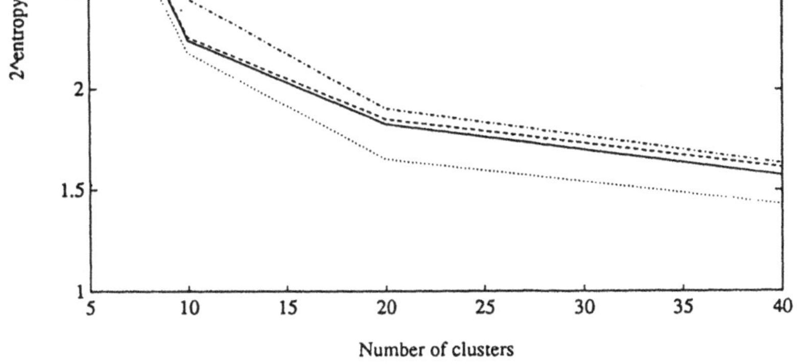

Fig. 4. Clustering entropies for random-size clusters.

the different clustering methods are about 30%.

The SOM and k-means have similar figures for classification errors in Table 1, which stems from the fact that k-means is effectively a batch-training version of the SOM without a neighborhood [9].

The experiments with the Isodata algorithm were performed with the Khoros data-processing and visualization tool[1]. The Isodata algorithm implemented in the Khoros system contains enhanced features to adapt the clusters to the local statistics of the data and to simplify selection of the split and merge parameters, but still it required much more manual experimenting than the almost automatic HSOM to find the optimal clustering.

Table 1. Classification errors in the clustering tests.

Method	Errors(%)
k-means	30.3
SOM	30.7
Isodata	28.0
HSOM	26.3

As a practical example we used HSOM to cluster sensory information from low-level feature detectors in a computer-vision system. The special feature of sensory information is that the signal space is often very high dimensional, but the actually occurring signals are implicitly rather low dimensional since the primary feature detectors tend to be orthogonal or otherwise mutually exclusive.

As the primary feature detectors we used Gabor filters (Gaussian band-pass filters in the frequency domain) in eight different orientations and two frequencies, $\pi/2$ and $\pi/4$, giving a total of 16 different spatial filters, which were applied to each point in the image. The ensuing 16-dimensional feature vectors at each pixel position were mapped with a 100-unit SOM to feature values. Another set of features were obtained by first mapping both resolutions separately by a 100-unit map and then clustering the map outputs by a second 100-unit map. For a complete description of the object-recognition system see [14].

[1] Khoros is software environment for data processing and visualization, a free software package copyrighted by the University of New Mexico, Albuguerque, NM 87106.

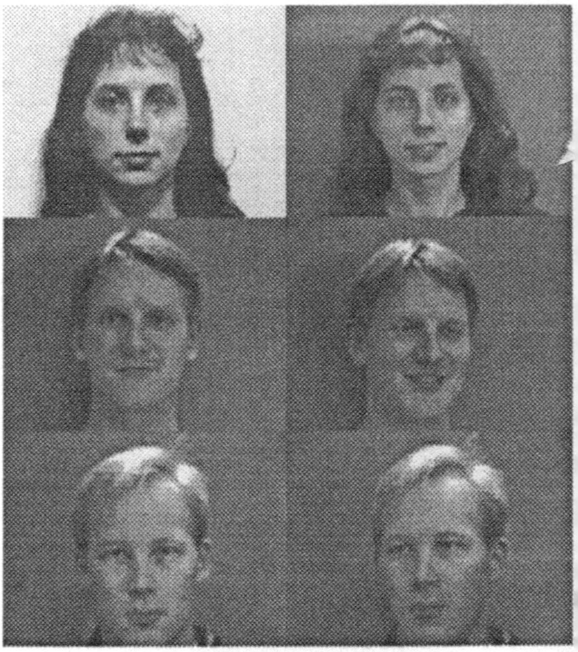

Fig. 5. Test images for table 2. from left to right and from top to bottom: T1, T2, J1, J2, P1, P2.

Distortions in the imaging (i.e., different lightings, contrasts, viewing angles) make the Gabor-filter responses move in the 16-dimensional space in a regular way. For example, increasing contrast makes all the edges sharper, increasing the responses of the high-frequency filters. If the clustering algorithm can find such regular trajectories and map them to the same cluster, the distortion tolerance of classification should be increased.

As a test problem we recognized human faces. According to our experiments the feature set produced by the HSOM is indeed clearly more distortion tolerant than that of the direct SOM. Examples of the tests are given in table 2. We compared the features by compiling normalized histograms of the features over each image in figure 5 and computing inner products of the histograms; this is a simplified form of the subspace classifier. One of the test images from each class was selected as the class prototype, and the other images were compared with the prototype.

In larger experiments we have been able to classify 19 out of 20 similar face images by using only the histograms of the 100 hierarchical features and the subspace classifier.

Table 2. Comparison of SOM and HSOM features (see text for details).

	Class Prototype Image					
	HSOM			SOM		
Test Image	T1	J1	P1	T1	J1	P1
T1	1.000	0.212	0.212	1.000	0.515	0.627
T2	0.550	0.295	0.508	0.790	0.504	0.698
J1	0.212	1.000	0.730	0.515	1.000	0.933
J2	0.291	0.814	0.717	0.572	0.984	0.980
P1	0.212	0.730	1.000	0.627	0.933	1.000
P2	0.231	0.703	0.860	0.568	0.985	0.971

Face recognition was selected as a test case because human faces clearly contain distinct features that characterize the face invariantly to changes in imaging conditions. The purpose was to verify whether the HSOM network can find such features in an unsupervised way. The classifier used in the experiments is very primitive since it loses all location information from the features and only the relative frequencies of the features in the image are considered. The successful tests show that the features extracted by the HSOM are rather robust and distortion tolerant.

References

1. R. Hecht-Nielsen, "Theory of backpropagation neural network," in *Proc. IEEE Int. Joint Conf. on Neural Networks*, vol. I, Washington, DC, 1989, pp. 593–611.
2. R. Hecht-Nielsen, *Neurocomputing*, Addison-Wesley: Reading, MA, 1990.
3. K. Hornik, M. Stinchcombe, and H. White, "Multi-layer feedforward networks are universal approximators," *Neural Net.*, vol. 2, pp. 359–366, 1989.
4. M.D. Richard and R.P. Lippmann, "Neural network classifiers estimate Bayesian *a posteriori* probabilities," *Neural Comput.*, vol. 3, 1991, pp. 461–483.
5. T. Kohonen, *Self-Organization and Associative Memory*, Springer-Verlag: Berlin, 1989.
6. T. Kohonen, "Self-organized formation of topologically correct feature maps," *Biol. Cybernet.*, vol. 43, pp. 59–69, 1982.
7. S. Amari, "Topographic organization of nerve fields," *Bull. Math. Biol.*, vol. 42, pp. 339–364, 1980.
8. H. Ritter and K. Schulten, "Kohonen's self-organizing maps: exploring their computational capabilities," in *Proc. IEEE Int. Joint Conf. on Neural Networks*, vol. 1, San Diego, CA, 1988, pp. 109–116.
9. S.P. Luttrell, "Self-organisation: A derivation from first principles of a class of learning algorithms," *Proc. IEEE Int. Joint Conf. on Neural Networks*, vol. 2, Washington, DC, 1989, pp. 495–498.
10. T. Kohonen, "Self-organizing maps: optimization approaches," in *Artificial Neural Networks*, vol. 2, T. Kohonen, K. Mäkisara, J. Kangas, and O. Simula, eds., North-Holland: Amsterdam, 1991, pp. 981–990.
11. P.A. Devijver and J. Kittler, *Pattern Recognition: A Statistical Approach*, Prentice Hall: London, 1982.
12. J.C Gower and G.J.S Ross, "Minimum spanning trees and single linkage cluster analysis," *Appl. Statist.*, vol. 18, pp. 54–64, 1969.
13. S.P. Luttrell, "Image compression using a multilayer neural network," *Pattern Recog. Lett.*, vol. 10, pp. 1–7, 1989.
14. J. Lampinen, "Distortion tolerant pattern recognition using invariant transformations and hierarchical SOFM clustering," in *Artificial Neural Networks*, vol. 1, T. Kohonen, K. Mäkisara, J. Kangas, and O. Simula, eds., North-Holland: Amsterdam, 1991, pp. 99–104.

Jouko Lampinen received the M.Sc. degree in applied physics and electronics from the University of Kuopio, Finland, in 1988. Currently he is finishing his doctoral thesis in computer science at Lappeenranta University of Technology, Finland, in the Department of Information Technology. His research interests are pattern recognition and neural networks, especially self-organizing models in feature extraction.

Erkki Oja is Professor of Computer Science at Lappeenranta University of Technology, Finland. He received his M.Sc. and Dr. Tech degrees from Helsinki University of Technology 1972 and 1977, respectively. He was visiting scientist at Brown University, Providence, in 1977-1978 and at Tokyo Institute of Technology in 1983-1984 where he also held the Toshiba Visiting Professor's Chair during the academic year 1990-1991. Professor Oja is the author of a number of articles on pattern recognition, computer vision, and neural computing and is the author of the book Subspace Methods of Pattern Recognition, which has been translated into Chinese and Japanese. His present research interests are in applying neural networks to computer vision and the study of subspace and PCA networks. Professor Oja is member of ACM, IEEE, INNS, ENNS and the Finnish Academy of Sciences, and he serves on the editorial boards of International Journal of Neural Systems, Neural Networks, Neural Computation, and IEEE Transactions on Neural Networks.

GPSR Compliance

The European Union's (EU) General Product Safety Regulation (GPSR) is a set of rules that requires consumer products to be safe and our obligations to ensure this.

If you have any concerns about our products, you can contact us on ProductSafety@springernature.com

In case Publisher is established outside the EU, the EU authorized representative is:

Springer Nature Customer Service Center GmbH
Europaplatz 3
69115 Heidelberg, Germany

Batch number: 10755809

Printed by Printforce, the Netherlands